Symbol	Meaning	Appendix
$\{a, b\}$	The set whose elements (or members) are a and b	A.1
$\in$	Is a member of, is an element of ($\notin$, is not a member of, is not an element of)	A.1
$\{x \mid . . .\}$	The set of all x such that . . .	A.1
$\cup$	The union of	A.1
$\cap$	The intersection of	A.1
$\varnothing$	The null set, the empty set	A.1
N	The set of natural numbers	A.2
J	The set of integers	A.2
Q	The set of rational numbers	A.2
H	The set of irrational numbers	A.2
R	The set of real numbers	A.2
$\sqrt{a}, \quad a \geq 0$	The nonnegative square root of a	A.2
$\approx$	Approximately equal to	A.2
$\lvert a \rvert$	The absolute value of a	A.2
$<$	Is less than	A.2
$\leq$	Is less than or equal to	A.2
$>$	Is greater than	A.2
$\geq$	Is greater than or equal to	A.2
(a, b)	The ordered pair of numbers whose first component is a and whose second component is b	A.3
f, F, g, etc.	Names a function or a relation	A.3
$f(x)$	f of x, or the value of f at x	A.3
$P(x, y)$	The name of a point whose coordinates are x and y	A.4
$\overline{P_1P_2}$	The line segment whose endpoints are designated P_1 and P_2	A.4
P_1P_2	The length of line segment $\overline{P_1P_2}$	A.4
$\mathcal{R}$	A relation	A.6
$\mathcal{R}^{-1}$	The inverse of relation $\mathcal{R}$	A.6
$\overrightarrow{AB}$	Ray with endpoints A and B	A.7
$\angle ABC$	Angle ABC (the vertex is at B)	A.7
$\alpha, \beta, \gamma, . . .$	Angles "alpha," "beta," "gamma," and so forth	A.7
α°	The measure in degrees of angle α	A.7
$\perp$	Is perpendicular to	A.7
$\triangle ABC$	A triangle with vertices A, B, C	A.8
$\cong$	Is congruent to	A.9
$\sim$	Is similar to	A.9

Essentials of Trigonometry

Essentials of Trigonometry

THIRD EDITION

Irving Drooyan
Walter Hadel
Charles C. Carico
Los Angeles Pierce College

Macmillan Publishing Co., Inc.
New York
Collier Macmillan Publishers
London

Macmillan Publishing Co., Inc.
866 Third Avenue, New York, New York 10022

Collier Macmillan Canada, Ltd.

Library of Congress Cataloging in Publication Data
Drooyan, Irving.
Essentials of trigonometry.

"A portion of this material has been adapted from Trigonometry: an analytical approach."
Includes index.
1. Trigonometry, Plane. I. Hadel, Walter, joint author. II. Carico, Charles C., joint author. III. Title.
QA533.D74 1981 516'.24 79-22994
ISBN 0-02-330270-4

Printing: 1 2 3 4 5 6 7 8 Year: 1 2 3 4 5 6 7

Preface

This third edition of *Essentials of Trigonometry* reflects ten years of classroom experience with the earlier editions. Like its predecessors, this edition is designed for use in a one-semester or one-quarter course in trigonometry. The organization of the material and the large number of exercises permit flexibility in courses that differ in the time available.

In this edition we have rewritten some textual material and made changes in the sequence of topics to improve the presentation. In particular, the material has been arranged so that instructors can easily accommodate their classroom procedures both to students who use a calculator and to those who do not. Answers to all examples and exercises that include degree measures of angles appear in decimal form to the nearest tenth of a degree. Table II in Appendix C has also been changed to provide such measures. A section on interpolation is provided as Appendix B for those instructors who want their students to use tables rather than a calculator.

The trigonometric functions and their applications to the solutions of triangles are discussed in Chapters 1 and 2. The presentation of these topics relies on geometric notions. The periodic properties of these functions are studied by considering physical models in applied problems in Chapter 2 and by investigating graphs of the functions and their inverses in Chapter 3. Identities and conditional equations are considered in detail in Chapter 4. Complex numbers and polar coordinates are treated in Chapter 5.

As in the previous editions, the subject matter of the text is continually reviewed through the use of Chapter Summaries and Review Exercises. Items in both the Summaries and Reviews are cross-referenced to the appropriate section of the text or Exercise Set, as is the list of symbols inside the front cover. A detailed summary of the important topics in the text that appears inside the back cover can serve as a convenient reference chart for students completing assignments or studying for examinations.

All Exercise Sets in this edition have been revised as necessary to insure that they are carefully graded. In addition, many new exercises have been added. All Exercise Sets include basic exercises designated

A; many also include more challenging exercises designated B, which provide the instructor with flexibility in making assignments, depending on the time available and the objectives of the course.

Answers, including graphs, for odd-numbered exercises in each Exercise Set and for all the problems in the Review Exercises follow the Appendixes.

A *Student Solution Key* containing *solutions* to the even-numbered exercises in each section is also available for this edition.

I. D.
W. H.
C. C. C.

Contents

1 Trigonometric Functions

Historically, the subject of trigonometry developed primarily as a means of finding the lengths of sides and the measures of angles of triangles for use in such practical areas as surveying and navigation. In more recent times, trigonometry has become important in describing phenomena that are periodic in nature. In this book we are concerned with presenting the necessary concepts and developing the necessary skills to solve problems in both areas.

In this chapter we first introduce some notions about angles and their measures. Then we shall consider six important functions whose domains are sets of angles.

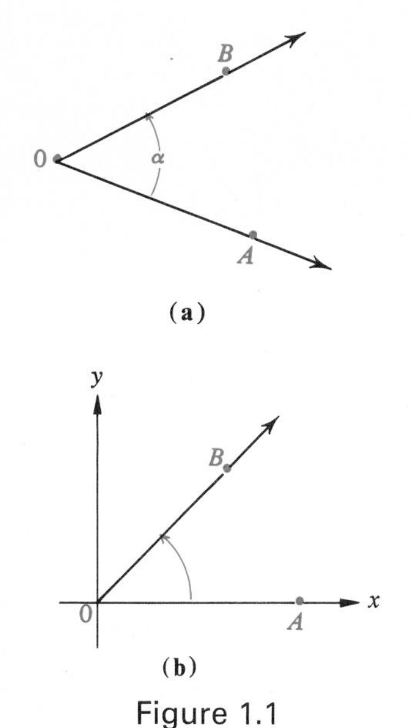

Figure 1.1

1.1 Angles and Their Radian Measures

In geometry, an angle is considered to be two rays with a common endpoint, such as $\overrightarrow{OA}$ and $\overrightarrow{OB}$ in Figure 1.1a (see Appendix A.7). Let us consider $\angle AOB$ and visualize it as being formed by rotating side $\overrightarrow{OA}$, with O as the pivotal point, to side $\overrightarrow{OB}$; side $\overrightarrow{OA}$ is called the **initial side,** and $\overrightarrow{OB}$ is called the **terminal side.** Angles viewed in this way, for which rotation is considered, are called **directed angles.**

Each angle in the plane can be made to coincide with an angle that has one of its sides along the positive half of the x-axis and its vertex at the origin, as shown in Figure 1.1b. Such an angle is said to be in *standard position*. The terminal side of an angle in standard position may lie in any quadrant or on any axis. The angle is said to be in that quadrant in which the terminal side lies. If the terminal side lies on an axis, the angle is called a **quadrantal angle.** For example, in Figure 1.2, α (alpha) is in Quadrant I, β (beta) is in Quadrant II, γ (gamma) is in Quadrant III, δ (delta) is in Quadrant IV, and θ (theta) is a quadrantal angle.

Measure of an Angle

The measure of an angle can be specified in different ways. The two kinds of angle measure most commonly used are **degree measure,** with which you are undoubtedly familiar (see Appendix A.7), and **radian measure.**

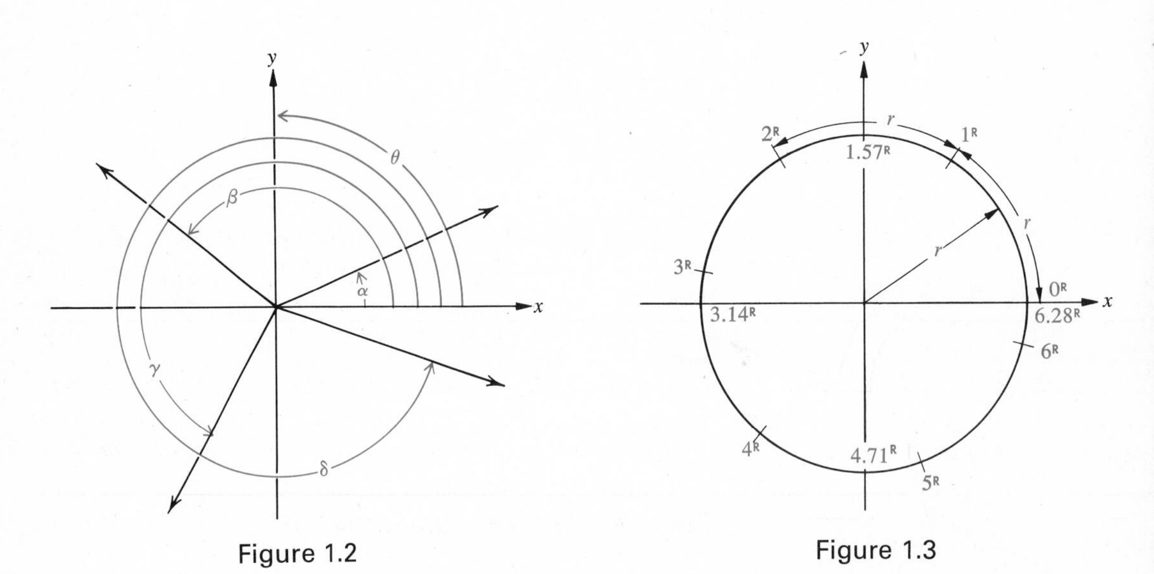

Figure 1.2

Figure 1.3

In radian measure, the circle is divided into unit arcs, each having a length equal to the radius of the circle, as shown in Figure 1.3. The circumference of a circle with radius r is $2\pi r$; hence, the circle will include 2π, or approximately 6.28, of these arcs. The number of such arcs intercepted by a central angle is the measure of the angle in radian units. For this measure we use the notation α^{R}.

Positive numbers are assigned for the measures of angles in which the rotation from the initial to the terminal side is in a counterclockwise direction, and negative numbers are assigned for the measure if the rotation is in a clockwise direction. In Figure 1.4a, the measuring circle is scaled in degree units and radian units for positive measures that are often used. In Figure 1.4b, the measuring circle is scaled for similar negative units.

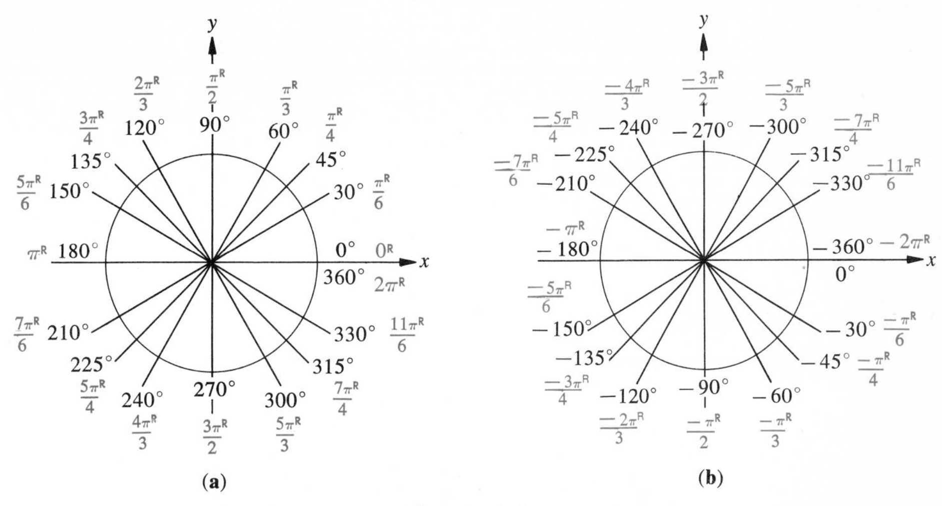

Figure 1.4

In Figure 1.5,

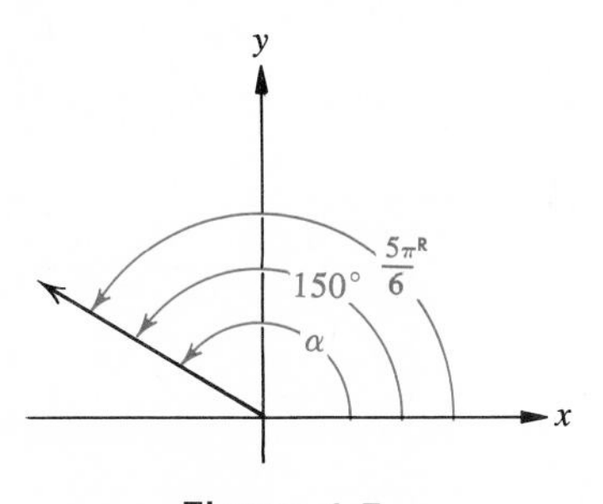

Figure 1.5

$$\alpha = 150^\circ \qquad \text{and} \qquad \alpha = \frac{5\pi^R}{6}.$$

As is customary, we shall use the terms "least positive angle" to mean that angle of a set of angles which has the least positive measure and "negative angle" to mean an angle with a negative measure.

Degree–Radian Conversion Formulas

The measure of an angle in radians can be related to the measure of the angle in degrees. Since the ratio of the measure of a given central angle to the measure of the central angle intercepting the entire circle is the same for any unit of measurement, we see that for degree measure and radian measure in particular

$$\frac{\alpha^\circ}{360} = \frac{\alpha^R}{2\pi}.$$

This equation is sometimes written in either of the following forms:

$$\textbf{I.} \qquad \alpha^\circ = \frac{180}{\pi}\,\alpha^R$$

$$\textbf{II.} \qquad \alpha^R = \frac{\pi}{180}\,\alpha^\circ$$

Examples Find the degree measure (to the nearest tenth of a degree) of each angle α with the given radian measure. Use $\pi \approx 3.14$.

a. $\alpha = \dfrac{3\pi^{R}}{5}$ b. $\alpha = 0.81^{R}$

Solutions

a. $\alpha^{\circ} = \left(\dfrac{180}{\pi} \cdot \dfrac{3\pi}{5}\right)^{\circ} = 108.0^{\circ}$ b. $\alpha^{\circ} = \left(\dfrac{180}{\pi} \cdot 0.81\right)^{\circ} \approx 46.4^{\circ}$*

Examples Find the radian measure (to the nearest hundredth of a radian) of each angle α with the given degree measure. Use $\pi \approx 3.14$.

a. $\alpha = 60^{\circ}$ b. $\alpha = 330^{\circ}$

Solutions

a. $\alpha^{R} = \left(\dfrac{\pi}{180} \cdot 60\right)^{R} = \dfrac{\pi^{R}}{3} \approx 1.05^{R}$

b. $\alpha^{R} = \left(\dfrac{\pi}{180} \cdot 330\right)^{R} = \dfrac{11\pi^{R}}{6} \approx 5.76^{R}$

Many exercises in this text, like the examples above, require some arithmetic computation. If you use a calculator for a closer approximation for the value of π than we used in the above examples, your results in the exercises may differ slightly from the answers in the answer section.

From conversion formulas I and II on page 3, we observe that the measure in degrees of an angle α with a measure of 1^{R} is given by

$$\alpha^{\circ} = \left(\frac{180}{\pi} \cdot 1\right)^{\circ} \approx 57.30^{\circ}$$

and that the measure in radians of an angle α with a measure of 1° is given by

$$\alpha^{R} = \left(\frac{\pi}{180} \cdot 1\right)^{R} \approx 0.01745^{R}.$$

These results can also be used to make conversions from radian units to degree units and from degree units to radian units, respectively.

* In any such chain of statements involving both the symbols = and ≈, it should be understood that either relationship is valid only for the expressions on either side of the particular symbol.

Examples

a. The degree measure of an angle $\alpha = 0.37^{\mathrm{R}}$ is given by

$$\alpha^{\circ} \approx (57.30 \cdot 0.37)^{\circ} \approx 21.2^{\circ}.$$

b. The radian measure of an angle $\alpha = 200^{\circ}$ is given by

$$\alpha^{\mathrm{R}} \approx (0.01745 \cdot 200)^{\mathrm{R}} \approx 3.49^{\mathrm{R}}.$$

Coterminal Angles

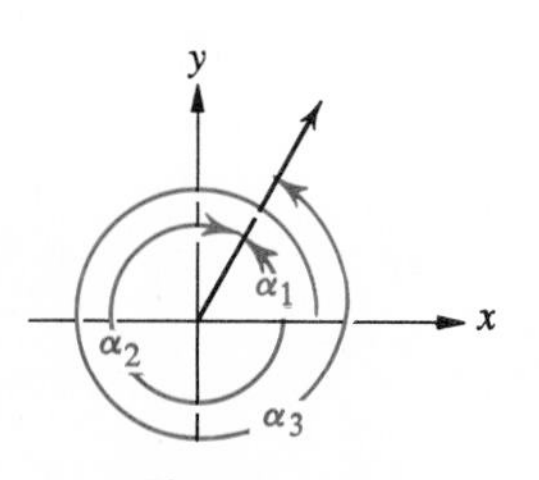

Figure 1.6

If an angle is viewed as being formed by rotation of the initial side to the terminal side, then it is possible to have an angle whose initial side is rotated more than 360°. For example, in Figure 1.6, $\alpha_3 > 360°$. Also, as indicated in the figure, angles may have the same initial side and the same terminal side but different measures.

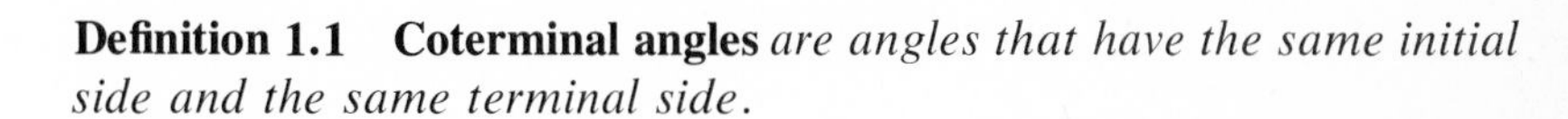

Definition 1.1 Coterminal angles *are angles that have the same initial side and the same terminal side.*

In Figure 1.6, α_1, α_2, and α_3 are coterminal.

Examples Find the least positive angle that is coterminal with each of the following.

a. 6.59^{R} b. 520°

Solutions

a. Sketch the angle, showing its initial and terminal sides. Since the angle generated in one revolution is $2\pi^{\mathrm{R}} \approx 6.28^{\mathrm{R}}$, the least positive angle having the same initial and same terminal sides is approximately $(6.59 - 6.28)^{\mathrm{R}} = 0.31^{\mathrm{R}}$.

b. Sketch the angle. In this case, the least positive angle having the same initial and terminal sides is $(520 - 360)^{\circ} = 160^{\circ}$.

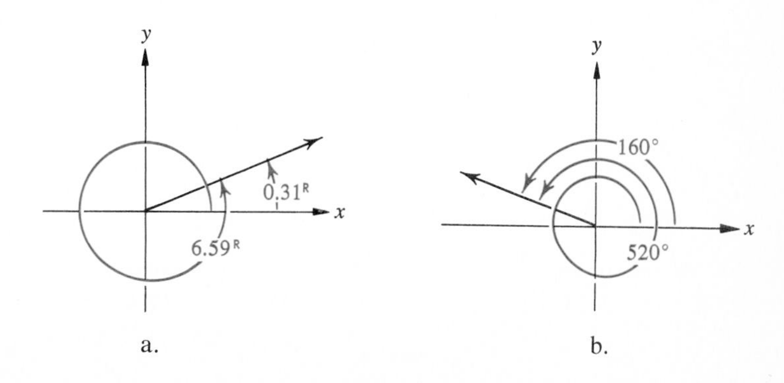

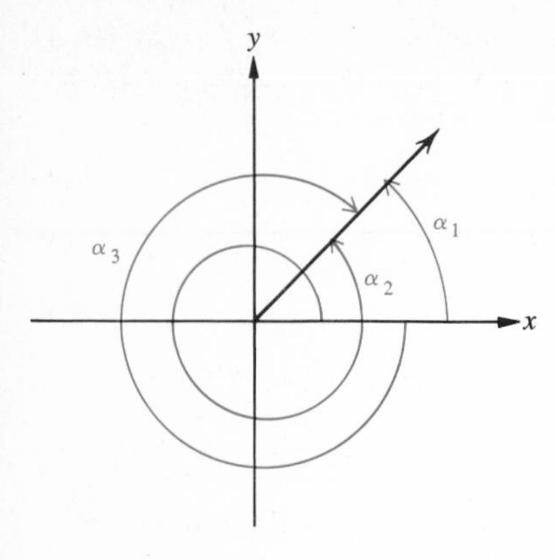

Example Write a general expression for all the angles that are coterminal with 45°.

Solution Since one revolution from a terminal side of any angle brings us back to that terminal side, then all angles coterminal with an angle of 45°, or $(\pi/4)^{R}$, are given by

$$(45 + k \cdot 360)^{\circ}, \quad k \in J \qquad \text{or} \qquad \left(\frac{\pi}{4} + k \cdot 2\pi\right)^{R}, \quad k \in J.^{*}$$

Note that

$$\text{for } k = 0, \qquad \alpha_1 = (45 + 0)^{\circ} = 45^{\circ} \quad \text{or} \quad \left(\frac{\pi}{4} + 0\right)^{R} = \frac{\pi}{4}^{R};$$

$$\text{for } k = 1, \qquad \alpha_2 = (45 + 360)^{\circ} = 405^{\circ} \quad \text{or} \quad \left(\frac{\pi}{4} + 2\pi\right)^{R} = \frac{9\pi}{4}^{R};$$

$$\text{for } k = -1, \quad \alpha_3 = (45 - 360)^{\circ} = -315^{\circ} \quad \text{or} \quad \left(\frac{\pi}{4} - 2\pi\right)^{R} = -\frac{7\pi}{4}^{R}.$$

EXERCISE SET 1.1

A

Complete the following degree–radian conversion wheels for special angles without referring to Figure 1.4. (These wheels are a handy reference.)

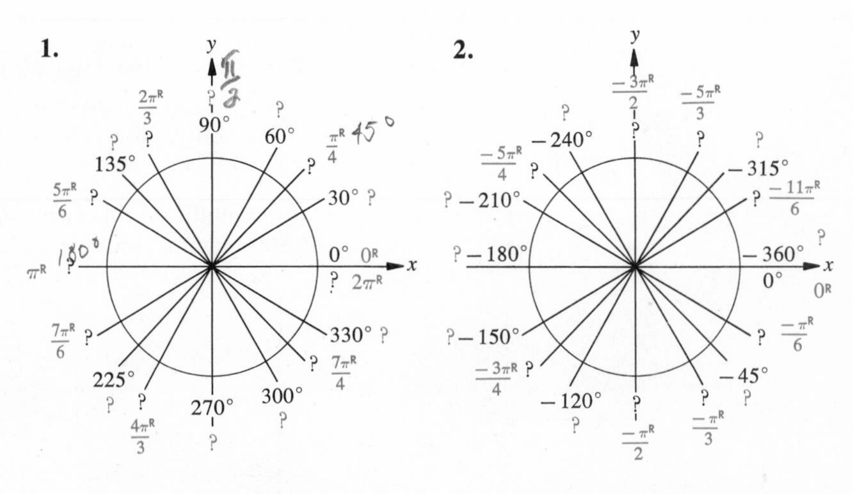

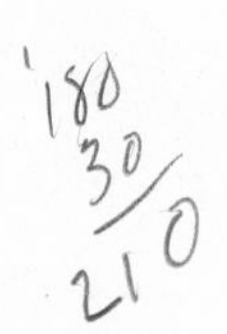

* The letter J is used to name the set of integers; the symbol $\in$ is read "is an element of" (see Appendix A.2).

Find the degree measure to the nearest tenth of a degree of an angle α with the given radian measure. (Use $\pi \approx 3.14$.) Specify the quadrant in which the terminal side of α lies.

Examples a. $\frac{6\pi}{5}^{R}$ b. -1.24^{R}

Solutions a. $= \left(\frac{180}{\pi} \cdot \frac{6\pi}{5}\right)^{\circ}$ b. $\alpha^{\circ} = (-1.24 \cdot 57.30)^{\circ}$

$= 216.0^{\circ}$; Quad. III $\approx -71.1^{\circ}$; Quad. IV

3. $\frac{\pi}{3}^{R}$ **4.** $\frac{2\pi}{3}^{R}$ **5.** $\frac{14\pi}{3}^{R}$ **6.** $\frac{15\pi}{4}^{R}$

7. $\frac{-8\pi}{5}^{R}$ **8.** $\frac{-11\pi}{8}^{R}$ **9.** $\frac{-18\pi}{8}^{R}$ **10.** $\frac{-12\pi}{5}^{R}$

11. 0.18^{R} **12.** 0.94^{R} **13.** 5.16^{R} **14.** 4.55^{R}

15. -0.27^{R} **16.** -0.86^{R} **17.** -4.04^{R} **18.** -3.61^{R}

Find the radian measure to the nearest hundredth of a radian of an angle α with the given degree measure. (Use $\pi \approx 3.14$.) Specify the quadrant in which the terminal side of α lies.

Example 118°

Solution $\alpha^{R} = \left(\frac{\pi}{180} \cdot 118\right)^{R} \approx 2.06^{R}$; Quad. II

Alternate Solution $\alpha^{R} = (0.01745 \cdot 118)^{R} \approx 2.06^{R}$

19. 30° **20.** 45° **21.** 75° **22.** 125°

23. 143° **24.** 137° **25.** 402° **26.** 560°

27. -37° **28.** -51° **29.** -80° **30.** -130°

31. -240° **32.** -120° **33.** -510° **34.** -400°

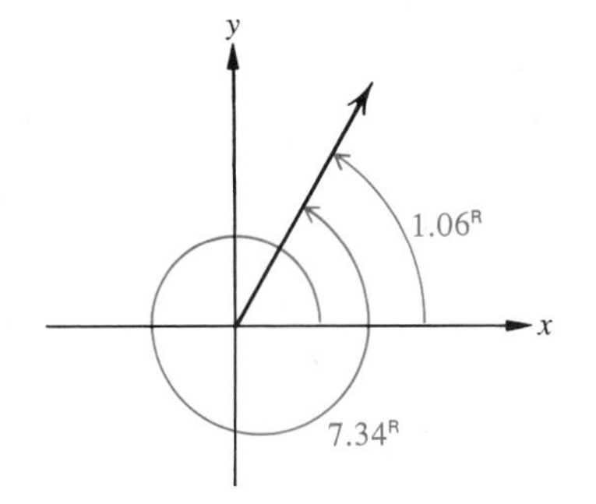

a.

Find the least positive angle that is coterminal with each of the following. Sketch each angle. (Use $\pi \approx 3.14$).

Examples a. 7.34^{R} b. -620°

(continued)

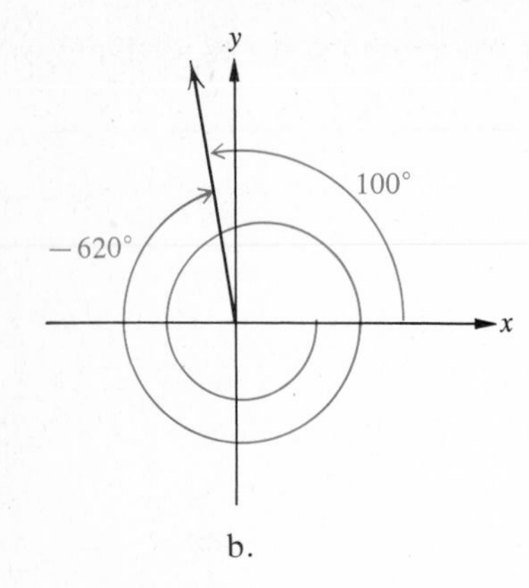

b.

Solutions a. $[7.34 + (-1) \cdot 2\pi]^{\text{R}}$ b. $(-620 + 2 \cdot 360)^\circ$

$\approx (7.34 - 6.28)^{\text{R}}$ $= (-620 + 720)^\circ$

$= 1.06^{\text{R}}$ $= 100^\circ$

35. 6.72^{R} **36.** 8.91^{R} **37.** 379° **38.** 423°

39. -11.11^{R} **40.** -13.48^{R} **41.** -792° **42.** -863°

Write a general expression for all angles coterminal with each of the following.

Examples a. 16° b. $\dfrac{\pi}{2}^{\text{R}}$

Solutions a. $(16 + k \cdot 360)^\circ$, $k \in J$ b. $\left(\dfrac{\pi}{2} + k \cdot 2\pi\right)^{\text{R}}$, $k \in J$

43. 21° **44.** 128° **45.** 242° **46.** 310°

47. $\dfrac{\pi}{6}^{\text{R}}$ **48.** $\dfrac{\pi}{3}^{\text{R}}$ **49.** $\dfrac{5\pi}{3}^{\text{R}}$ **50.** $\dfrac{7\pi}{6}^{\text{R}}$

1.2 Trigonometric Ratios

Figure 1.7

Consider the three real numbers x, y, and $\sqrt{x^2 + y^2}$ that can be associated with any point on the terminal side of an angle in standard position, as shown in Figure 1.7. The components of the ordered pair (x, y) are the coordinates of the point, and from the distance formula (see Appendix A.4) or the Pythagorean theorem, $\sqrt{x^2 + y^2}$ is the length of the line segment from the origin to the point. For each angle α, six possible ratios can be formed using the numbers x, y, and $\sqrt{x^2 + y^2}$. These ratios are called **trigonometric ratios** and are named as follows.

Definition 1.2 *For all angles α in standard position, if x and y are the coordinates of any point on the terminal side of α [where $(x, y) \neq (0, 0)$] and $r = \sqrt{x^2 + y^2}$,*

$$\textbf{sine } \alpha = \frac{y}{r}, \qquad \textbf{cosecant } \alpha = \frac{r}{y} \quad (y \neq 0),$$

$$\textbf{cosine } \alpha = \frac{x}{r}, \qquad \textbf{secant } \alpha = \frac{r}{x} \quad (x \neq 0),$$

$$\textbf{tangent } \alpha = \frac{y}{x} \quad (x \neq 0), \qquad \textbf{cotangent } \alpha = \frac{x}{y} \quad (y \neq 0).$$

The names of the trigonometric ratios are abbreviated, respectively, as

$$\sin \alpha, \qquad \csc \alpha,$$
$$\cos \alpha, \qquad \sec \alpha,$$
$$\tan \alpha, \qquad \cot \alpha.$$

Notice that $\csc \alpha$, $\sec \alpha$, and $\cot \alpha$ are the reciprocals of $\sin \alpha$, $\cos \alpha$, and $\tan \alpha$, respectively. That is,

$$\boldsymbol{\csc \alpha = \frac{1}{\sin \alpha}, \quad \sec \alpha = \frac{1}{\cos \alpha}, \quad \text{and} \quad \cot \alpha = \frac{1}{\tan \alpha}.}$$

These three equations are **identities;** each is true for all the replacements of α for which each member is defined. They are basic to the study of trigonometry. You will be introduced to other important identities in the work that follows.

Also, note in Definition 1.2 that r is the positive square root of $x^2 + y^2$. Thus, r is never negative or zero.

It can be shown by using similar triangles, and this will be left as an exercise, that if (x_1, y_1) and (x_2, y_2) correspond to any two points (excluding the origin) on the terminal side of an angle in standard position, as shown in Figure 1.8, then

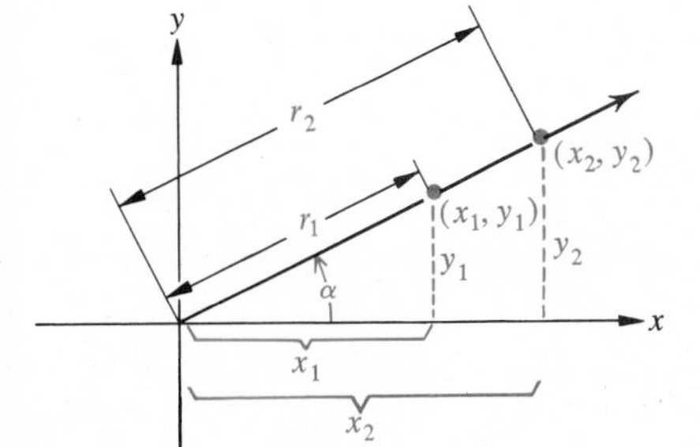

Figure 1.8

$$\frac{y_1}{r_1} = \frac{y_2}{r_2}, \quad \frac{x_1}{r_1} = \frac{x_2}{r_2}, \quad \text{and} \quad \frac{y_1}{x_1} = \frac{y_2}{x_2}.$$

Similar equalities for the reciprocals of these ratios also hold. Thus, the six trigonometric ratios are determined only by the position of the terminal side of an angle and not by the point selected on the terminal side.

Trigonometric Functions

Notice that each equation in Definition 1.2 defines a function because any angle α is paired with only one number (see Appendix A.3). These functions are called **trigonometric functions.** If an element in the range of any one of the six trigonometric functions is denoted by **trig α,**

$$\text{trig } \alpha \in \{\sin \alpha, \cos \alpha, \tan \alpha, \csc \alpha, \sec \alpha, \cot \alpha\},$$

the functions can be specified by $\{(\alpha, \text{trig } \alpha)\}$. The **domain** of each function is a set of angles determined by Definition 1.2. Note that, since $r \neq 0$, $\sin \alpha$ and $\cos \alpha$ are defined for all angles. However, $\tan \alpha$

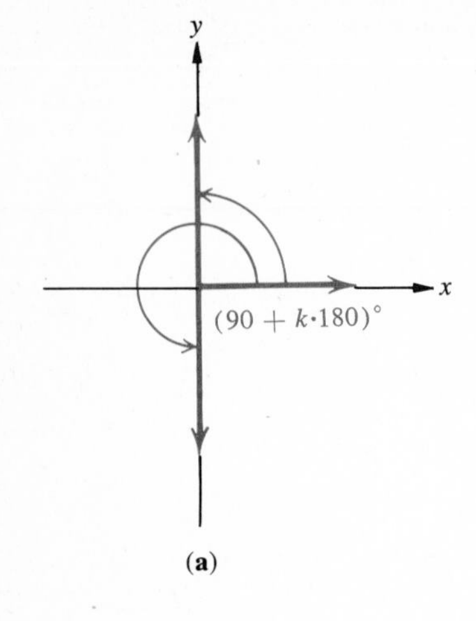

(a)

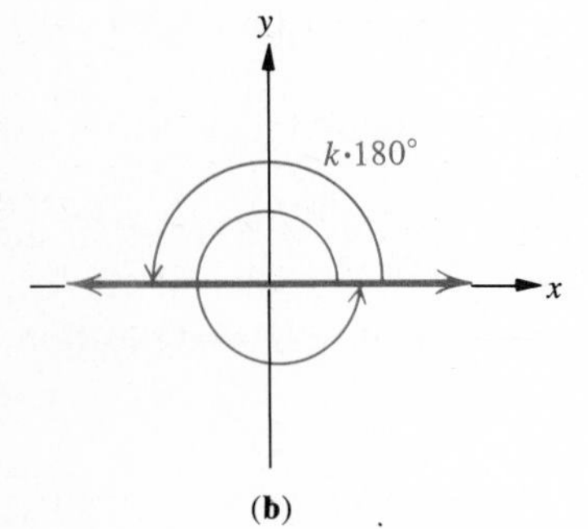

(b)

Figure 1.9

and sec α are undefined if $x = 0$. From this it follows that the domains of the tangent and the secant functions are the set of all angles α such that

$$\alpha \neq (90 + k \cdot 180)^\circ, \quad k \in J.$$

Furthermore, because cot α and csc α are undefined if $y = 0$, the domains of the cotangent and the cosecant functions are the set of all angles α such that

$$\alpha \neq (0 + k \cdot 180)^\circ = k \cdot 180^\circ, \quad k \in J.$$

Hence, if α is in standard position and the terminal side is on the y-axis (see Figure 1.9a), then tan α and sec α are undefined; if the terminal side is on the x-axis (see Figure 1.9b), then cot α and csc α are undefined.

Since $r = \sqrt{x^2 + y^2}$, then $r \geq |x|$ and $r \geq |y|$ for *all* (x, y) on the terminal side of an angle in standard position. Thus, from Definition 1.2, the *ranges* of the trigonometric functions are given by the following:

$$-1 \leq \sin \alpha \leq 1, \qquad \csc \alpha \leq -1 \quad \text{or} \quad \csc \alpha \geq 1,$$

$$-1 \leq \cos \alpha \leq 1, \qquad \sec \alpha \leq -1 \quad \text{or} \quad \sec \alpha \geq 1,$$

$$-\infty < \tan \alpha < \infty, \qquad -\infty < \cot \alpha < \infty.$$

Signs of Function Values

The trigonometric ratios sin α, cos α, etc., that are elements in the ranges of the functions are called **function values.**

Since, for an angle α in standard position, $r = \sqrt{x^2 + y^2}$ is always positive, the *sign* of a trigonometric ratio depends on the values of x and y, and hence on the quadrant in which the terminal side of the angle lies. The signs of function values of nonquadrantal angles are shown in Table 1.1. The signs of function values for quadrantal angles are considered in Section 1.3.

EXERCISE SET 1.2

A

Find each of the six trigonometric function values of α if the terminal side of α (in standard position) contains the given point. Sketch each angle.

TABLE 1.1

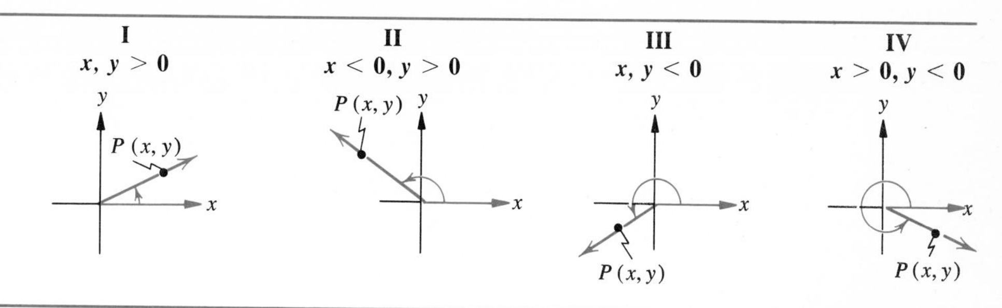

Quadrant	I $x, y > 0$	II $x < 0, y > 0$	III $x, y < 0$	IV $x > 0, y < 0$
$\sin\alpha$ or $\csc\alpha$	+	+	−	−
$\cos\alpha$ or $\sec\alpha$	+	−	−	+
$\tan\alpha$ or $\cot\alpha$	+	−	+	−

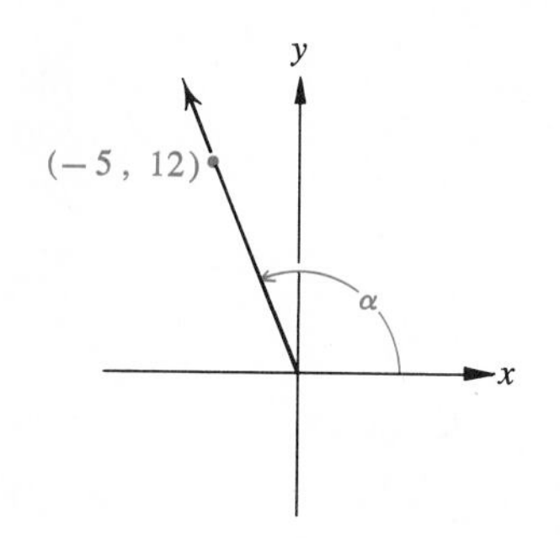

Example $(-5, 12)$

Solution By Definition 1.2,

$$r = \sqrt{x^2 + y^2} = \sqrt{(-5^2 + (12)^2} = 13.$$

Hence,

$$\sin\alpha = \frac{y}{r} = \frac{12}{13}, \qquad \csc\alpha = \frac{r}{y} = \frac{13}{12},$$

$$\cos\alpha = \frac{x}{r} = \frac{-5}{13}, \qquad \sec\alpha = \frac{r}{x} = \frac{13}{-5} = \frac{-13}{5},$$

$$\tan\alpha = \frac{y}{x} = \frac{12}{-5} = \frac{-12}{5}, \qquad \cot\alpha = \frac{x}{y} = \frac{-5}{12}.$$

1. $(4, 3)$ **2.** $(12, 5)$ **3.** $(8, -15)$ **4.** $(2, -1)$

5. $(-1, 3)$ **6.** $(5, 2)$ **7.** $(-2, 2)$ **8.** $(-\sqrt{3}, 1)$

9. $(-2, -2\sqrt{3})$ **10.** $(0, 2)$ **11.** $(0, -4)$ **12.** $(-2, 0)$

Determine the quadrant in which the terminal side of each angle α lies.

Example $\sin\alpha < 0$ and $\tan\alpha > 0$

Solution For $\sin\alpha < 0$, the terminal side of α lies in Quadrant III or IV, and for $\tan\alpha > 0$, the terminal side of α lies in Quadrant I or III. Therefore, the terminal side of α must lie in Quadrant III, the intersection of {III, IV} and {I, III}.

13. $\sin \alpha < 0$ and $\cos \alpha > 0$ **14.** $\sin \alpha < 0$ and $\tan \alpha < 0$

15. $\sin \alpha > 0$ and $\sec \alpha > 0$ **16.** $\sin \alpha > 0$ and $\cot \alpha < 0$

17. $\cos \alpha < 0$ and $\tan \alpha > 0$ **18.** $\cos \alpha < 0$ and $\csc \alpha < 0$

19. $\cos \alpha < 0$ and $\cot \alpha < 0$ **20.** $\tan \alpha > 0$ and $\csc \alpha > 0$

Given one function value of an angle α, determine the other five trigonometric function values of α. In Exercises 21–32, $0° \leq \alpha \leq 90°$.

Example $\sin \alpha = \frac{3}{4}$

Solution A sketch of a right triangle as shown in the figure is helpful. The acute angle α and $\overline{QP}$ and $\overline{OP}$ are sketched so that the lengths of these sides are approximately in the ratio of 3 to 4. From the Pythagorean theorem,

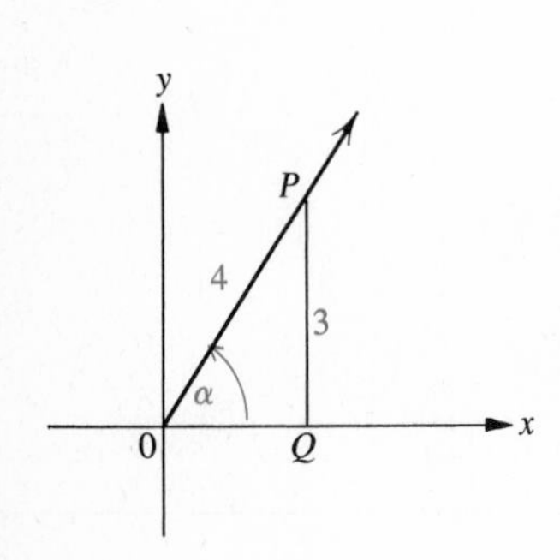

$$(OQ)^2 + 3^2 = 4^2,$$
$$(OQ)^2 = 16 - 9 = 7,$$
$$OQ = \pm\sqrt{7}.$$

Since α is in Quadrant I, we use the positive number, $\sqrt{7}$. The other five ratios can now be specified directly from Definition 1.2. Thus,

$$\sin \alpha = \frac{3}{4}, \qquad \csc \alpha = \frac{4}{3},$$
$$\cos \alpha = \frac{\sqrt{7}}{4}, \qquad \sec \alpha = \frac{4}{\sqrt{7}},$$
$$\tan \alpha = \frac{3}{\sqrt{7}}, \qquad \cot \alpha = \frac{\sqrt{7}}{3}.$$

21. $\tan \alpha = \frac{4}{3}$ **22.** $\cot \alpha = \frac{5}{12}$ **23.** $\sin \alpha = \frac{2}{5}$ **24.** $\cos \alpha = \frac{1}{4}$

25. $\sec \alpha = \frac{7}{3}$ **26.** $\csc \alpha = 2$ **27.** $\cos \alpha = \frac{1}{2}$ **28.** $\sin \alpha = \frac{2}{3}$

29. $\csc \alpha = 4$ **30.** $\sec \alpha = \frac{5}{4}$ **31.** $\cot \alpha = 1$ **32.** $\tan \alpha = 2$

Given one function value and the sign of another, determine the other five trigonometric values of α, $0° \leq \alpha < 360°$.

Example $\sin \alpha = \frac{3}{5}$, if $\cos \alpha < 0$

Solution From Table 1.1, if $\sin \alpha > 0$ and $\cos \alpha < 0$, then α must be in Quadrant II. From the Pythagorean theorem,

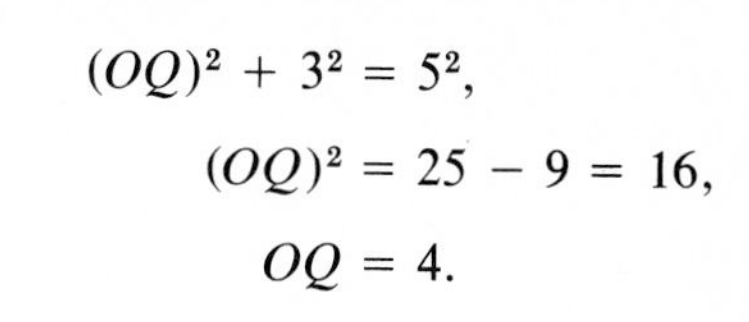

$$(OQ)^2 + 3^2 = 5^2,$$
$$(OQ)^2 = 25 - 9 = 16,$$
$$OQ = 4.$$

Hence the coordinates of point P are $(-4, 3)$, and the six ratios are

$$\sin \alpha = \frac{3}{5}, \qquad \csc \alpha = \frac{5}{3},$$

$$\cos \alpha = -\frac{4}{5}, \qquad \sec \alpha = -\frac{5}{4},$$

$$\tan \alpha = -\frac{3}{4}, \qquad \cot \alpha = -\frac{4}{3}.$$

33. $\tan \alpha = -\frac{5}{12}$, if $\sin \alpha > 0$

34. $\cos \alpha = -\frac{3}{4}$, if $\sin \alpha > 0$

35. $\sin \alpha = -\frac{2}{3}$, if $\tan \alpha > 0$

36. $\tan \alpha = \frac{8}{15}$, if $\cos \alpha > 0$

37. $\cos \alpha = \frac{4}{5}$, if $\cot \alpha < 0$

38. $\sin \alpha = -\frac{1}{3}$, if $\tan \alpha < 0$

39. $\tan \alpha = -\frac{1}{2}$, if $\cos \alpha < 0$

40. $\sin \alpha = -\frac{3}{5}$, if $\cot \alpha > 0$

41. $\cot \alpha = -\frac{2}{3}$, if $\sin \alpha < 0$

42. $\cos \alpha = -\frac{2}{5}$, if $\tan \alpha < 0$

B

Give the domain and range of each function.

43. $\{(\alpha, \sin \alpha)\}$

44. $\{(\alpha, \cos \alpha)\}$

45. $\{(\alpha, \tan \alpha)\}$

46. $\{(\alpha, \csc \alpha)\}$

47. $\{(\alpha, \sec \alpha)\}$

48. $\{(\alpha, \cot \alpha)\}$

49. Prove that if (x_1, y_1) and (x_2, y_2) correspond to any two points (excluding the origin) on the terminal side of an angle α in the first quadrant, then for $r_1 = \sqrt{x_1^2 + y_1^2}$ and $r_2 = \sqrt{x_2^2 + y_2^2}$,

(*continued*)

$$\frac{x_1}{y_1} = \frac{x_2}{y_2}, \quad \frac{x_1}{r_1} = \frac{x_2}{r_2}, \quad \text{and} \quad \frac{y_1}{r_1} = \frac{y_2}{r_2}.$$

50. Show that the statement in Problem 49 is valid for an angle α in Quadrant II.

1.3 Function Values for Special Angles

Function values for each of the six trigonometric functions for 30°, 45°, and 60° angles can be obtained from geometric considerations of right triangles.

Function Values for 30°, 45°, and 60°

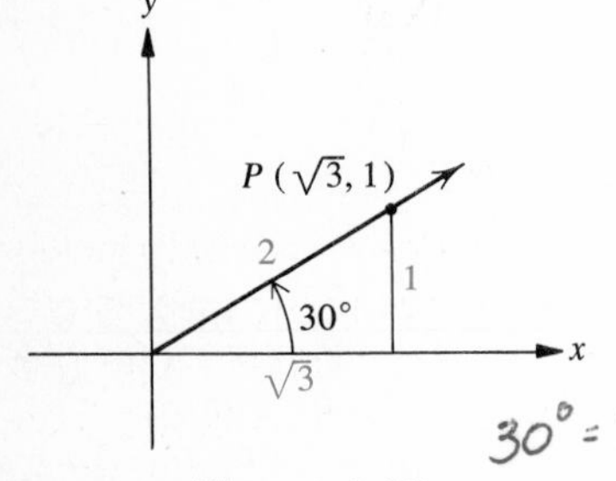

Figure 1.10

First let us recall from geometry (see Appendix A.8) that the lengths of the sides of a 30°–60° right triangle are proportional to the numbers 1, $\sqrt{3}$, and 2, as shown in Figure 1.10. Since any point on the terminal side of the 30° angle may be chosen to determine the trigonometric ratios, it is convenient to select $P\ (\sqrt{3}, 1)$. From Definition 1.2,

$$\sin 30^\circ = \frac{1}{2}, \qquad \csc 30^\circ = \frac{2}{1} = 2,$$

$$\cos 30^\circ = \frac{\sqrt{3}}{2}, \qquad \sec 30^\circ = \frac{2}{\sqrt{3}},$$

$$\tan 30^\circ = \frac{1}{\sqrt{3}}, \qquad \cot 30^\circ = \frac{\sqrt{3}}{1} = \sqrt{3}.$$

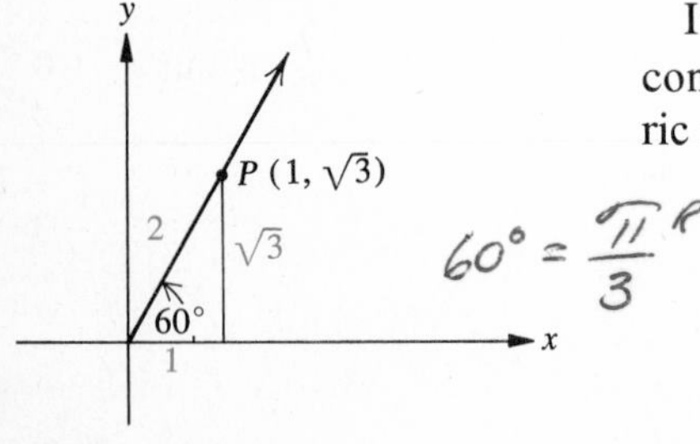

Figure 1.11

In Figure 1.11, a 60° angle is shown in standard position. Here it is convenient to choose the point $P\ (1, \sqrt{3})$ to determine the trigonometric ratios associated with this angle. From Definition 1.2,

$$\sin 60^\circ = \frac{\sqrt{3}}{2}, \qquad \csc 60^\circ = \frac{2}{\sqrt{3}},$$

$$\cos 60^\circ = \frac{1}{2}, \qquad \sec 60^\circ = \frac{2}{1} = 2,$$

$$\tan 60^\circ = \frac{\sqrt{3}}{1} = \sqrt{3}, \qquad \cot 60^\circ = \frac{1}{\sqrt{3}}.$$

Let us also recall from geometry (see Appendix A.8) that the lengths of the sides of a 45°–45° right triangle are proportional to the numbers

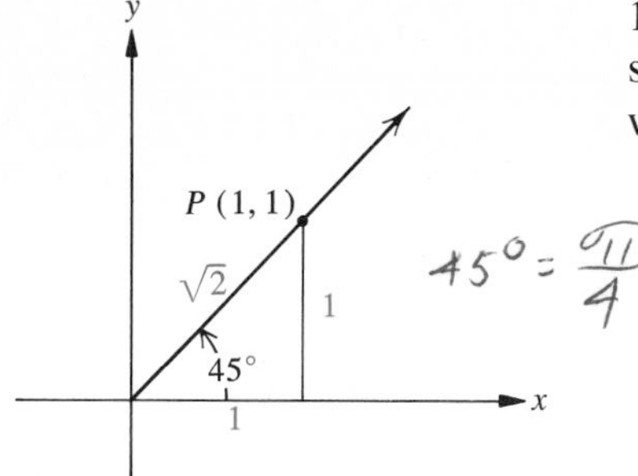

Figure 1.12

1, 1, and $\sqrt{2}$, as shown in Figure 1.12. In this case it is convenient to select the point $P(1, 1)$ to determine the trigonometric ratios associated with a 45° angle. From Definition 1.2,

$$\sin 45° = \frac{1}{\sqrt{2}}, \qquad \csc 45° = \frac{\sqrt{2}}{1} = \sqrt{2},$$

$$\cos 45° = \frac{1}{\sqrt{2}}, \qquad \sec 45° = \frac{\sqrt{2}}{1} = \sqrt{2},$$

$$\tan 45° = 1, \qquad \cot 45° = 1.$$

Note that some of the function values above for 30°, 45°, and 60° are shown in fraction form in which the denominator is a radical expression. Such forms are often more convenient to use than fractions in which the denominators have been rationalized. In our work we will in general use the forms shown.

Function Values for Quadrantal Angles

As we have noted, the terminal side of a quadrantal angle coincides with a coordinate axis, and one of the coordinates of any point on the terminal side is zero. Hence, some of the trigonometric ratios for quadrantal angles are not defined. For convenience in finding the trigonometric ratios that are defined, we select the point on the terminal side with the nonzero coordinate equal to 1 or -1, as shown in Figure 1.13. For 0°, 90°, 180°, and 270°,

$$r = \sqrt{x^2 + y^2} = \sqrt{1^2} = 1.$$

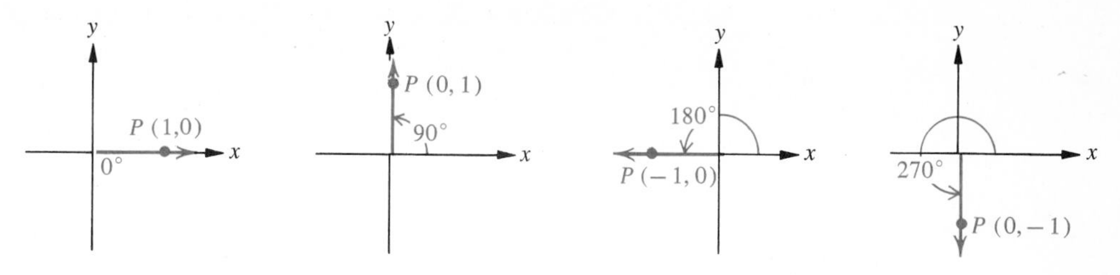

Figure 1.13

The values of the trigonometric ratios that are defined for the angles shown can be determined directly from Definition 1.2.

The function values (trigonometric ratios) for the special angles that have been discussed above are summarized in Table 1.2 on page 16.

TABLE 1.2

α°	α^R	sin α	cos α	tan α	csc α	sec α	cot α
0°	0^R	0	1	0	undef.	1	undef.
30°	$\frac{\pi}{6}^R$	$\frac{1}{2}$	$\frac{\sqrt{3}}{2}$	$\frac{1}{\sqrt{3}}$	2	$\frac{2}{\sqrt{3}}$	$\sqrt{3}$
45°	$\frac{\pi}{4}^R$	$\frac{1}{\sqrt{2}}$	$\frac{1}{\sqrt{2}}$	1	$\sqrt{2}$	$\sqrt{2}$	1
60°	$\frac{\pi}{3}^R$	$\frac{\sqrt{3}}{2}$	$\frac{1}{2}$	$\sqrt{3}$	$\frac{2}{\sqrt{3}}$	2	$\frac{1}{\sqrt{3}}$
90°	$\frac{\pi}{2}^R$	1	0	undef.	1	undef.	0
180°	π^R	0	−1	0	undef.	−1	undef.
270°	$\frac{3\pi}{2}^R$	−1	0	undef.	−1	undef.	0

You may want to refer directly to the entries in the table, or you may want to construct a figure (as we did to obtain these values) each time you want a particular trigonometric ratio for one of the special angles.

Example Find the value of (csc 30°)(cos 30°)(tan 30°)².

Solution From Table 1.2, or from the construction of appropriate triangles,

$$(\csc 30^\circ)(\cos 30^\circ)(\tan 30^\circ)^2 = (2)\left(\frac{\sqrt{3}}{2}\right)\left(\frac{1}{\sqrt{3}}\right)^2 = \frac{\sqrt{3}}{3}.$$

Calculators with trigonometric function value capability make it possible to obtain the table entries quite readily in decimal form. However, the definition of the trigonometric ratios will be more meaningful to you if you obtain these values in fractional form directly from the triangles in Figures 1.10, 1.11, and 1.12, using radicals rather than decimal approximations. *In our work in this text we shall express the trigonometric ratios for these special angles in such form.*

A symbol such as (tan 30°)², as in the example above, is usually written in more concise form as tan² 30°. Thus,

$$\sin^2 \alpha \text{ means } (\sin \alpha)^2,$$

$$\cos^2 \alpha \text{ means } (\cos \alpha)^2,$$

$$\vdots$$

Example Show that $\sin^2 \alpha + \cos^2 \alpha = 1$ for $\alpha = 30°$.

Solution From Table 1.2, or from an appropriate triangle, we have $\sin 30° = 1/2$ and $\cos 30° = \sqrt{3}/2$. Hence,

$$\sin^2 30° + \cos^2 30° = \left(\frac{1}{2}\right)^2 + \left(\frac{\sqrt{3}}{2}\right)^2$$
$$= \frac{1}{4} + \frac{3}{4} = 1.$$

Some Basic Identities

The equation $\sin^2 \alpha + \cos^2 x = 1$ in the above example is in fact an identity and is true for all replacements of α. This equation follows directly from Definition 1.2. For any point $P(x, y)$ on the terminal side of an angle α in standard position,

$$y^2 + x^2 = r^2.$$

Multiplying each member by $1/r^2$ $(r \neq 0)$ yields

$$\left(\frac{y}{r}\right)^2 + \left(\frac{x}{r}\right)^2 = 1,$$

from which we obtain

$$\mathbf{\sin^2 \alpha + \cos^2 \alpha = 1.} \tag{1}$$

Several other useful identities also follow directly from Definition 1.2. Since $\sin \alpha = y/r$ and $\cos \alpha = x/r$ $(r \neq 0)$,

$$\frac{\sin \alpha}{\cos \alpha} = \frac{\frac{y}{r}}{\frac{x}{r}} = \frac{y}{x} \quad (x \neq 0).$$

By Definition 1.2, $\tan \alpha = y/x$. Hence,

$$\mathbf{\tan \alpha = \frac{\sin \alpha}{\cos \alpha}} \tag{2}$$

for those values of α for which $\cos \alpha \neq 0$. Because $\cot \alpha = x/y$ and

$$\frac{\cos \alpha}{\sin \alpha} = \frac{\frac{x}{r}}{\frac{y}{r}} = \frac{x}{y} \quad (y \neq 0),$$

$$\boldsymbol{\cot \alpha = \frac{\cos \alpha}{\sin \alpha},} \tag{3}$$

for those values of α for which $\sin \alpha \neq 0$.

EXERCISE SET 1.3

A

Find each function value, if it exists, by making an appropriate sketch and reading the ratio directly from the figure. Do not use Table 1.2.

Example $\cos 60°$

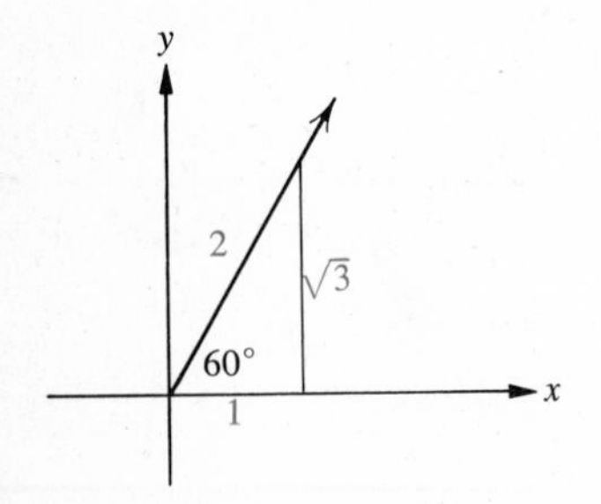

Solution A 30°–60° right triangle is sketched with the 60° angle in standard position and with sides of length, 1, $\sqrt{3}$, and 2. From Definition 1.2,

$$\cos 60° = \frac{x}{r} = \frac{1}{2}.$$

1. $\sin 60°$	**2.** $\tan 30°$	**3.** $\cos 45°$	**4.** $\csc 60°$
5. $\cos 30°$	**6.** $\cot 45°$	**7.** $\cot 60°$	**8.** $\sin 45°$
9. $\csc 45°$	**10.** $\cot 30°$	**11.** $\sec 30°$	**12.** $\sec 60°$
13. $\sin 90°$	**14.** $\sec 180°$	**15.** $\cos 270°$	**16.** $\tan 0°$
17. $\csc 180°$	**18.** $\tan 90°$	**19.** $\cos 180°$	**20.** $\sin 270°$
21. $\sin 0°$	**22.** $\csc 0°$	**23.** $\sec 90°$	**24.** $\cos 0°$

Compute each expression if the expression represents a real number.

Example $\sin 45° + (\cos^2 30°)(\tan 60°)$

Solution The function values can be obtained from Table 1.2 or by sketching the appropriate triangles. By using either method, we find

$$\sin 45° = \frac{1}{\sqrt{2}}, \quad \cos^2 30° = \left(\frac{\sqrt{3}}{2}\right)^2 = \frac{3}{4}, \quad \text{and} \quad \tan 60° = \sqrt{3}.$$

Hence,

$$\sin 45° + (\cos^2 30°)(\tan 60°) = \frac{1}{\sqrt{2}} + \left(\frac{3}{4}\right)(\sqrt{3})$$

$$= \frac{1(4)}{\sqrt{2}(4)} + \frac{3\sqrt{3}(\sqrt{2})}{4(\sqrt{2})}$$

$$= \frac{4 + 3\sqrt{6}}{4\sqrt{2}}.$$

25. $\cos^2 30°$

26. $\tan^2 30°$

27. $\sin^2 45°$

28. $\csc^2 270°$

29. $(\cos 180°)(\tan^2 60°)$

30. $(\sin 90°)(\cos^2 45°)$

31. $(\sin^2 30°)(\cot^2 60°)$

32. $(\tan^2 30°)(\sec^2 0°)$

33. $\sec 60° - \cos^2 45°$

34. $\csc 30° - \tan^2 30°$

35. $\tan^2 45° - \cot^2 45°$

36. $\tan^2 60° - \cot^2 30°$

37. $(\sin 45°)(\cos 30°)(\tan^2 60°)$

38. $(\sec^2 180°)(\sin 270°)(\tan 0°)$

39. $(\cos 0°)(\sin 90°) + \tan 45°$

40. $\cot 45° + (\tan 60°)(\cot 60°)$

41. $\cos^2 60° - (\sec 270°)(\tan 45°)$

42. $2(\csc^2 180°)(\sin^2 45°) + \tan 45°$

43. $\dfrac{1 + \tan^2 30°}{\sec^2 30°}$

44. $\dfrac{1 - \cos^2 45°}{\sin^2 45°}$

45. $\dfrac{(\sin^2 30°)(\cos 180°)}{\sin 30°}$

46. $\dfrac{\tan 60°}{(\sin 60°)(\tan^2 60°)}$

47. $\dfrac{(\sin 90°)(\cos 45°)}{\cos 90°}$

48. $\dfrac{\tan^2 30° + \sec^2 45°}{\cos 270°}$

Use the identities on pages 9 and 17 to write each expression in terms of $\sin \alpha$ *and/or* $\cos \alpha$.

Examples a. $\tan \alpha \cdot \csc \alpha$ b. $\dfrac{\cot^2 \alpha}{\cos \alpha}$

Solutions

a. Since $\tan \alpha = \sin \alpha / \cos \alpha$ and $\csc \alpha = 1/\sin \alpha$, we have that

$$\tan \alpha \cdot \csc \alpha = \frac{\sin \alpha}{\cos \alpha} \cdot \frac{1}{\sin \alpha} = \frac{1}{\cos \alpha}.$$

(continued)

b. Since $\cot \alpha = \cos \alpha / \sin \alpha$, we have that

$$\frac{\cot^2 \alpha}{\cos \alpha} = \frac{\left(\frac{\cos \alpha}{\sin \alpha}\right)^2}{\cos \alpha} = \frac{\cos^2 \alpha}{\sin^2 \alpha} \cdot \frac{1}{\cos \alpha} = \frac{\cos \alpha}{\sin^2 \alpha}.$$

49. $\sec \alpha \cdot \cot \alpha$

50. $\csc \alpha \cdot \sec \alpha$

51. $\tan \alpha \cdot \csc^2 \alpha$

52. $\cot^2 \alpha \cdot \sec \alpha$

53. $\sin^2 \alpha \cdot \cot^2 \alpha$

54. $\tan^2 \alpha \cdot \cos^2 \alpha$

55. $\dfrac{\cot^2 \alpha}{\sec \alpha}$

56. $\dfrac{\tan^2 \alpha}{\csc \alpha}$

57. $\dfrac{\tan^2 \alpha + 1}{\sec \alpha}$

58. $\dfrac{\cot^2 \alpha + 1}{\csc \alpha}$

59. $\dfrac{\csc^2 \alpha - 1}{\cot \alpha}$

60. $\dfrac{\sec^2 \alpha - 1}{\tan \alpha}$

61. For which angles are the tangent function values undefined? For which are they equal to 0?

62. For which angles are the cotangent function values undefined? For which are they equal to 0?

63. For which angles are the secant function values undefined? Are there any angles for which the secant function value is 0?

64. For which angles are the cosecant function values undefined? Are there any angles for which the cosecant function value is 0?

B

65. Given that $\sin \alpha = y$, find expressions in terms of y for the other five trigonometric ratios. *Hint:* Consider $\sin \alpha = y/1$.

66. Given that $\tan \alpha = y$, find expressions in terms of y for the other five trigonometric ratios.

67. Show that $1 + \tan^2 \alpha = \sec^2 \alpha$ is an identity.

68. Show that $1 + \cot^2 \alpha = \csc^2 \alpha$ is an identity.

1.4 Function Values for Acute Angles

In Section 1.3 we obtained function values for the six trigonometric functions for selected angles in their respective domains. In this section we shall find ways to approximate function values for other acute angles by using specially prepared tables or by using a calculator. We shall first consider the use of tables.

Using Tables

Table II in Appendix C lists rational number approximations for function values of selected angles over the interval $0° \leq \alpha \leq 90°$. The table is graduated in intervals of six minutes (6′), where one minute is equal to one-sixtieth of a degree, and in intervals of tenths of a degree. Observe that the table reads from top to bottom in the left margin with function values identified at the tops of the columns. It reads from bottom to top in the right margin with function values identified at the bottoms of the columns. Function values are, in general, irrational numbers, and the four-place decimals shown in the table are mostly rational number approximations.

Examples Find approximations for

a. $\sin 23° 12'$ b. $\cot 57.7°$

Solutions From Table II,

a. $\sin 23° 12' \approx 0.3939$ b. $\cot 57.7° \approx 0.6322$

Table III in Appendix C lists function values for angles whose measures are given in radian units in intervals of 0.01^R.

Examples Find approximations for

a. $\cos 0.53^R$ b. $\tan 1.40^R$

Solutions From Table III,

a. $\cos 0.53^R \approx 0.8628$ b. $\tan 1.40^R \approx 5.798$

Notice that in both tables the function values $\sin \alpha$, $\tan \alpha$, and $\sec \alpha$ increase, while $\cos \alpha$, $\cot \alpha$, and $\csc \alpha$ decrease, for increasing measures of α where $0° \leq \alpha \leq 90°$.

Tables II and III consist of function values for angles with the specified measures. However, Definition 1.2 implies that, with certain restrictions, function values exist for every $\alpha \in A$.* To approximate function values for angle measures not listed in the tables, we can read the entry in the table for that angle measure nearest the angle measure with which we are concerned. Thus, from Table III,

$$\sin 0.714^R \approx \sin 0.71^R \approx 0.6518.$$

* The letter A names the set of all angles.

If we want a closer approximation, we can either refer to tables with greater detail or we can use a method called **linear interpolation,** which is considered in Appendix B. However, for our purpose *we shall use the nearest reading in Table II or Table III as an adequate approximation for function values.*

Although at this time you may not have the mathematical background to understand completely how the function values for Table II or Table III are obtained, it may be of interest to you to become acquainted with the kinds of formulas, developed in calculus, used to produce these tables. Two such formulas are introduced in the exercise set.

Using Calculators

Finding the function values for trigonometric functions by using a calculator is a simple process. However, because the steps to be taken vary somewhat for different types of calculators, we will not consider specific procedures. The instruction booklet for your particular calculator can provide this information.

Calculators vary in the number of digits that are shown in the display; however, we shall round off all readings so that they will be consistent with values obtained from the tables.

Example Set Degree/Radian switch on RAD:

$$\sin 0.714^{R} \approx 0.65486.$$

Rounding off the result to four decimal places gives

$$\sin 0.714^{R} \approx 0.6549.$$

Example Set Degree/Radian switch on DEG:

$$\sin 21^{\circ}\, 42' = \sin 21.7^{\circ} \approx 0.36974.$$

Rounding off the result to four decimal places gives

$$\sin 21^{\circ}\, 42' \approx 0.3697.$$

Note that in the above example it was necessary first to express the angle $21^{\circ}\, 42'$ in decimal form as 21.7° before using the calculator.

Values for $\sin x$, $\cos x$, and $\tan x$ can be obtained directly on most scientific calculators. Values for $\sec x$, $\csc x$, and $\cot x$ can be obtained by using the reciprocal relationships

$$\sec x = \frac{1}{\cos x}, \quad \csc x = \frac{1}{\sin x}, \quad \text{and} \quad \cot x = \frac{1}{\tan x}$$

introduced in Section 1.2.

Example Set Degree/Radian switch on DEG:

$$\begin{aligned} \sec 14^\circ\ 18' &= \sec 14.3^\circ \\ &= \frac{1}{\cos 14.3^\circ} \approx 1.03197. \end{aligned}$$

Rounding off to the nearest four places gives

$$\sec 14^\circ\ 18' \approx 1.0320.$$

Function values obtained using a calculator may sometimes differ from the values obtained using the tables.

Finding Acute Angles

Given a positive element in the range of any of the six trigonometric functions, we can use Table 1.2, Table II, Table III, or a calculator to find the corresponding element in the domain. For example, if

$$\sin \alpha = \sqrt{3}/2,$$

we can obtain $\alpha = 60^\circ$ or $\alpha = (\pi/3)^R$ from Table 1.2. If

$$\sin \alpha = 0.5564,$$

we can obtain $\alpha = 33.8^\circ$ from Table II and $\alpha = 0.59^R$ from Table III. A calculator can also be used to find these angles. However, we shall first consider a special notation that is usually printed on calculator keys that are used for this purpose.

Inverse Notation

Special notations called **inverse notations** are often used to express the element α in the domain of a trigonometric function explicitly in terms of the corresponding element in its range. For example,

$$\sin \alpha = 0.5388$$

can be expressed equivalently as

$$\alpha = \operatorname{Sin}^{-1} 0.5388 \qquad \text{or} \qquad \alpha = \operatorname{Arcsin} 0.5388,$$

which are both read as "α equals Arcsin 0.5388." The phrase "inverse sine of 0.5388" is also used in this context. In each case we can simply view each of the expressions

$$\text{Sin}^{-1}\,0.5388 \qquad \text{and} \qquad \text{Arcsin}\,0.5388$$

as the *angle* whose sine is 0.5388. Note that a capital letter is used for the first letter in each symbol.

More generally, we have the following:

Definition 1.3 *For $y > 0$, the inverse notation* **Trig^{-1} y**, *or* **Arctrig y**, *names an acute angle α such that* trig $\alpha = y$.*

Example Express $\tan \alpha = 1.4882$ using inverse notation.

Solution $\alpha = \text{Tan}^{-1}\,1.4882$, or
$\alpha = \text{Arctan}\,1.4882$, or
$\alpha =$ Inverse tangent of 1.4882.

Example Express $\alpha = \text{Cos}^{-1}\,0.9461$ without using inverse notation and find the value of the acute angle.

Solution $\cos \alpha = 0.9461$. Using Table II or a calculator (set on DEG), we find that

$$\alpha \approx 18.9°.$$

Using Table III or a calculator (set on RAD), we find that

$$\alpha \approx 0.33^{R}.$$

EXERCISE SET 1.4

A

Use Table II or a calculator to find an approximation for each function value.

Examples a. $\cos 63.7°$ b. $\cot 17°\,6'$

Solutions (Using Table II)

a. $\cos 63.7° \approx 0.4431$ b. $\cot 17°\,6' \approx 3.2506$

* A more general definition for Trig^{-1} y is given in Section 3.6 in conjunction with a discussion of inverse functions.

Alternate Solutions (Using a calculator)

a. $\cos 63.7° \approx 0.4431$

b. $\cot 17° 6' = \cot 17.1°$

$$= \frac{1}{\tan 17.1°} \approx 3.2506$$

1. $\sin 16.4°$ **2.** $\cos 21.5°$ **3.** $\tan 69.2°$ **4.** $\sin 82.6°$

5. $\cos 18° 48'$ **6.** $\sin 27° 6'$ **7.** $\tan 6° 12'$ **8.** $\cos 63° 30'$

9. $\cot 14.3°$ **10.** $\sec 79.6°$ **11.** $\csc 12° 24'$ **12.** $\cot 62° 12'$

Use Table III or a calculator to find an approximation for each function value.

Examples a. $\tan 0.43^R$ b. $\sec 1.07^R$

Solutions (Using Table III)

a. $\tan 0.43^R \approx 0.4586$ b. $\sec 1.07^R \approx 2.083$

Alternate Solutions (Using a calculator)

a. $\tan 0.43^R \approx 0.4586$ b. $\sec 1.07^R = \dfrac{1}{\cos 1.07^R} \approx 2.083$

13. $\sin 0.91^R$ **14.** $\cos 1.23^R$ **15.** $\tan 0.42^R$ **16.** $\sin 1.05^R$

17. $\sec 0.63^R$ **18.** $\csc 1.42^R$ **19.** $\cot 0.31^R$ **20.** $\sec 1.15^R$

Express each of the following using the inverse notation $\text{Trig}^{-1}(\alpha)$.

Examples a. $\cos \alpha = 0.4231$ b. $\csc \alpha = 2.4123$

Solutions a. $\alpha = \text{Cos}^{-1} 0.4231$ b. $\alpha = \text{Csc}^{-1} 2.4123$

21. $\sin \alpha = 0.2169$ **22.** $\tan \alpha = 3.1242$ **23.** $\cot \alpha = 2.0941$

24. $\cos \alpha = 0.6929$ **25.** $\sec \alpha = 2.2013$ **26.** $\csc \alpha = 1.0491$

Express each of the following without using inverse notation.

Examples a. $\alpha = \text{Sin}^{-1} 0.2314$ b. $\alpha = \text{Arctan}\ 1.4321$

Solutions a. $\sin \alpha = 0.2314$ b. $\tan \alpha = 1.4321$

27. $\alpha = \text{Cot}^{-1}\ 2.1431$

28. $\alpha = \text{Cos}^{-1}\ 0.2169$

29. $\alpha = \text{Arcsec}\ 1.4214$

30. $\alpha = \text{Arccsc}\ 2.0031$

31. $\alpha = \text{Arcsin}\ 0.4112$

32. $\alpha = \text{Arctan}\ 1.0341$

Find an approximation to the nearest tenth for the measure of angle α, *in degrees, where* $0° \le \alpha \le 90°$. *Use Table II or a calculator.*

Example $\sec \alpha = 1.1937$

Solution (Using Table II)

$$\alpha = \text{Sec}^{-1}\ 1.1937 \approx 33.1°.$$

Alternate Solution (Using a calculator)

$$\cos \alpha = \frac{1}{\sec \alpha} = \frac{1}{1.1937};$$

$$\alpha = \text{Cos}^{-1} \frac{1}{1.1937} \approx 33.1°.$$

33. $\sin \alpha = 0.3616$

34. $\cos \alpha = 0.9164$

35. $\tan \alpha = 0.4877$

36. $\sin \alpha = 0.8695$

37. $\sec \alpha = 1.7655$

38. $\cot \alpha = 1.6775$

39. $\csc \alpha = 1.7348$

40. $\sec \alpha = 1.5590$

Find an approximation for the measure of angle α, *in radians, where* $0^R \le \alpha \le 1.57^R$. *Use Table III or a calculator.*

Example $\cot \alpha = 1.560$

Solution (Using Table III)

$$\alpha = \text{Cot}^{-1}\ 1.560 \approx 0.57^R.$$

Alternate Solution (Using a calculator)

$$\tan \alpha = \frac{1}{\cot \alpha} = \frac{1}{1.560};$$

$$\alpha = \text{Tan}^{-1} \frac{1}{1.560} \approx 0.57^R.$$

41. $\sin \alpha = 0.2667$

42. $\cos \alpha = 0.9211$

43. $\tan \alpha = 0.5334$

44. $\tan \alpha = 1.592$

45. $\cot \alpha = 0.2669$

46. $\sec \alpha = 1.022$

47. $\csc \alpha = 1.355$

48. $\cot \alpha = 0.0910$

In Exercises 49–52 use the fact that the path of a light ray changes direction as it passes from one medium to another in accordance with the following law:

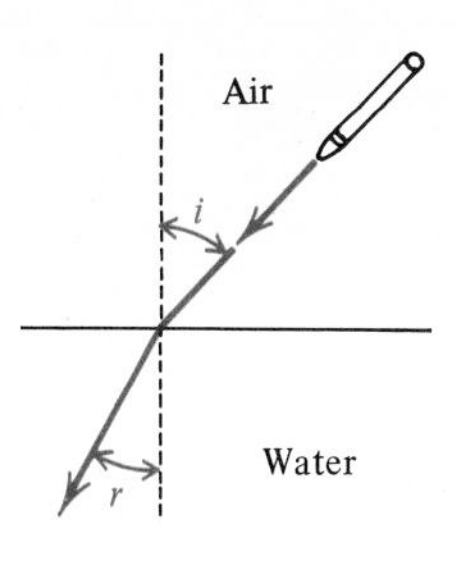

$$n_i \sin i = n_r \sin r,$$

where i and r are angles with the vertical of the incident (initial) and refracted ray, and n_i and n_r are the respective constants for specific media (see figure). In particular, $n_i = 1.00$ for air and $n_r = 1.33$ for water.

49. Find the angle of refraction if the angle of incidence is 28.1°.

50. Find the angle of incidence if the angle of refraction is 14.7°.

51. Find the angle of incidence if the angle of refraction is 42° 24′.

52. Find the angle of refraction if the angle of incidence is 36° 18′.

B

Use Table 1.2 to find the value of each of the following.

Example $\sin\left(\text{Cos}^{-1}\frac{1}{2}\right)$

Solution $\text{Cos}^{-1}\frac{1}{2} = 60°$; hence,

$$\sin\left(\text{Cos}^{-1}\frac{1}{2}\right) = \sin 60° = \frac{\sqrt{3}}{2}.$$

53. $\tan\left(\text{Sin}^{-1}\frac{1}{\sqrt{2}}\right)$ **54.** $\cos(\text{Tan}^{-1}\sqrt{3})$ **55.** $\sin(\text{Tan}^{-1} 1)$

56. $\tan\left(\text{Cos}^{-1}\frac{\sqrt{3}}{2}\right)$ **57.** $\cos\left(\text{Cos}^{-1}\frac{1}{\sqrt{2}}\right)$ **58.** $\sin\left(\text{Sin}^{-1}\frac{1}{2}\right)$

In Exercises 59–64, use Table III to find the value of each expression.

59. $\sin(\text{Cos}^{-1} 0.9916)$ **60.** $\tan(\text{Sin}^{-1} 0.6594)$ **61.** $\cos(\text{Tan}^{-1} 1.210)$

62. $\cos(\text{Sin}^{-1} 0.9246)$ **63.** $\tan(\text{Cos}^{-1} 0.7038)$ **64.** $\sin(\text{Tan}^{-1} 12.35)$

65. Simplify $\cos(\text{Cos}^{-1} x)$, where x is a positive real number less than 1.

66. Simplify $\sin(\text{Sin}^{-1} x)$, where x is a positive real number less than 1.

In calculus it is shown that for all $x = \alpha^{R}$ and $n \in N$,

$$\cos x = 1 - \frac{x^2}{2!} + \frac{x^4}{4!} + \cdots + (-1)^{n-1}\frac{x^{2n-2}}{(2n-2)!} + \cdots, \qquad (1)$$

$$\sin x = x - \frac{x^3}{3!} + \frac{x^5}{5!} + \cdots + (-1)^{n-1}\frac{x^{2n-1}}{(2n-1)!} + \cdots, \qquad (2)$$

where

$$2! = 1 \cdot 2, \quad 3! = 1 \cdot 2 \cdot 3, \quad 4! = 1 \cdot 2 \cdot 3 \cdot 4, \quad \text{etc.}$$

The expression in the right member of each equation is called an **alternating infinite power series in** x. *If any finite number of terms in either of these power series is used to obtain an approximation for the function value, the error that exists is less than the absolute value of the next term in the series.*

Example Find $\cos 0.3^{R}$ correct to four decimal places.

Solution Replacing x in Equation (1) by 0.3 yields

$$\cos 0.3^{R} = 1 - \frac{(0.3)^2}{2!} + \frac{(0.3)^4}{4!} - \frac{(0.3)^6}{6!} + \cdots$$

$$= 1 - \frac{0.09}{2} + \frac{0.0081}{24} - \frac{0.000729}{720} + \cdots$$

$$= 1 - 0.045 + 0.00034 - 0.000001 + \cdots.$$

Since we wish accuracy to four decimal places, and since the error introduced by taking the first three terms is less than the absolute value of the fourth term (0.000001), then correct to at least four decimal places,

$$\cos 0.3^{R} \approx 1 - 0.045 + 0.00034 = 0.9553.$$

Note that this is the value for $\cos 0.3^{R}$ shown in Table III.

Use Equation (1) or (2) above to find each function value correct to four decimal places.

67. $\sin 0.01^{R}$ **68.** $\cos 0.50^{R}$ **69.** $\sin 1.00^{R}$

70. $\cos 0.02^{R}$ **71.** $\sin 0.60^{R}$ **72.** $\cos 1.10^{R}$

1.5 Function Values for Angles α, $0° < \alpha < 360°$

In Section 1.4, we obtained trigonometric function values for *acute* angles by using Tables II and III in Appendix C or by using a calculator. In this section we shall obtain function values for angles greater than 90° and less than 360° by using certain relationships from geometry and appropriate tables. We shall first introduce a notion that will be helpful in obtaining such function values.

Reference Angles

Definition 1.4 *The* **reference angle** *of any nonquadrantal angle α in standard position is the positive acute angle $\tilde{\alpha}$ whose sides are the terminal side of α and a ray of the x-axis.*

The reference angles for angles in each of the four quadrants are shown in Figure 1.14.

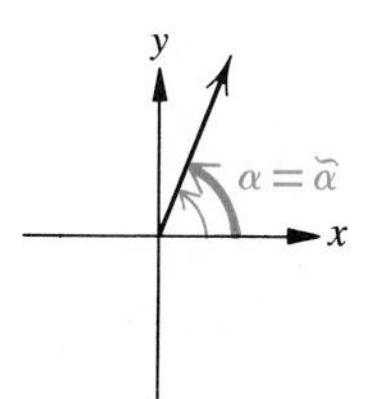

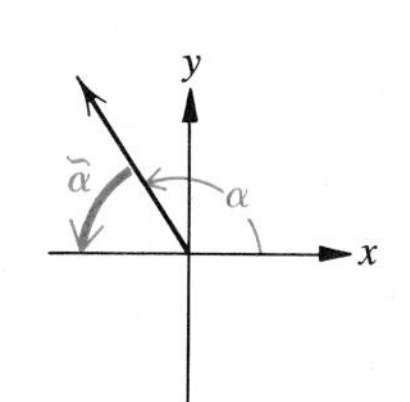

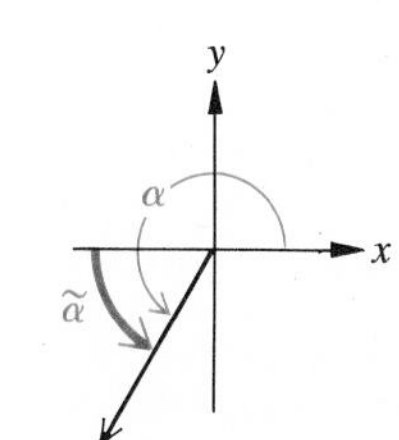

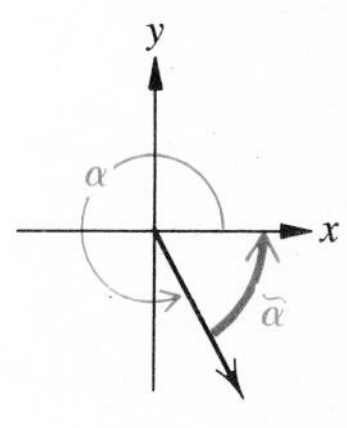

Figure 1.14

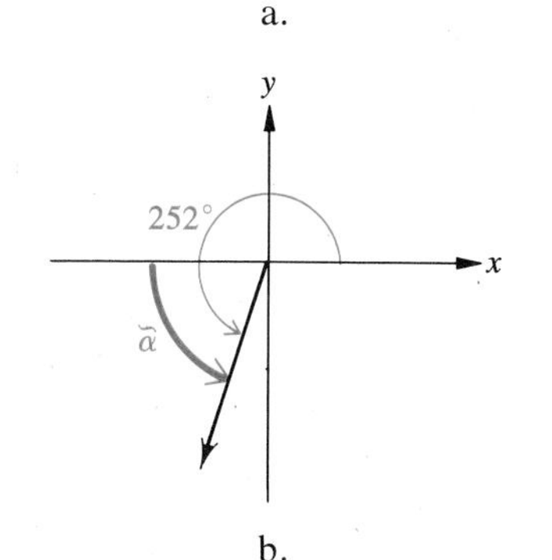

Examples Sketch $\tilde{\alpha}$ for each angle and find $\tilde{\alpha}°$.

a. $\alpha = 139°$

b. $\alpha = 252°$

Solutions

a. $\tilde{\alpha} = (180 - 139)°$

$= 41°$

b. $\tilde{\alpha} = (252 - 180)°$

$= 72°$

Using Reference Angles

Using geometric considerations, we can establish that *the absolute values of the trigonometric ratios of any angle α are equal, respectively, to the values of the corresponding trigonometric ratios for $\tilde{\alpha}$.*

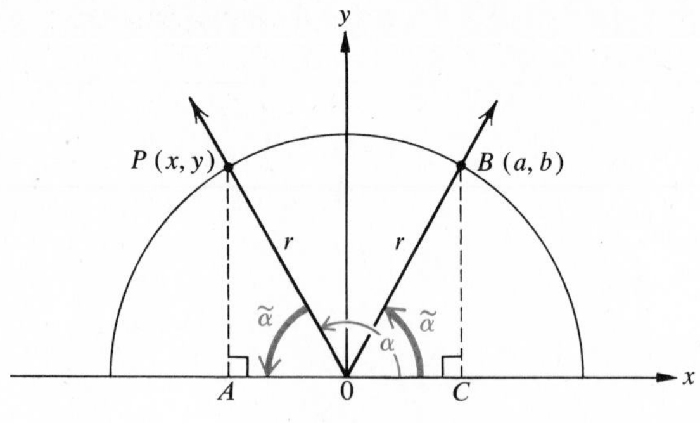

Figure 1.15

For the case in which α is in Quadrant II, as shown in Figure 1.15, we can construct:

1. $\overline{PA}$ perpendicular to the x-axis for any point P on the terminal side of α.
2. Semicircle with radius $\overline{OP}$.
3. Radius $\overline{OB}$, making an angle $\tilde{\alpha}$ with the x-axis.
4. $\overline{BC}$ perpendicular to the x-axis.

Because the hypotenuse and an acute angle of triangle POA are congruent, respectively, to the hypotenuse and an acute angle of triangle BOC, the triangles are congruent, and hence $PA = BC$ and $OA = OC$. Thus, $x = -a$ and $y = b$. We then have

$$|\sin\alpha| = \left|\frac{y}{r}\right| = \frac{b}{r} = \sin\tilde{\alpha}, \qquad |\csc\alpha| = \left|\frac{r}{y}\right| = \frac{r}{b} = \csc\tilde{\alpha},$$

$$|\cos\alpha| = \left|\frac{x}{r}\right| = \frac{a}{r} = \cos\tilde{\alpha}, \qquad |\sec\alpha| = \left|\frac{r}{x}\right| = \frac{r}{a} = \sec\tilde{\alpha},$$

$$|\tan\alpha| = \left|\frac{y}{x}\right| = \frac{b}{a} = \tan\tilde{\alpha}, \qquad |\cot\alpha| = \left|\frac{x}{y}\right| = \frac{a}{b} = \cot\tilde{\alpha}.$$

The above relationships hold for an angle in any quadrant, as can be shown by using congruent triangles as above. Hence, for all α for which trig α is defined,

$$|\text{trig }\alpha| = \text{trig }\tilde{\alpha}.$$

The notion of a reference angle facilitates finding a function value of nonquadrantal angles with measures greater than 90°. For such an angle α, we proceed as follows:

1. Find trig $\tilde{\alpha}$.

2. Prefix the algebraic sign, + or −, depending on whether the original function value, trig α, is positive or negative; use Table 1.1 as necessary.

For the special angles given in Table 1.2, we shall continue to show function values in the fractional and/or radical form shown in the table, both in examples and in answers to exercises.

Example Find tan 210°.

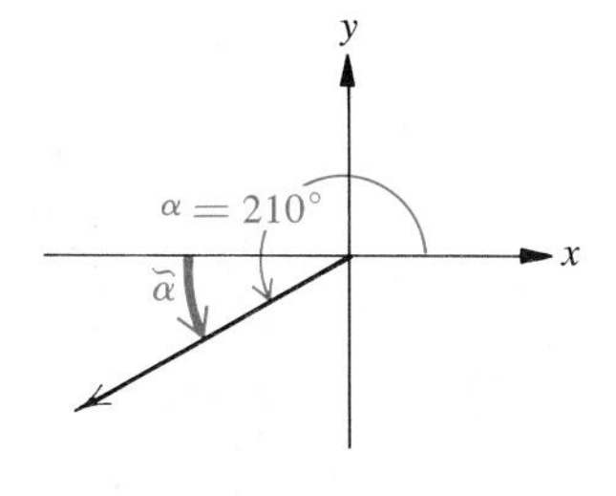

Solution A sketch showing the reference angle is helpful. Observe that $\tilde{\alpha} = (210 - 180)° = 30°$. Since α is in Quadrant III, tan 210° is positive (see Table 1.1). Therefore,

$$\tan 210° = \tan \tilde{\alpha}$$
$$= \tan 30° = \frac{1}{\sqrt{3}}.$$

Using a Calculator

Most calculators are programmed to give function values directly for values of x outside the interval $0 \leq x \leq \pi/2$. For example,

$$\sin 5.84^{\mathrm{R}} \approx -0.4288192 \approx -0.4288.$$

Some calculators are programmed so that *only* function values in which the domains are between $-\pi^{\mathrm{R}}$ and π^{R} can be obtained directly. Thus, it is sometimes necessary to first use a reference angle before using a calculator. In the above example, we would first have

$$\tilde{\alpha} \approx 6.28 - 5.84 = 0.44$$

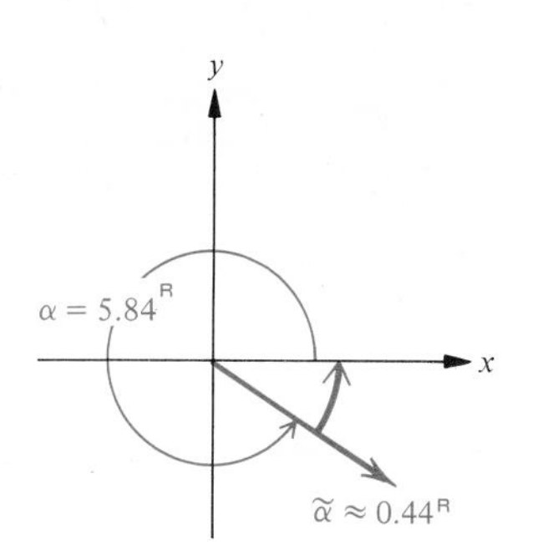

Figure 1.16

(see Figure 1.16), and then using a calculator we would obtain

$$\sin 5.84 \approx -\sin 0.44$$
$$\approx -0.42593 \approx -0.4259.$$

Note that the value -0.4288 that we obtained without using a reference angle differs from the value -0.4259 that we obtained by using a reference angle. This variation occurs because we used 6.28 as the approximation for 2π when we used the reference angle, while the calculator used a different approximation. In fact, if the element in the domain is a relatively large number, the difference in the values obtained by the two methods can be significant.

There are times, other than when finding trigonometric function values, that the notion of a reference angle is useful. To reinforce this idea, the examples in the text and the answers to exercises are based on the use of reference angles with the approximation 3.14 for π. *If you use your calculator to find function values directly, you should expect to obtain slightly different approximations than those given in the text.*

Example Find an approximate value for $\cos 1.95^{R}$.

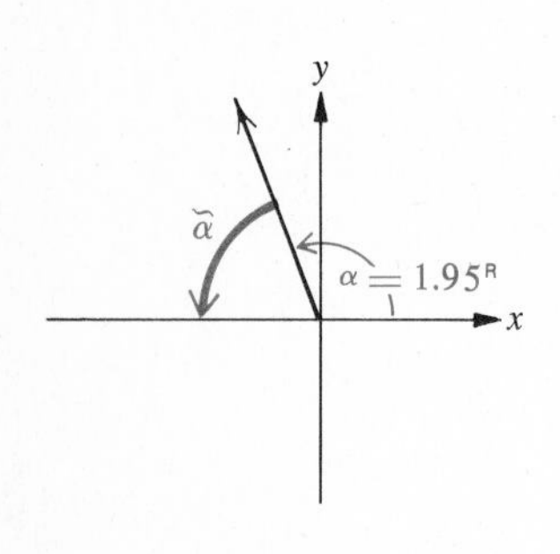

Solution (Using Table III or a calculator) A sketch showing the reference angle is helpful. Using $\pi \approx 3.14$, we have

$$\tilde{\alpha} \approx (3.14 - 1.95)^{R} = 1.19^{R}.$$

Since α is in Quadrant II, $\cos 1.95^{R}$ is negative. Therefore,

$$\begin{aligned} \cos 1.95^{R} &= -\cos \tilde{\alpha} \\ &\approx -\cos 1.19^{R} = -0.3717. \end{aligned}$$

Example Find an approximation for $\csc 334^\circ 36'$.

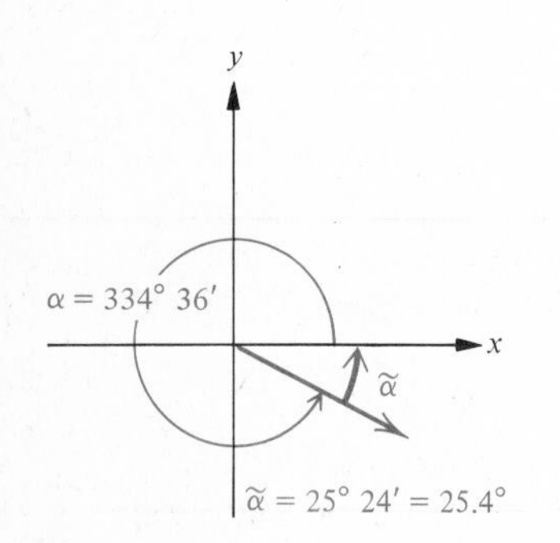

Solution (Using Table II) A sketch showing the reference angle is helpful. Observe

$$\tilde{\alpha} = 360^\circ - 334^\circ 36' = 25^\circ 24'.$$

Also, since α is in Quadrant IV, $\csc 334^\circ 36'$ is negative. Therefore,

$$\begin{aligned} \csc 334^\circ 36' &= -\csc \tilde{\alpha} \\ &\approx -\csc 25^\circ 24' \approx -2.3314. \end{aligned}$$

Alternate Solution (Using a Calculator)

$$\begin{aligned} \csc 334^\circ 36' &\approx -\csc 25^\circ 24' = -\csc 25.4^\circ \\ &\approx -\frac{1}{\sin 25.4^\circ} \approx -2.3314. \end{aligned}$$

EXERCISE SET 1.5

A

Sketch $\tilde{\alpha}$ and find $\tilde{\alpha}$ in the same units in which each α is given.

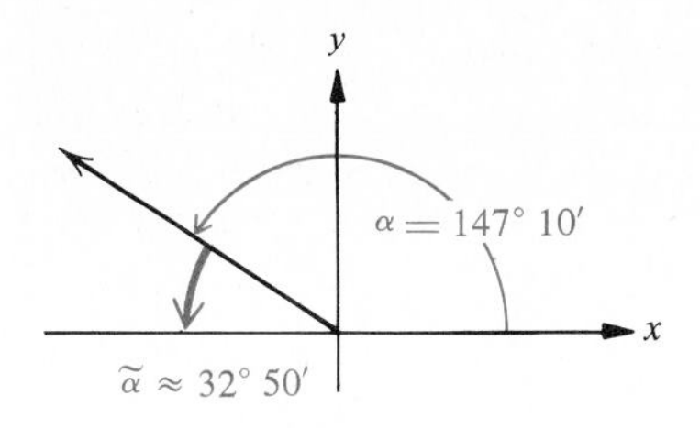

Example 147° 10′

Solution $\tilde{\alpha} = 180° - 147°\,10'$

$= 32°\,50'$

1. 152° **2.** 146° **3.** 327° 20′

4. 215° 40′ **5.** 262° 12′ **6.** 341° 28′

Examples a. $\frac{11\pi^R}{6}$ b. 4.28^R

Solutions a. $\tilde{\alpha} = 2\pi^R - \frac{11\pi^R}{6}$ b. $\tilde{\alpha} \approx 4.28^R - 3.14^R$

$= \frac{\pi^R}{6}$ $= 1.14^R$

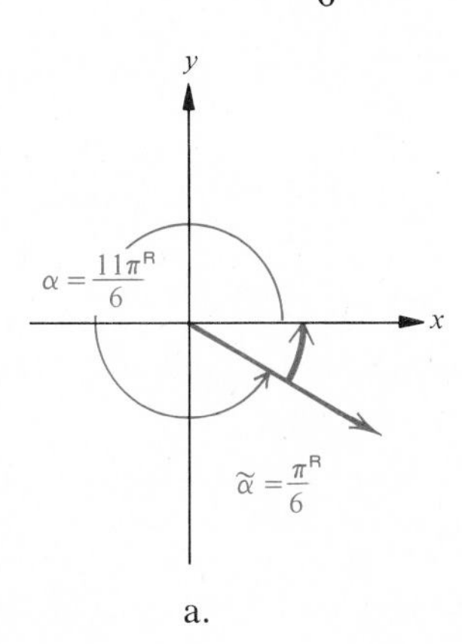

a.

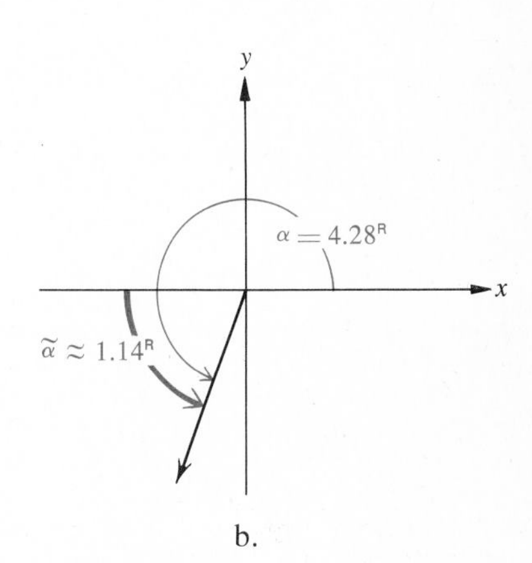

b.

7. $\frac{5\pi^R}{6}$ **8.** $\frac{7\pi^R}{6}$ **9.** $\frac{5\pi^R}{4}$ **10.** $\frac{7\pi^R}{4}$

11. $\frac{5\pi^R}{3}$ **12.** $\frac{4\pi^R}{3}$ **13.** 1.96^R **14.** 2.05^R

15. 3.27^R **16.** 4.11^R **17.** 5.95^R **18.** 4.90^R

Find each function value. Leave answers in fractional and/or radical form as shown in Table 1.2.

Examples a. sin 240° b. $\tan \frac{7\pi^R}{4}$

(continued)

$\alpha = 240°$

$\tilde{\alpha} = 60°$

a.

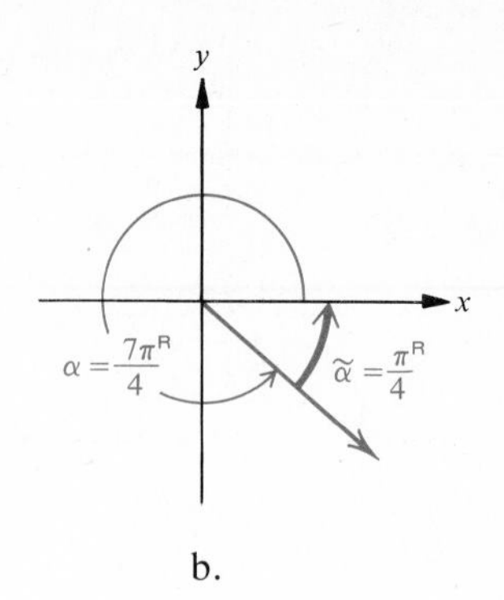

b.

Solutions

a. The reference angle

$$\tilde{\alpha} = (240 - 180)° = 60°.$$

Since 240° is in Quadrant III, sin 240° is negative. Therefore,

$$\sin 240° = -\sin 60° = -\frac{\sqrt{3}}{2}.$$

b. The reference angle

$$\tilde{\alpha} = 2\pi^{R} - \frac{7\pi}{4}^{R} = \frac{\pi}{4}^{R}.$$

Since $(7\pi/4)^{R}$ is in Quadrant IV, $\tan\,(7\pi/4)^{R}$ is negative. Therefore,

$$\tan \frac{7\pi}{4}^{R} = -\tan \frac{\pi}{4}^{R} = -1.$$

19. sin 330° **20.** cot 120° **21.** cos 300° **22.** cos 240°

23. tan 225° **24.** sin 315° **25.** sec 330° **26.** cos 120°

27. cot 225° **28.** csc 150° **29.** csc 300° **30.** tan 330°

31. $\sin \frac{5\pi}{4}^{R}$ **32.** $\cos \frac{3\pi}{4}^{R}$ **33.** $\tan \frac{2\pi}{3}^{R}$ **34.** $\cot \frac{5\pi}{3}^{R}$

35. $\sec \frac{7\pi}{6}^{R}$ **36.** $\csc \frac{5\pi}{6}^{R}$

Find an approximation for each function value. Use Table II or Table III in Appendix C, or use a calculator.

Example sec 128° 12′

Solution (Using Table II)

$$\tilde{\alpha} = 180° - 128° \, 12' = 51° \, 48'.$$

Since 128° 12′ is in Quadrant II, sec 128° 12′ is negative. Therefore,

$$\sec 128° \, 12' = -\sec 51° \, 48' = -1.6171.$$

Alternate Solution (Using a Calculator)

$$\sec 128° \, 12' = \sec 128.2° = -\sec 51.8$$

$$= -\frac{1}{\cos 51.8} = -1.6171.$$

37. cos 310° **38.** sin 230° **39.** tan 263.4° **40.** cos 138.7°

41. cot 342.5° **42.** csc 273.6° **43.** sin 195° 12′ **44.** cos 240° 30′

45. tan 184° 18′ **46.** tan 153° 48′ **47.** csc 322° 24′ **48.** sec 269° 42′

Example cot 5.36^{R}

Solution (Using Table III)

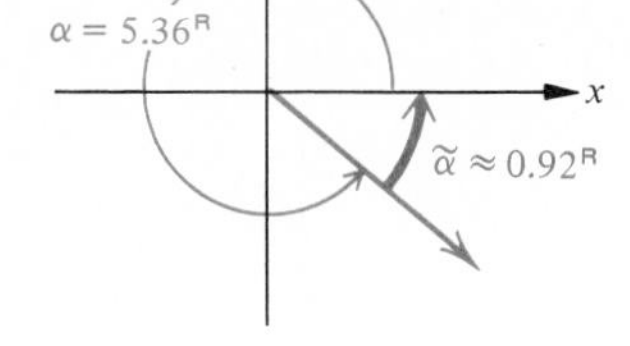

$$\tilde{\alpha} \approx 6.28^{\mathrm{R}} - 5.36^{\mathrm{R}} = 0.92^{\mathrm{R}}.$$

Since 5.36^{R} is in Quadrant IV, cot 5.36^{R} is negative. Hence,

$$\cot 5.36^{\mathrm{R}} \approx -\cot 0.92^{\mathrm{R}} \approx -0.7615.$$

Alternate Solution (Using a Calculator)

$$\cot 5.36^{\mathrm{R}} \approx -\cot 0.92^{\mathrm{R}}$$

$$= -\frac{1}{\tan 0.92^{\mathrm{R}}} \approx -0.7615.$$

49. tan 2.10^{R} **50.** sin 4.22^{R} **51.** cos 1.62^{R} **52.** tan 3.48^{R}

53. sec 4.87^{R} **54.** csc 1.84^{R} **55.** cos 2.93^{R} **56.** sec 6.20^{R}

57. sin 2^{R} **58.** tan 4^{R} **59.** cos 3^{R} **60.** sin 5^{R}

1.6 Finding Values for Angles α, $0° \le \alpha < 360°$

For most elements in the range of a trigonometric function, there are two corresponding elements in its domain over the interval 0° to 360°. For example, if

$$\sin \alpha = \frac{1}{2},$$

then, over the interval 0° to 360°, either

$$\alpha_1 = 30° \qquad \text{or} \qquad \alpha_2 = 150°$$

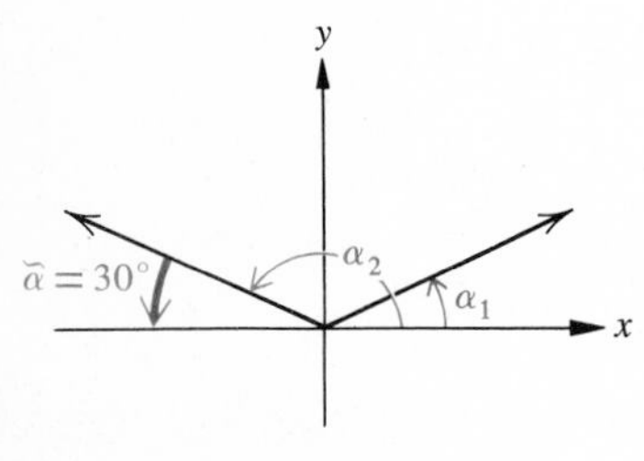

Figure 1.17

satisfies the equation. For α_2, as shown in Figure 1.17, $\tilde{\alpha} = 30°$.

The following examples illustrate how the reference angle can be used to find all elements in the domain of a trigonometric function over the interval 0° to 360° or 0^R to $2\pi^R$ that satisfies a given condition. Note that *for any positive function value, the angle obtained from any table, or calculator, is the reference angle.*

Example Given that $\tan \alpha = 1$, find the following.

a. The least nonnegative angle α in degrees.

b. $\{\alpha \mid 0° \le \alpha < 360°\}$ for which the statement is true.

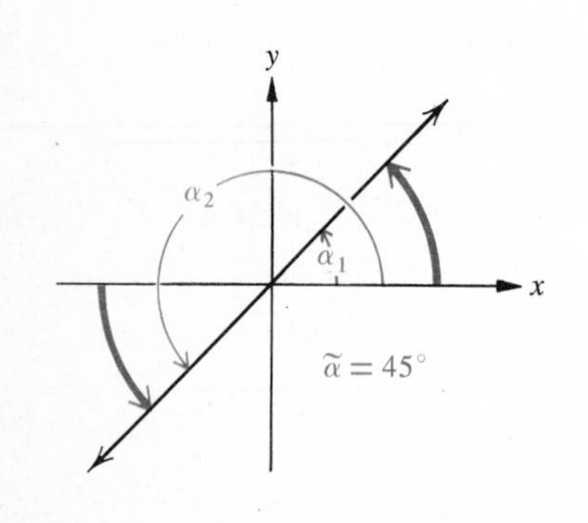

Solution The reference angle $\tilde{\alpha}$, for which

$$\tan \tilde{\alpha} = |\tan \alpha| = 1,$$

is 45° (see Table 1.2). Since $\tan \alpha$ (equal to 1) is positive, α is in Quadrant I or Quadrant III. A sketch showing the reference angle in each of these quadrants is helpful.

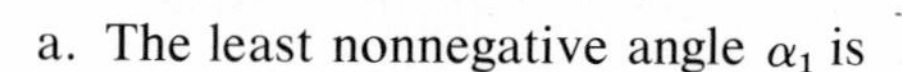

a. The least nonnegative angle α_1 is

$$\alpha_1 = \tilde{\alpha} = 45°.$$

b. In Quadrant III,

$$\alpha_2 = 180° + \tilde{\alpha} = (180 + 45)° = 225°.$$

Hence,

$$\{\alpha \mid \tan \alpha = 1,\ 0° \le \alpha < 360°\} = \{45°, 225°\}.$$

Example Given that $\sin \alpha = -0.6494$, find the following.

a. The least nonnegative angle α in degrees.

b. $\{\alpha \mid 0° \le \alpha < 360°\}$ for which the statement is true.

Solution The reference angle $\tilde{\alpha}$, for which

$$\sin \tilde{\alpha} = |\sin \alpha| = |-0.6494| = 0.6494,$$

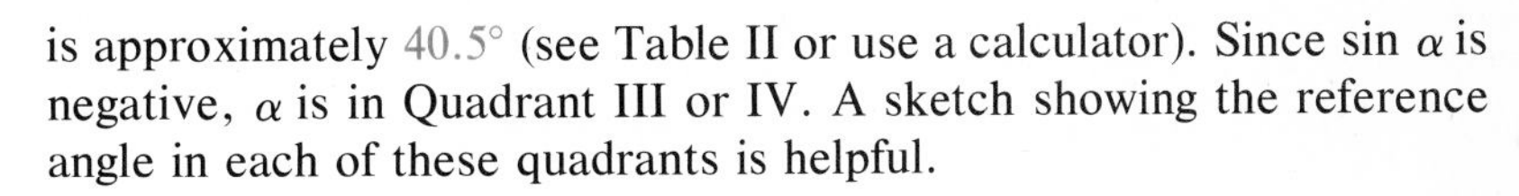

is approximately $40.5°$ (see Table II or use a calculator). Since $\sin \alpha$ is negative, α is in Quadrant III or IV. A sketch showing the reference angle in each of these quadrants is helpful.

a. The least nonnegative angle is

$$\alpha_1 = 180° + \tilde{\alpha}$$
$$\approx 180° + 40.5° = 220.5°.$$

b. In Quadrant IV,

$$\alpha_2 = 360° - \tilde{\alpha}$$
$$\approx 360° - 40.5° = 319.5°.$$

Hence,

$$\{\alpha \mid \sin \alpha = -0.6494,\ 0° \le \alpha < 360°\} = \{220.5°, 319.5°\}.$$

Example Given that $\cos \alpha = -0.7518$, find the following.

a. The least nonnegative angle α in radians.

b. $\{\alpha \mid 0^R \le \alpha < 2\pi^R\}$ for which the statement is true.

Solution The reference angle $\tilde{\alpha}$, for which

$$\cos \tilde{\alpha} = |\cos \alpha| = |-0.7518| = 0.7518,$$

is approximately 0.72^R (see Table III or use a calculator). Since $\cos \alpha$ is negative, α is in Quadrant II or Quadrant III. A sketch showing the reference angle in each of these quadrants is helpful.

a. The least nonnegative angle is

$$\alpha_1 = \pi^R - \tilde{\alpha}$$
$$\approx (3.14 - 0.72)^R = 2.42^R.$$

(*continued*)

b. In Quadrant III,

$$\alpha_2 = \pi^{\mathrm{R}} + \tilde{\alpha}$$
$$\approx (3.14 + 0.72)^{\mathrm{R}} = 3.86^{\mathrm{R}}.$$

Hence,

$$\{\alpha \mid \cos \alpha = -0.7518,\ 0^{\mathrm{R}} \leq \alpha < 2\pi^{\mathrm{R}}\} = \{2.42^{\mathrm{R}}, 3.86^{\mathrm{R}}\}.$$

EXERCISE SET 1.6

For each function value, find (**a**) *the least nonnegative angle* α *in degrees and* (**b**) $\{\alpha \mid 0° \leq \alpha < 360°\}$ *for which the statement is true. Leave answer in fractional and/or radical form as shown in Table 1.2.*

Example $\sin \alpha = -\dfrac{\sqrt{3}}{2}$

Solution The reference angle $\tilde{\alpha}$, for which

$$\sin \tilde{\alpha} = |\sin \alpha| = \left| -\frac{\sqrt{3}}{2} \right| = \frac{\sqrt{3}}{2},$$

is 60° (see Table 1.2). Since $\sin \alpha$ (equal to $-\sqrt{3}/2$) is negative, α is in Quadrant III or Quadrant IV. A sketch showing the reference angle in each of these quadrants is helpful.

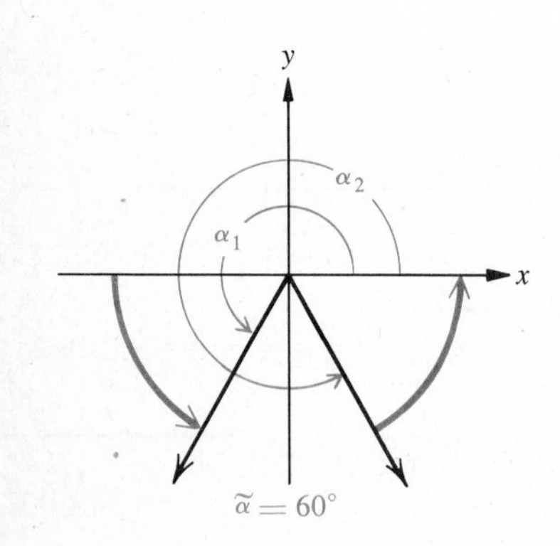

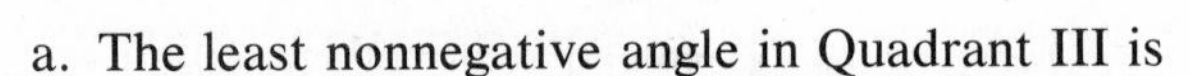

a. The least nonnegative angle in Quadrant III is

$$\alpha_1 = 180° + \tilde{\alpha} = 180° + 60° = 240°.$$

b. The angle in Quadrant IV is

$$\alpha_2 = 360° - 60° = 300°.$$

Hence,

$$\{\alpha \mid \sin \alpha = -\frac{\sqrt{3}}{2},\ 0° \leq \alpha < 360°\} = \{240°, 300°\}.$$

1. $\cos \alpha = \dfrac{\sqrt{3}}{2}$ **2.** $\sin \alpha = \dfrac{1}{\sqrt{2}}$ **3.** $\tan \alpha = -\dfrac{1}{\sqrt{3}}$

4. $\cos \alpha = \dfrac{1}{2}$ **5.** $\cot \alpha = \sqrt{3}$ **6.** $\tan \alpha = 1$

7. $\sec \alpha = \sqrt{2}$ **8.** $\cot \alpha = -1$ **9.** $\csc \alpha = -2$

10. $\sec \alpha = 2$ **11.** $\sin \alpha = -\frac{1}{2}$ **12.** $\csc \alpha = \frac{2}{\sqrt{3}}$

13. $\cos \alpha = -\frac{1}{\sqrt{2}}$ **14.** $\sin \alpha = -\frac{\sqrt{3}}{2}$ **15.** $\csc \alpha = \sqrt{2}$

16. $\cos \alpha = -1$ **17.** $\sin \alpha = 0$ **18.** $\tan \alpha = -\sqrt{3}$

19. $\sec \alpha = -\frac{2}{\sqrt{3}}$ **20.** $\cot \alpha = 0$ **21.** $\csc \alpha$ undefined

22. $\sec \alpha$ undefined **23.** $\cot \alpha$ undefined **24.** $\tan \alpha$ undefined

For each function value, find an approximation in degrees for (**a**) *the least nonnegative angle* α *and* (**b**) $\{\alpha \mid 0° \le \alpha < 360°\}$ *for which each statement is true.*

Example $\tan \alpha = -0.8693$

Solution The reference angle $\tilde{\alpha}$, for which

$$\tan \tilde{\alpha} = |\tan \alpha| = |-0.8693| = 0.8693,$$

is approximately 41° (see Table II or use a calculator). Since $\tan \alpha$ (equal to -0.8693) is negative, α is in Quadrant II or Quadrant IV. A sketch showing the reference angle in each of these quadrants is helpful.

a. The least nonnegative angle in Quadrant II is

$$\alpha_1 = 180° - \tilde{\alpha}$$
$$\approx 180° - 41° = 139°.$$

b. The angle in Quadrant IV is

$$\alpha_2 \approx 360° - 41° = 319°.$$

Hence,

$$\{\alpha \mid \tan \alpha = -0.8693,\ 0° \le \alpha < 360°\} = \{139°, 319°\}.$$

25. $\sin \alpha = 0.2419$ **26.** $\cos \alpha = 0.5150$ **27.** $\tan \alpha = -2.6051$

28. $\cot \alpha = -2.1445$ **29.** $\sec \alpha = 1.0320$ **30.** $\csc \alpha = 1.1570$

31. $\cos \alpha = -0.9085$ **32.** $\sin \alpha = -0.3535$ **33.** $\cot \alpha = 2.1060$

34. $\tan \alpha = 0.3134$ **35.** $\csc \alpha = -1.2796$ **36.** $\sec \alpha = -3.8140$

37. $\sin \alpha = -0.5476$ **38.** $\cot \alpha = 0.6032$ **39.** $\cos \alpha = 0.3955$

40. $\csc \alpha = -2.3400$ **41.** $\tan \alpha = 1.0212$ **42.** $\sec \alpha = 1.6316$

For each function value, find an approximation in radians for (**a**) *the least nonnegative angle* α *and* (**b**) $\{\alpha \mid 0^R \le \alpha < 2\pi^R\}$ *for which each statement is true. Use* $\pi \approx 3.14$.

Example $\cos \alpha = -0.9131$

Solution The reference angle $\tilde{\alpha}$, for which

$$\cos \tilde{\alpha} = |\cos \alpha| = |-0.9131| = 0.9131,$$

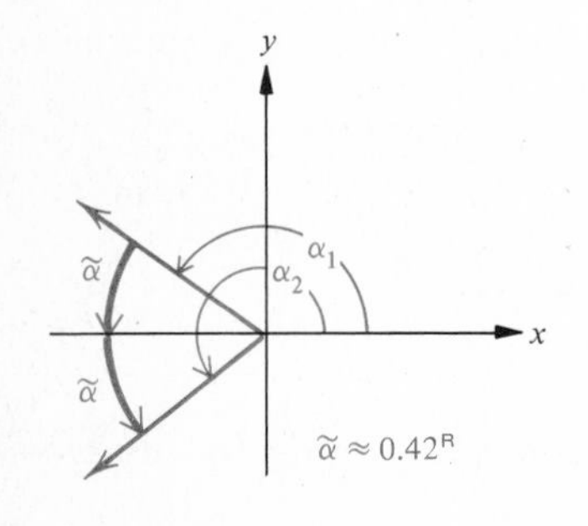

is approximately 0.42^R (see Table III or use a calculator). Since $\cos \alpha$ (equal to -0.9131) is negative, α is in Quadrant II or Quadrant III. A sketch showing the reference angle in each of these quadrants is helpful.

a. The least nonnegative angle in Quadrant II is

$$\alpha_1 \approx 3.14 - \tilde{\alpha} = 3.14^R - 0.42^R = 2.72^R.$$

b. The angle in Quadrant III is

$$\alpha_2 \approx 3.14^R + 0.42^R = 3.56^R.$$

43. $\sin \alpha = 0.6442$ **44.** $\cos \alpha = 0.5898$ **45.** $\tan \alpha = 0.2341$

46. $\sin \alpha = 0.2280$ **47.** $\sec \alpha = 1.795$ **48.** $\csc \alpha = 1.467$

49. $\cos \alpha = -0.9689$ **50.** $\sin \alpha = -0.5480$ **51.** $\sin \alpha = -0.1197$

52. $\tan \alpha = -1.050$ **53.** $\csc \alpha = -2.125$ **54.** $\sec \alpha = -5.273$

1.7 Function Values for All $\alpha \in A$

The trigonometric function values are ratios of numbers associated with a point on the terminal side of an angle in standard position. Therefore, the trigonometric function values for an angle α are equal to the analogous function values for any angle coterminal with α and can be obtained for any α, an element of the set A of all angles, by using the methods we used in Section 1.5 to obtain function values for $0° \le \alpha < 360°$.

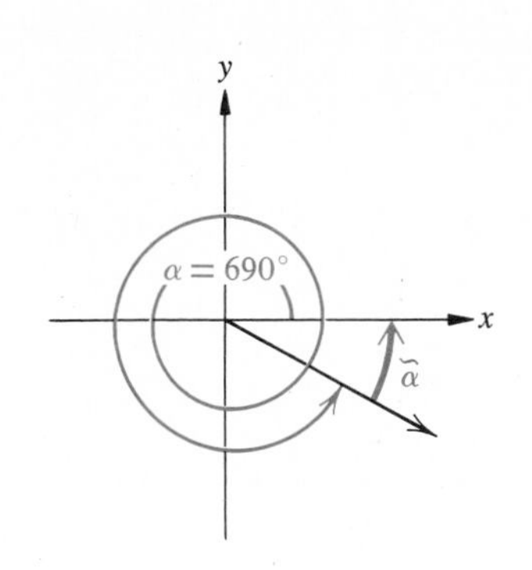

Example Find $\tan 690°$.

Solution Since $690° - 360° = 330°$, the reference angle

$$\tilde{\alpha} = 360° - 330° = 30°.$$

Because 690° is in Quadrant IV, we have, using Table 1.2,

$$\tan 690° = -\tan 30° = -\frac{1}{\sqrt{3}}.$$

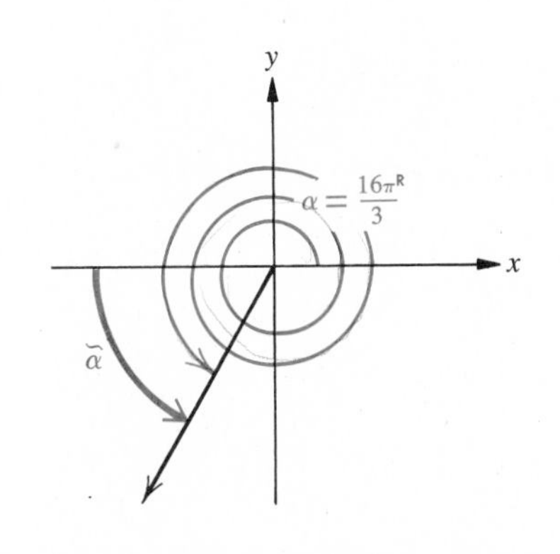

Example Find $\cos \dfrac{16\pi^R}{3}$.

Solution Since

$$\frac{16\pi^R}{3} - 4\pi^R = \frac{16\pi^R}{3} - \frac{12\pi^R}{3} = \frac{4\pi^R}{3},$$

the reference angle

$$\tilde{\alpha} = \frac{4\pi^R}{3} - \pi^R = \frac{\pi^R}{3}.$$

Because $(16\pi/3)^R$ is in Quadrant III, we have, using Table 1.2,

$$\cos \frac{16\pi^R}{3} = -\cos \frac{\pi^R}{3} = -\frac{1}{2}.$$

Function values for angles with negative measures can also be found using similar procedures.

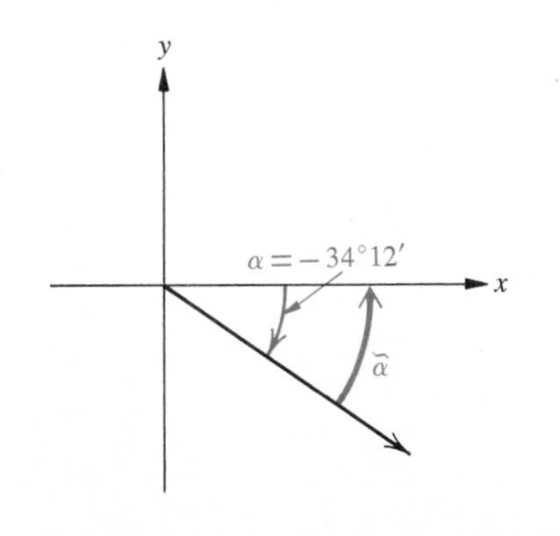

Example Find $\sin(-34°\ 12')$.

Solution The reference angle is

$$\tilde{\alpha} = 34°\ 12'.$$

Because $-34°\ 12'$ is in Quadrant IV, we have, using Table II or a calculator,

$$\begin{aligned}\sin(-34°\ 12') &= -\sin 34°\ 12' \\ &= -\sin 34.2° \approx -0.5621.\end{aligned}$$

We can obtain the infinite set of values for α for a given trigonometric function value by first finding the values of α in the interval $0° \leq \alpha < 360°$ by the methods that we considered in Section 1.6. An example of this procedure is shown in the Exercise Set.

EXERCISE SET 1.7

A

Find each function value. Leave answers in one of the forms shown in Table 1.2.

Example $\tan 480°$

Solution Since $480° - 360° = 120°$, the reference angle

$$\tilde{\alpha} = 180° - 120° = 60°.$$

Because 480° is in Quadrant II, tan 480° is negative. Using Table 1.2 or by constructing an appropriate triangle, we have that

$$\tan 480° = -\tan 60° = -\sqrt{3}.$$

1. $\tan 420°$ **2.** $\sin 570°$ **3.** $\cos 675°$ **4.** $\sec 510°$

5. $\cot 930°$ **6.** $\tan 870°$ **7.** $\sin 630°$ **8.** $\cos 765°$

9. $\csc 495°$ **10.** $\cot 960°$ **11.** $\sin 4\pi^R$ **12.** $\tan 5\pi^R$

13. $\cos \frac{11\pi}{4}^R$ **14.** $\sec \frac{8\pi}{3}^R$ **15.** $\tan \frac{10\pi}{3}^R$ **16.** $\csc \frac{17\pi}{6}^R$

17. $\cot \frac{25\pi}{6}^R$ **18.** $\cos \frac{11\pi}{3}^R$

Find an approximation for each function value. Use Table II or Table III as appropriate, or use a calculator.

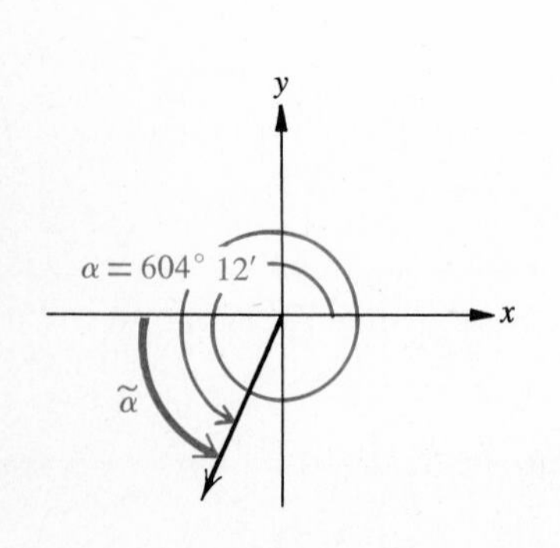

Example $\cos 604°\ 12'$

Solution Since $604°\ 12' - 360° = 244°\ 12'$, the reference angle

$$\tilde{\alpha} = 244°\ 12' - 180° = 64°\ 12'.$$

Because 604° 12′ is in Quadrant III, cos 604° 12′ is negative, and we have, using Table II or a calculator,

$$\cos 604° \, 12' = -\cos 64° \, 12'$$
$$= -\cos 64.2° \approx -0.4352.$$

19. $\sin 380.2°$ **20.** $\tan 780.4°$ **21.** $\cos 910.3°$ **22.** $\cos 590.5°$

23. $\sec 551° \, 30'$ **24.** $\cot 842° \, 48'$ **25.** $\cos 7.14^{R}$ **26.** $\sin 8.21^{R}$

27. $\tan 15.01^{R}$ **28.** $\cos 12.86^{R}$ **29.** $\sec 10.43^{R}$ **30.** $\cot 16.23^{R}$

Find each function value. Leave answers in one of the forms shown in Table 1.2.

Example $\cot\left(-\frac{3\pi}{4}^{R}\right)$

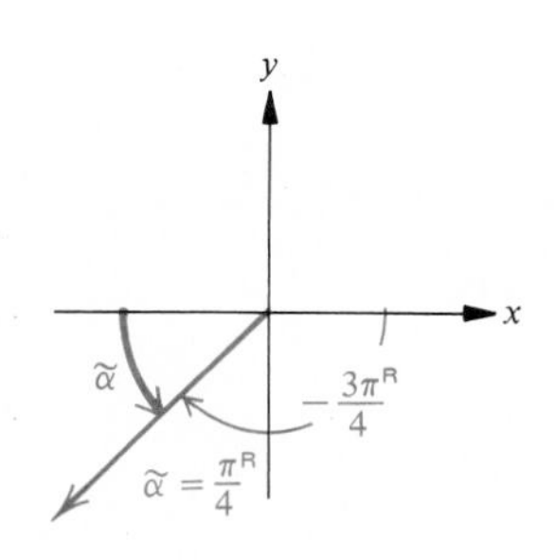

Solution The reference angle

$$\tilde{\alpha} = \pi^{R} - \frac{3\pi}{4}^{R} = \frac{\pi}{4}^{R}.$$

Because $(-3\pi/4)^{R}$ is in Quadrant III, we have, using Table 1.2,

$$\cot\left(-\frac{3\pi}{4}^{R}\right) = \cot \frac{\pi}{4}^{R} = 1.$$

31. $\sin(-30°)$ **32.** $\cos(-45°)$ **33.** $\tan(-210°)$

34. $\tan(-300°)$ **35.** $\cos\left(-\frac{\pi}{6}^{R}\right)$ **36.** $\sin\left(-\frac{\pi}{2}^{R}\right)$

37. $\sin\left(-\frac{11\pi}{4}^{R}\right)$ **38.** $\cos\left(-\frac{17\pi}{3}^{R}\right)$

Find an approximation for each function value. Use Table II or Table III as appropriate, or use a calculator.

Example $\cos(-422° \, 30')$

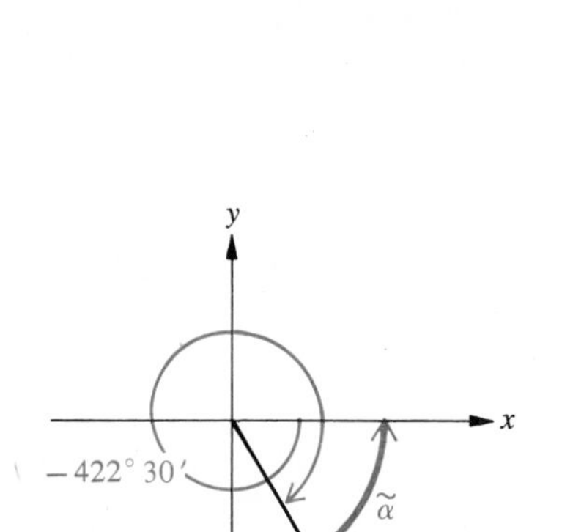

Solution The reference angle

$$\tilde{\alpha} = |360° - 422° \, 30°| = 62° \, 30'.$$

Because $-422° \, 30'$ is in Quadrant IV, we have, using Table II or a calculator,

$$\cos(-422°\ 30') = \cos 62°\ 30'$$
$$= \cos 62.5° \approx 0.4617.$$

39. $\cos(-648°)$ **40.** $\sin(-800°)$ **41.** $\tan(-430°\ 18')$

42. $\tan(-542°\ 42')$ **43.** $\sin(-780°\ 30')$ **44.** $\cos(-844°\ 48')$

45. $\tan(-7.43^R)$ **46.** $\tan(-8.14^R)$ **47.** $\cos(-12.32^R)$

48. $\sin(-14.66^R)$

Find the set of all angles for which each statement is true. Specify α *in degree measure.*

Example $\tan \alpha = -0.2162$

Solution The reference angle $\tilde{\alpha}$, for which

$$\tan \tilde{\alpha} = |\tan \alpha| = |-0.2162| = 0.2162,$$

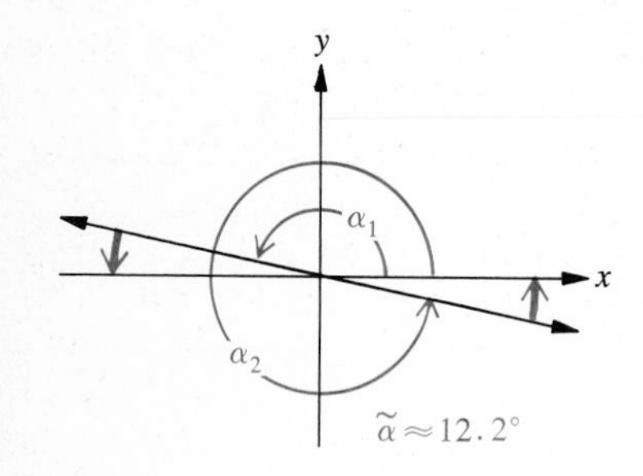

is approximately 12.2° (see Table II or use a calculator). Since $\tan \alpha$ (equal to -0.2162) is negative, α is in Quadrant II or IV. The figure shows the reference angles in each of these quadrants. The least non-negative angle is

$$\alpha_1 = 180° - \tilde{\alpha}$$
$$\approx 180° - 12.2° = 167.8°.$$

In Quadrant IV,

$$\alpha_2 = 360° - \tilde{\alpha}$$
$$\approx 360° - 12.2° = 347.8°.$$

Hence,

$$\{\alpha \mid \tan \alpha = -0.2162\}$$
$$= \{\alpha \mid \alpha \approx 167.8° + k \cdot 360° \text{ or } \alpha \approx 347.8° + k \cdot 360°\}^*$$
$$= \{\alpha \mid \alpha \approx 167.8° + k \cdot 180°, \quad k \in J\}.$$

* This set can be written in an alternate form using set notation as

$$\{\alpha \mid \alpha = 167.8° + k \cdot 360°\} \cup \{\alpha \mid \alpha = 347.8° + k \cdot 360°\}.$$

See Appendix A.1 for a discussion of set operations.

49. $\sin \alpha = 0.2790$

50. $\cos \alpha = 0.9311$

51. $\cot \alpha = -1.2799$

52. $\tan \alpha = -1.4882$

53. $\csc \alpha = -1.2813$

54. $\sec \alpha = -1.1164$

55. $\tan \alpha = 0.3019$

56. $\sin \alpha = -0.9426$

Chapter Summary

1.1 An **angle** is the union of two rays that have a common end point. The rays are called the **sides of the angle.**

An angle that has its initial side along the positive x-axis and its vertex at the origin is said to be in **standard position.** Such an angle is said to be in that quadrant in which its terminal side lies.

Two kinds of angle measure most commonly used are called **degree measure** and **radian measure.**

The measure of an angle in radians is related to the measure of the angle in degrees by the equations

$$\alpha^\circ = \frac{180}{\pi}\alpha^{\mathrm{R}} \quad \text{and} \quad \alpha^{\mathrm{R}} = \frac{\pi}{180}\alpha^\circ.$$

In particular, $1^{\mathrm{R}} \approx 57.30^\circ$ and $1^\circ \approx 0.01745^{\mathrm{R}}$.

Angles having the same initial side and the same terminal side are called **coterminal angles.**

1.2 For all angles α in standard position, if x and y are the coordinates of any point on the terminal side of α $[(x, y) \neq (0, 0)]$ and $r = \sqrt{x^2 + y^2}$,

$$\text{sine } \alpha = \frac{y}{r}, \qquad \text{cosecant } \alpha = \frac{r}{y} \quad (y \neq 0),$$

$$\text{cosine } \alpha = \frac{x}{r}, \qquad \text{secant } \alpha = \frac{r}{x} \quad (x \neq 0),$$

$$\text{tangent } \alpha = \frac{y}{x} \quad (x \neq 0), \qquad \text{cotangent } \alpha = \frac{x}{y} \quad (y \neq 0).$$

The following identities follow from the definitions of the trigonometric ratios.

$$\csc \alpha = \frac{1}{\sin \alpha}, \quad \sec \alpha = \frac{1}{\cos \alpha}, \quad \cot \alpha = \frac{1}{\tan \alpha}.$$

The sign (plus or minus) of a trigonometric ratio, also called a **function value,** depends on the quadrant in which the terminal side of the angle lies. These ratios are elements in the ranges of the trigonometric functions and are restricted as follows:

$$-1 \le \sin\alpha \le 1, \qquad \csc\alpha \le -1 \quad \text{or} \quad \csc\alpha \ge 1,$$

$$-1 \le \cos\alpha \le 1, \qquad \sec\alpha \le -1 \quad \text{or} \quad \sec\alpha \ge 1,$$

$$-\infty < \tan\alpha < \infty, \qquad -\infty < \cot\alpha < \infty.$$

1.3 Function values for each of the six trigonometric functions for certain angles can be obtained from geometric considerations of right triangles.

The following identities follow directly from the definition of the trigonometric ratios.

$$\sin^2\alpha + \cos^2\alpha = 1, \quad \tan\alpha = \frac{\sin\alpha}{\cos\alpha}, \quad \cot\alpha = \frac{\cos\alpha}{\sin\alpha}.$$

1.4 Tables II and III in Appendix C can be used to find rational number approximations for function values of selected angles over the intervals $0^\circ \le \alpha \le 90^\circ$ and $0^R \le \alpha \le 1.57^R$, respectively. Alternatively, a calculator can be used to obtain such function values.

The tables or a calculator can also be used to find a positive acute angle α, where trig α is given. The **inverse notation** $\text{Trig}^{-1}\,y$ can be used to name an angle α such that trig $\alpha = y$.

1.5 The **reference angle** of any nonquadrantal angle α in standard position is the positive acute angle $\tilde{\alpha}$ whose sides are the terminal side of α and a ray of the x-axis. The absolute values of the trigonometric ratios of any angle α are equal, respectively, to the analogous trigonometric ratios for $\tilde{\alpha}$. The use of a reference angle facilitates finding function values of nonquadrantal angles with measures greater than 90°.

1.6 The use of a reference angle helps finding angles α, $0^\circ \le \alpha < 360^\circ$ for a given trigonometric function value.

1.7 The trigonometric function values for an angle α are equal to the analogous function values for *any* angle coterminal with α, including angles with negative measures.

Review Exercises

1.1 *Find the degree measure of α to the nearest tenth of a degree.*

1. $\alpha = \frac{2\pi^{R}}{5}$

2. $\alpha = 2.41^{R}$

Find the radian measure of α to the nearest hundredth of a radian.

3. $\alpha = 75°$

4. $\alpha = 257°$

Find the least positive angle coterminal with each angle α.

5. $\alpha = 938°$

6. $\alpha = 7.23^{R}$

Write a general expression for all angles coterminal with α.

7. $\alpha = 131°$

8. $\alpha = \frac{7\pi^{R}}{4}$

1.2 *Find the element in the range of each of the six trigonometric functions of α if the terminal side of α (in standard position) contains the given point.*

9. $(-4, 3)$

10. $(-1, 2)$

11. $(\sqrt{3}, -1)$

12. $(-8, -15)$

Determine the quadrant in which the terminal side of each angle α lies if α is in standard position.

13. $\sin \alpha < 0$ and $\sec \alpha < 0$

14. $\cos \alpha < 0$ and $\cot \alpha > 0$

Given one function value of an angle α, determine the other five trigonometric function values of α.

15. $\cos \alpha = \frac{3}{5}$, if $\sin \alpha > 0$

16. $\tan \alpha = \frac{4}{7}$, if $\cos \alpha > 0$

17. $\sin \alpha = -\frac{1}{3}$, if $\cot \alpha > 0$

18. $\sec \alpha = -\frac{5}{2}$, if $\tan < 0$

1.3 *Find each function value, if it exists, by making an appropriate sketch and reading the ratio directly from the figure.*

19. $\csc 60°$

20. $\sec 45°$

21. $\tan 270°$

22. $\sin 180°$

Compute each expression if the expression represents a real number.

23. $\sec^2 30°$

24. $(\cos 180°)(\tan^2 45°)$

25. $(\sec 45°)(\sin 30°)(\cot^2 60°)$

26. $\dfrac{2(\tan^2 60°) + \cos^2 45°}{\sin^2 90°}$

Use the identities on pages 9 and 17 to write each expression in terms of $\sin \alpha$ *and/or* $\cos \alpha$.

27. $\dfrac{\cos^2 \alpha}{\cot \alpha}$

28. $\csc^2 \alpha \cdot \tan \alpha$

29. $\dfrac{\sec^2 \alpha - \tan^2 \alpha}{\tan \alpha}$

30. $\dfrac{\csc^2 \alpha - \cot^2 \alpha}{\cot \alpha}$

1.4 *Find an approximation for each function value.*

31. $\tan 79° 30'$

32. $\sec 44° 18'$

33. $\csc 0.34^R$

34. $\cos 0.73^R$

35. Express $\tan \alpha = 1.4321$ equivalently using inverse notation.

36. Express $\alpha = \text{Cos}^{-1} 0.4232$ equivalently without inverse notation.

Find an approximation for the measure of each angle α *in degree measure, where* $0° \leq \alpha \leq 90°$.

37. $\tan \alpha = 1.4770$

38. $\sec \alpha = 1.6502$

Find an approximation for the measure of each angle α *in radian measure, where* $0^R \leq \alpha \leq 1.57^R$.

39. $\csc \alpha = 4.582$

40. $\cos \alpha = 0.7712$

1.5 *Sketch* $\tilde{\alpha}$ *and find the measure of* $\tilde{\alpha}$ *in the same units in which each* α *is given. Use* $\pi \approx 3.14$.

41. $231° 20'$ **42.** $301° 43'$ **43.** 2.45^R **44.** 4.76^R

Find each function value (or an approximation). Use $\pi \approx 3.14$.

45. $\sin 150°$

46. $\cot 331° 48'$

47. $\tan 3.27^R$

48. $\sec 2^R$

1.6 *Find the set of all angles* α *(or approximations),* $0° \leq \alpha < 360°$, *for each function value.*

49. $\sin \alpha = 0.3584$

50. $\cot \alpha = -1.7045$

Find the set of all angles α *(or approximations),* $0^{R} \leq \alpha < 2\pi^{R}$, *for each function value. Use* $\pi \approx 3.142$.

51. $\tan \alpha = 2.176$

52. $\sec \alpha = 1.823$

1.7 *Find each function value (or an approximation). Use* $\pi \approx 3.14$.

53. $\csc 450°$

54. $\sin 472° \, 24'$

55. $\cos 6\pi^{R}$

56. $\cot 9.37^{R}$

Find each function value.

57. $\tan(-60°)$

58. $\sec(-330°)$

59. $\csc\left(-\frac{\pi}{3}^{R}\right)$

60. $\cos(-543°)$

61. $\sin(-472° \, 12')$

62. $\cot(-9.17^{R})$

Find the set of all angles for which each statement is true. Specify α *in degree measure.*

63. $\sin \alpha = 0.3040$

64. $\cos \alpha = 0.4848$

65. $\cot \alpha = 1.9542$

66. $\csc \alpha = -1.4242$

2 Applications

When we use given information about the lengths of one, two, or three sides of a triangle and the measure of one or two angles to find the remaining, or unknown, lengths and measures, we are said to be *solving a triangle*. In this chapter you will first solve different kinds of triangles. Then, you will be introduced to the notion of a geometric vector, an important prerequisite for the study of advanced topics in mathematics. You may want to review some pertinent topics of geometry in Appendices A.7–A.9 before you begin this chapter.

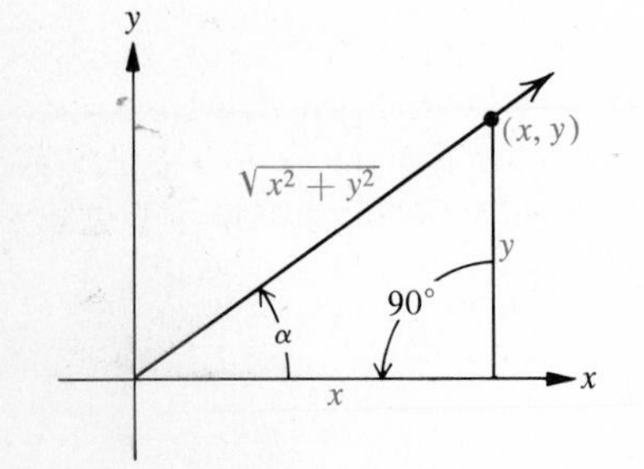

Figure 2.1

2.1 Right Triangles

Right triangles can be solved by using the trigonometric ratios of the acute angles of the triangle. Figure 2.1 shows that, in the first quadrant, any point (x, y) on the terminal side of an angle α in standard position determines a right triangle with sides of length x and y and with hypotenuse of length $\sqrt{x^2 + y^2}$. Therefore, in the special case where α is an acute angle in a right triangle,

$$\sin \alpha = \frac{\text{length of side opposite } \alpha}{\text{length of hypotenuse}}, \qquad \csc \alpha = \frac{\text{length of hypotenuse}}{\text{length of side opposite } \alpha},$$

$$\cos \alpha = \frac{\text{length of side adjacent to } \alpha}{\text{length of hypotenuse}}, \qquad \sec \alpha = \frac{\text{length of hypotenuse}}{\text{length of side adjacent to } \alpha},$$

$$\tan \alpha = \frac{\text{length of side opposite } \alpha}{\text{length of side adjacent to } \alpha}, \qquad \cot \alpha = \frac{\text{length of side adjacent to } \alpha}{\text{length of side opposite } \alpha}.$$

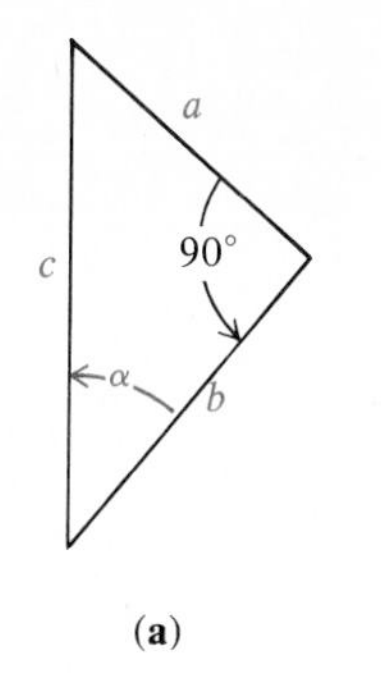

(a)

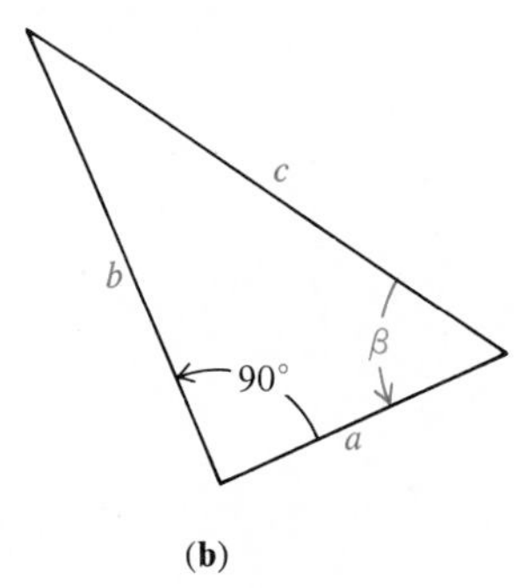

(b)

Figure 2.2

It is not necessary that α be in standard position. The above relations involving the ratios of the lengths of the sides of a right triangle are equally applicable to a right triangle in any position. For example, in Figure 2.2a, the six trigonometric ratios of the angle α are

$$\sin \alpha = \frac{a}{c}, \qquad \csc \alpha = \frac{c}{a},$$

$$\cos \alpha = \frac{b}{c}, \qquad \sec \alpha = \frac{c}{b},$$

$$\tan \alpha = \frac{a}{b}, \qquad \cot \alpha = \frac{b}{a}.$$

In Figure 2.2b, the trigonometric ratios of the angle β are

$$\sin \beta = \frac{b}{c}, \qquad \csc \beta = \frac{c}{b},$$

$$\cos \beta = \frac{a}{c}, \qquad \sec \beta = \frac{c}{a},$$

$$\tan \beta = \frac{b}{a}, \qquad \cot \beta = \frac{a}{b}.$$

Generally, the sides of a triangle are labeled with their lengths, a, b, and c, and the angles opposite a, b, and c are labeled α, β, and γ, or A, B, and C, respectively, where γ or C names the right angle in a right triangle. The measures of angles in triangles are usually given in degree units.

Cofunctions

Two functions f and g are called **cofunctions** if

$$f(x) = g(y) \qquad \text{and} \qquad x + y = 90°.$$

In the trigonometric ratios above, note that $\sin \alpha = \cos \beta$ and $\cos \alpha = \sin \beta$, where $\alpha + \beta = 90°$. Hence, the sine and cosine are cofunctions. Furthermore, for the same reason, the secant and cosecant are cofunctions, as are the tangent and cotangent. Note that these relationships are used in the construction of Table II. For example,

$$\tan 32° = 0.6249 \quad \text{and} \quad \cot(90° - 32°) = \cot 58° = 0.6249.$$

Solution of Triangles

Right triangles can be solved by using the trigonometric ratios on page 50 and the fact that *the sum of the measures of the angles of any triangle equals* 180°.

Example If one angle of a right triangle measures 34° and the side opposite this angle has a length of 12 inches, solve the triangle and find its area.

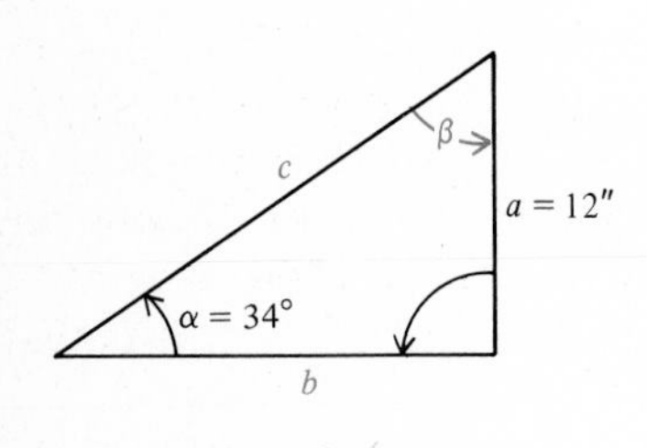

Solution Sketch the right triangle, showing the given information. First, from the fact that $\alpha + \beta + \gamma = 180°$ and $\gamma = 90°$,

$$\beta = 90° - \alpha = 90° - 34° = 56°.$$

By definition,

$$\cot 34° = \frac{b}{a} = \frac{b}{12} \quad \text{or} \quad b = 12 \cot 34°,$$

and

$$\csc 34° = \frac{c}{a} = \frac{c}{12} \quad \text{or} \quad c = 12 \csc 34°.$$

From Table II or by using a calculator, we have $\cot 34° \approx 1.4826$ and $\csc 34° \approx 1.7883$. Thus,

$$b \approx 12(1.4826) \approx 17.8,$$

$$c \approx 12(1.7883) \approx 21.5.$$

The area $\mathscr{A}$ of a triangle equals one-half the product of the length of the altitude and the length of the base. Thus,

$$\mathscr{A} \approx \tfrac{1}{2}(12)(17.8) \approx 107.$$

Note that the ratio selected in each case in the above example for cot 34° and csc 34° was one in which the desired length, b or c, appeared in the numerator of the ratio. With these choices, b and c were found by multiplication. If the ratios

$$\tan 34° = \frac{12}{b} \quad \text{and} \quad \sin 34° = \frac{12}{c}$$

had been used, then, equivalently,

$$b = \frac{12}{\tan 34^\circ} \quad \text{and} \quad c = \frac{12}{\sin 34^\circ},$$

and the operation of division would have to be performed to find b and c. If a calculator is used, there would be no advantage in using a ratio that leads to a multiplication operation over a ratio that leads to a division operation.

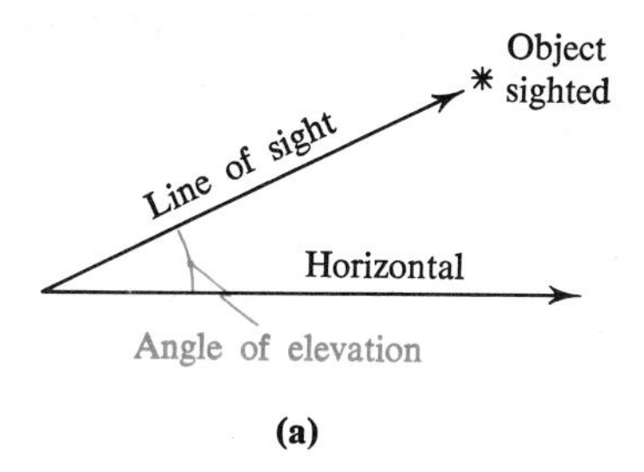

(a)

(b)

Figure 2.3

In the foregoing example the lengths b and c and the area $\mathscr{A}$ were expressed by an approximation using *three significant digits.** Unless otherwise stated, we shall continue to specify lengths of line segments and areas of triangles to this degree of accuracy when solutions are given in decimal notation, and we shall specify measures of angles to the *nearest tenth of a degree or nearest six minutes*. Furthermore, we shall assume that all data have similar accuracy. Thus, $a = 12$ in the above example implies that $a = 12.0$ and $\alpha = 34^\circ$ implies that $\alpha = 34.0^\circ = 34^\circ\ 00'$.

Certain angles are given special names in the fields of surveying and navigation. For example, the angle formed by a horizontal ray and an observer's "line of sight" to any object above the horizontal is called an **angle of elevation** (Figure 2.3a). The angle formed by a horizontal ray and an observer's "line of sight" to any object below the horizontal is called an **angle of depression** (Figure 2.3b).

Example On level ground at point A the measure of the angle of elevation of the top of a tower is $57^\circ\ 24'$. The distance from A to the base of the tower is 50 meters. How high is the tower?

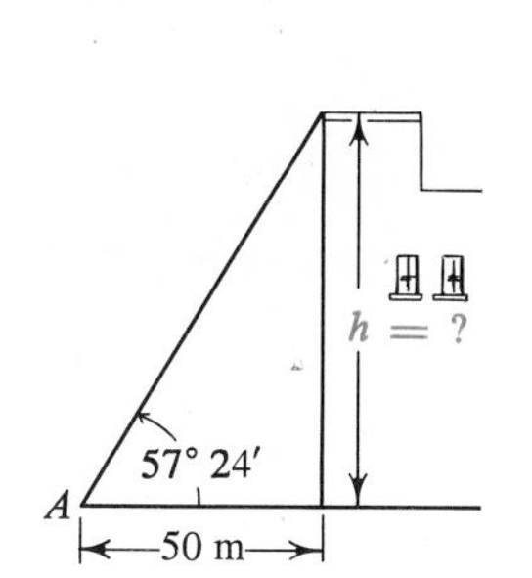

Solution By definition,

$$\tan 57^\circ\ 24' = \frac{h}{50},$$

where h represents the height of the tower. Equivalently,

$$h = 50 \tan 57^\circ\ 24' = 50 \tan 57.4^\circ$$
$$\approx 50(1.5637) \approx 78.2.$$

Thus, the height of the tower is approximately 78.2 meters.

* Significant digits are the digits required for specifying n when a number is written in scientific notation as $n \cdot 10^k$, $k \in J$; $1 \le n < 10$. For example, since $0.030552 = 3.0552 \times 10^{-2}$, an approximation for 0.030552 to *three* significant digits is 0.0306, where the third significant digit was arbitrarily changed from 5 to 6 because the following digit was 5.

EXERCISE SET 2.1

A

Express each of the following in terms of a function value of an angle between 0° and 45°.

Examples

a. $\sin 83.6°$

b. $\sec 63° 24'$

Solutions

a. $\sin 83.6° = \cos(90° - 83.6°)$
$= \cos 16.4°$

b. $\sec 63° 24' = \csc(90° - 63° 24')$
$= \csc 26° 36'$

1. $\cos 62°$

2. $\sec 51°$

3. $\tan 79.3°$

4. $\sin 59.6°$

5. $\csc 88° 12'$

6. $\cot 48° 48'$

Solve each right triangle. In each case, $\gamma = 90°$.

7. $a = 2$ and $b = 3$

8. $c = 41$ and $b = 9$

9. $c = 17$ and $a = 15$

10. $c = 14$ and $\alpha = 17°$

11. $c = 23.5$ and $\beta = 51°$

12. $a = 6.1$ and $\alpha = 73°$

13. $a = 11$ and $\beta = 22° 12'$

14. $b = 5$ and $\alpha = 64° 42'$

15. $b = 15$ and $\alpha = 50° 18'$

16. $c = 20$ and $\beta = 27° 30'$

17. $b = 30$ and $\beta = 41° 36'$

18. $a = 32$ and $\beta = 75° 6'$

Find the area of the indicated triangle.

19. The triangle in Exercise 9.

20. The triangle in Exercise 10.

21. The triangle in Exercise 13.

22. The triangle in Exercise 14.

23. The measure of an angle of elevation from a surveyor's position (ignoring the surveyor's height) to the top of a hill is 37° 24′. The top of the hill is 275 meters above a level line through the surveyor's position. How far is the surveyor from a point directly below the top of the hill?

24. From the top of a vertical cliff, 80 meters above the surface of the ocean, the measure of the angle of depression to a marker on the surface of the ocean is 18° 12′. How far is the marker from the foot of the cliff?

25. A 40-foot ladder is used to reach the top of a 24-foot wall. If the ladder extends 10 feet past the top of the wall, find the measure of the angle that the ladder forms with the horizontal ground.

26. The lengths of the sides of an isosceles triangle are 17, 17, and 30 centimeters. Find the measures of the angles in the triangle. *Hint:* Use the altitude of the triangle.

27. A ladder 30 decimeters long is leaning against the side of a building. If the ladder makes an angle of $25° \ 18'$ with the side of the building, how far up from the ground does the ladder touch the building?

28. A tower 120 feet high casts a shadow 60 feet long. Find the angle of elevation of the sun.

29. A railroad track makes an angle of $4° \ 48'$ with the horizontal. How far must a train go up the track to gain 20 meters in altitude?

30. A rectangle is 20 centimeters long and 14 centimeters wide. Find the angles that a diagonal makes with the sides.

B

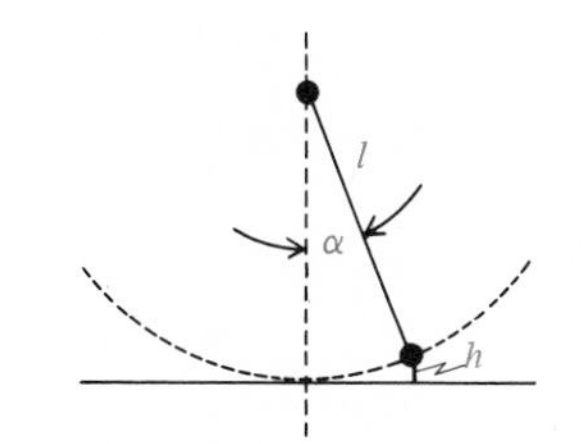

Ex. 31

31. Prove that when a simple pendulum of length l is inclined at an angle α with the vertical, the bob is at a height h above its lowest position, where $h = l(1 - \cos \alpha)$.

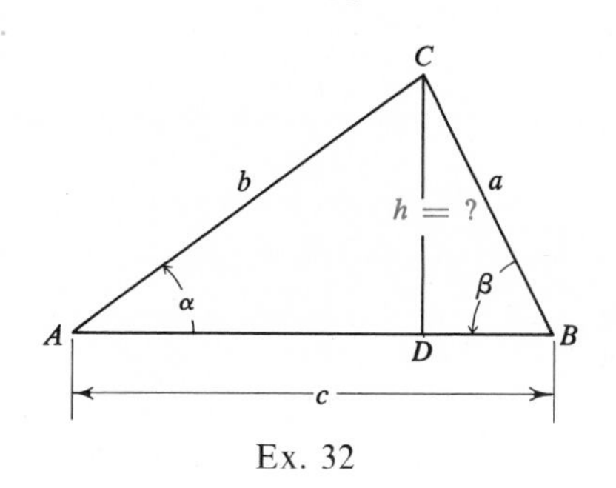

Ex. 32

32. Show that the length h of the perpendicular from C to the side $\overline{AB}$ of any triangle ABC can be found by

$$h = \frac{c}{\cot \alpha + \cot \beta}.$$

Hint: $AD = h \cot \alpha$.

2.2 The Law of Sines

Any side and two angles: use the law of sines

In Section 2.1 we solved right triangles using trigonometric ratios expressed in terms of the lengths of the sides of the triangle. In this section and in the next, special formulas will be derived to enable us to solve any triangle, right, acute, or obtuse.

Consider $\triangle ABC$ in Figure 2.4 (β may be either acute or obtuse), and introduce the altitude $\overline{CD}$. Thus, $\triangle ADC$ and $\triangle BDC$ are right triangles. In either figure on page 56, in $\triangle ADC$,

$$\sin \alpha = \frac{h}{b}, \qquad \text{from which} \qquad h = b \sin \alpha.$$

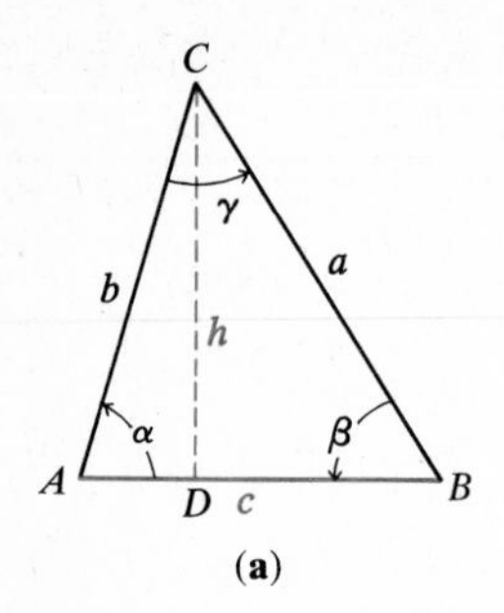

(a)

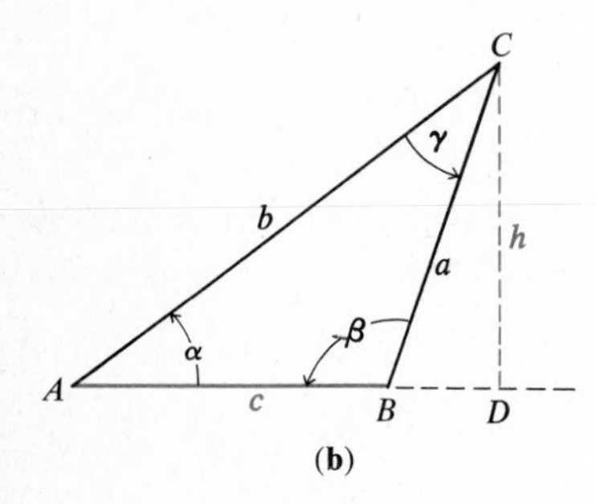

(b)

Figure 2.4

In $\triangle BDC$ in Figure 2.4a,

$$\sin \beta = \frac{h}{a}, \qquad \text{from which} \qquad h = a \sin \beta.$$

In $\triangle BDC$ in Figure 2.4b, $\angle CBD$ is equal to the reference angle for β; therefore,

$$\sin \beta = \sin \angle CBD = \frac{h}{a},$$

$$h = a \sin \beta.$$

Hence, we have in each case

$$h = b \sin \alpha = a \sin \beta.$$

Dividing each member of $b \sin \alpha = \alpha \sin \beta$ by $a \cdot b$, we have

$$\frac{\sin \alpha}{a} = \frac{\sin \beta}{b}. \tag{1}$$

If in Figure 2.4 we use the altitude from vertex A and a similar argument, we will have

$$\frac{\sin \beta}{b} = \frac{\sin \gamma}{c}. \tag{2}$$

Comparing Equations (1) and (2), we find that

$$\boldsymbol{\frac{\sin \alpha}{a} = \frac{\sin \beta}{b} = \frac{\sin \gamma}{c}.}$$

In many applications this relationship, called the **Law of Sines,** is conveniently used in the form

$$\boldsymbol{\frac{a}{\sin \alpha} = \frac{b}{\sin \beta} = \frac{c}{\sin \gamma}.}$$

Two Angles and a Side

The Law of Sines can be used to solve triangles when several kinds of data are given. We consider first the case in which the measures of two angles and the length of one side of a triangle are given.

Example Solve the triangle for which $\alpha = 30°$, $\beta = 45°$, and $a = 20$.

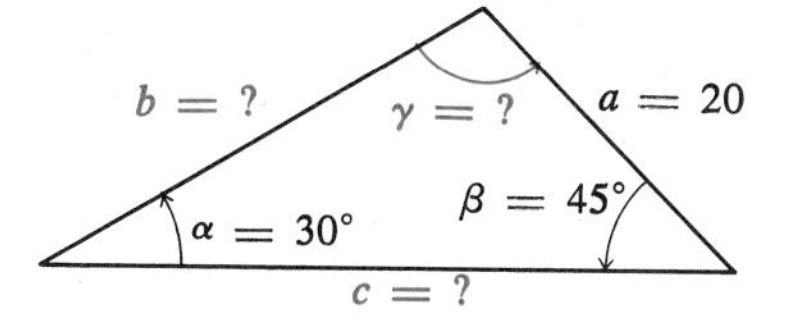

Solution A sketch of the triangle showing the given information is helpful. First, from the fact that $\alpha + \beta + \gamma = 180°$, it follows that

$$\begin{aligned}\gamma &= 180° - \alpha - \beta \\ &= 180° - 30° - 45° = 105°.\end{aligned}$$

From the Law of Sines,

$$\frac{20}{\sin 30°} = \frac{b}{\sin 45°},$$

or, equivalently,

$$b = \frac{20 \sin 45°}{\sin 30°} \approx \frac{20(0.7071)}{0.5000} \approx 28.3.$$

Then, again from the Law of Sines,

$$\frac{20}{\sin 30°} = \frac{c}{\sin 105°},$$

or, equivalently,

$$\begin{aligned}c &= \frac{20 \sin 105°}{\sin 30°} = \frac{20 \sin 75°}{\sin 30°} \approx \frac{20(0.9659)}{0.5000} \\ &= 40(0.9659) \approx 38.6.\end{aligned}$$

Hence,

$$b \approx 28.3, \quad c \approx 38.6, \quad \text{and} \quad \gamma = 105°.$$

Two Sides and an Opposite Angle

A second case in which the Law of Sines is used is one in which the lengths of two sides of a triangle, say a and b, and the measure of an angle opposite one of them, say α, are given. Ambiguity may arise here, depending on a in relation to b and the measure of α. Figure 2.5, in which α is an *acute angle* and h is the length of a perpendicular segment, illustrates the following possibilities that may occur.

1. If $a < h$, the side with length a cannot intersect the side opposite C and no triangle is possible, as shown in Figure 2.5a. In this case, since $h = b \sin \alpha$, we have $a < b \sin \alpha$.

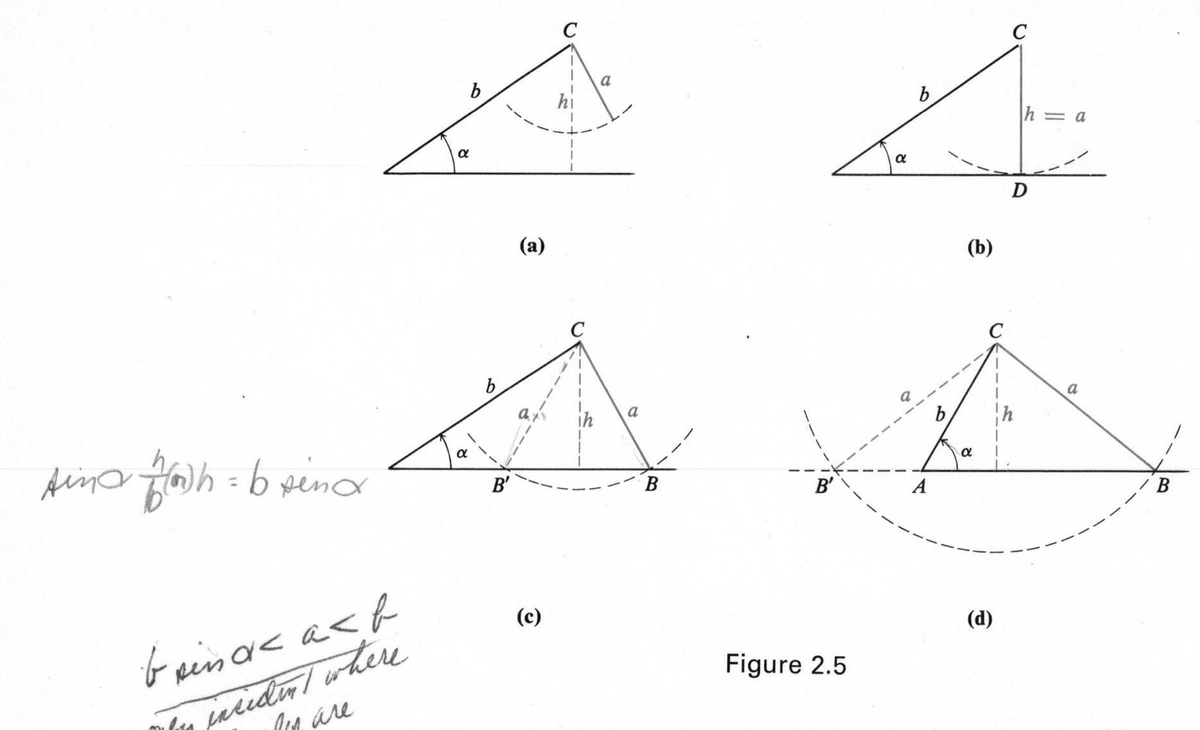

Figure 2.5

2. If $a = h = b \sin \alpha$, the side with length a intersects the side opposite C at D. In this case exactly one right triangle is possible, as shown in Figure 2.5b.
3. If $a > h$ (or $b \sin \alpha$) and $a < b$, or $b \sin \alpha < a < b$, two triangles are possible, as shown in Figure 2.5c.
4. If $a \geq b$, the triangle $AB'C$, shown in Figure 2.5d, does not contain the given angle α and therefore does not lead to a solution. Hence, in this case, only one triangle is possible. In particular, if $a = b$, the triangle is isosceles.

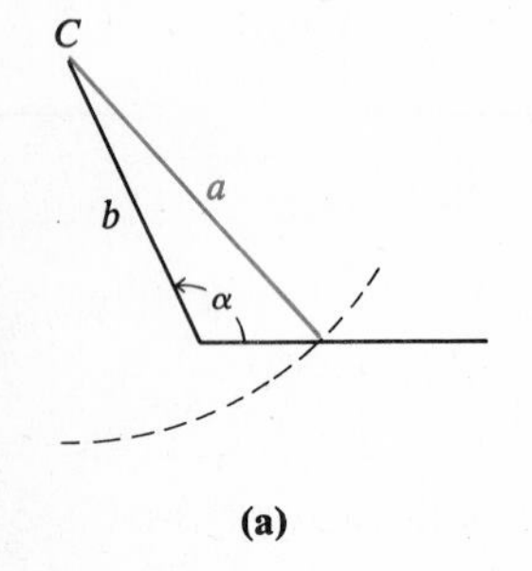

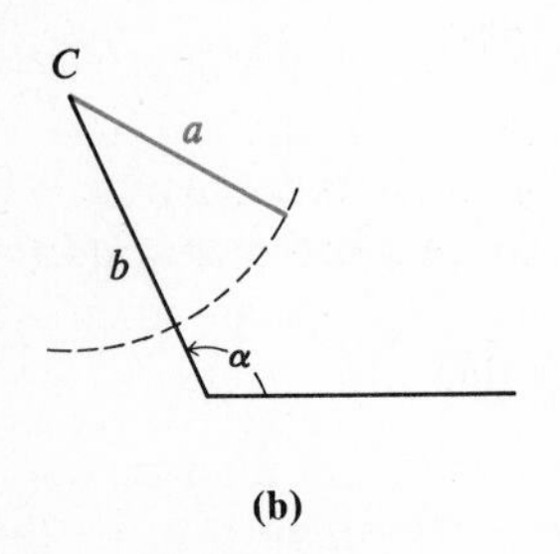

Figure 2.6

Figure 2.6, in which α is an *obtuse angle,* illustrates the following two possibilities.

1. If $a > b$, then one triangle is possible, as shown in Figure 2.6a.
2. If $a \leq b$, then no triangle is possible as shown in Figure 2.6b.

While we have concerned ourselves only with the variables a, b, and α in the above discussion, similar results are valid when other variables are involved in the given information. A sketch of the triangle using the given information will help to indicate the possibilities that exist. In particular, sketches of triangles like those shown in Figures 2.5 and 2.6, in which the given angle is drawn on the left, are usually the most helpful in deciding which possibility exists for given data.

Example Solve the triangle(s) for which $a = 7$, $b = 5$, and $\beta = 30°$.

Solution. Sketch the triangle(s) using the given information. Since the lengths of two sides and the measure of an angle β, opposite the side with measure b, are given, a check must be made to see how many triangles are possible. Since

$$\begin{aligned} h &= a \sin \beta = 7 \sin 30° \\ &= 7(0.5000) = 3.5, \end{aligned}$$

it follows that $h < b < a$ and that two triangles are possible, as indicated in the figure.

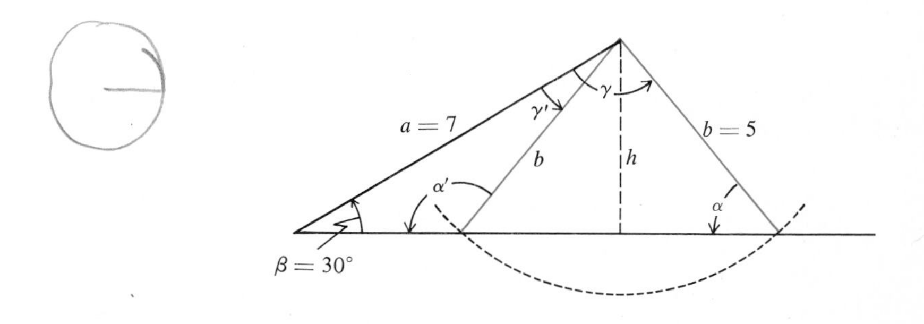

From the Law of Sines,

$$\frac{\sin \alpha}{7} = \frac{\sin 30°}{5}$$

$$\sin \alpha = \frac{7 \sin 30°}{5} = \frac{7(0.5000)}{5} = 0.7000.$$

Since $\sin \alpha > 0$ in Quadrants I and II, to the nearest tenth,

$$\alpha \approx 44.4° \quad \text{or} \quad \alpha' \approx 180° - 44.4° = 135.6°.$$

For $\alpha \approx 44.4°$, the solution of one of the two possible triangles is completed as follows. Since the sum of the measures of the angles of a triangle equals 180°,

$$\begin{aligned} \gamma &= 180° - \alpha - \beta \\ &\approx 180° - 44.4° - 30° = 105.6°. \end{aligned}$$

(*continued*)

From the Law of Sines and the fact that $\sin 105.6° = \sin 74.4°$,

$$\frac{5}{\sin 30°} = \frac{c}{\sin 105.6°};$$

$$c = \frac{5 \sin 74.4°}{\sin 30°} \approx \frac{5(0.9632)}{0.5000} \approx 9.63.$$

For $\alpha' \approx 135.6°$, the solution of the second possible triangle can be completed. Since

$$\gamma' \approx 180° - 135.6° - 30° = 14.4°,$$

then by the Law of Sines

$$\frac{5}{\sin 30°} = \frac{c'}{\sin 14.4°};$$

$$c' \approx \frac{5(0.2487)}{0.5000} \approx 2.49.$$

Thus, the two solutions are

$$\alpha \approx 44.4°, \quad \gamma \approx 105.6°, \quad \text{and} \quad c \approx 9.63;$$

$$\alpha' \approx 135.6°, \quad \gamma' \approx 14.4°, \quad \text{and} \quad c' \approx 2.49.$$

EXERCISE SET 2.2

A

Solve each triangle.

1. $a = 5$, $\alpha = 20°$, and $\beta = 75°$

2. $a = 32$, $\alpha = 26.3°$, and $\gamma = 81.8°$

3. $a = 58.4$, $\beta = 37.2°$, and $\gamma = 100°$

4. $b = 1.9$, $\alpha = 111.7°$, and $\beta = 5.1°$

5. $b = 0.42$, $\alpha = 35° \, 36'$, and $\gamma = 91° \, 30'$

6. $b = 0.88$, $\beta = 63° \, 54'$, and $\gamma = 34° \, 12'$

7. $c = 13.6$, $\alpha = 30° \, 24'$, and $\beta = 72° \, 6'$

8. $c = 1.4$, $\alpha = 135° \, 12'$, and $\gamma = 34° \, 48'$

Determine the number of triangles that satisfy each set of conditions and solve each triangle.

9. $a = 7$, $b = 5$, and $\alpha = 30°$

10. $a = 10$, $b = 9$, and $\beta = 60°$

11. $a = 4.2$, $c = 6.1$, and $\alpha = 32.2°$

12. $a = 38$, $c = 45$, and $\gamma = 35.6°$

13. $b = 16$, $c = 32$, and $\beta = 30°$

14. $b = 0.3$, $c = 0.4$, and $\gamma = 62°$

15. $b = 3.9$, $a = 5$, and $\beta = 42° 42'$

16. $b = 37$, $a = 51$, and $\alpha = 135° 30'$

Find the area of each triangle.

17. $a = 3.4$, $\beta = 13°$, and $\alpha = 50°$

18. $c = 3.8$, $\beta = 15° 6'$, and $\gamma = 50°$

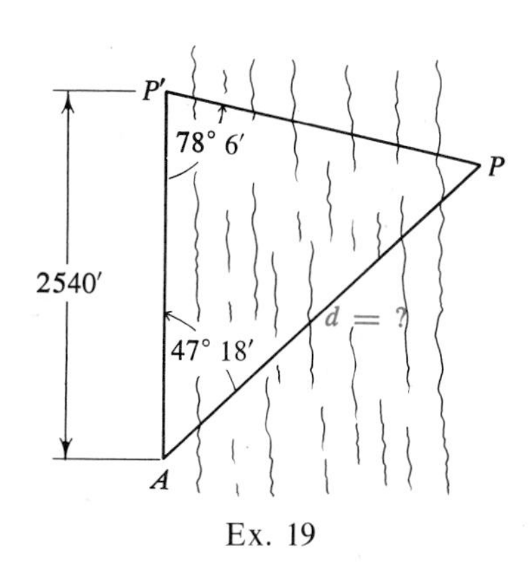

Ex. 19

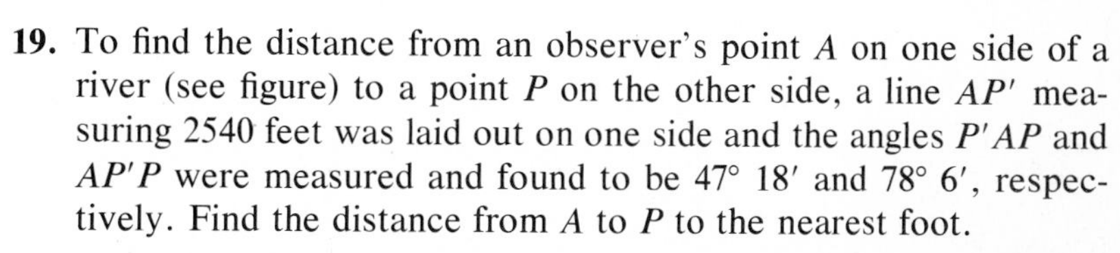

19. To find the distance from an observer's point A on one side of a river (see figure) to a point P on the other side, a line AP' measuring 2540 feet was laid out on one side and the angles $P'AP$ and $AP'P$ were measured and found to be 47° 18′ and 78° 6′, respectively. Find the distance from A to P to the nearest foot.

20. Two men, 400 feet apart, observe a balloon between them that is in a vertical plane with them. The respective angles of elevation of the balloon are observed by the men to be 75° 24′ and 49° 30′. Find the height of the balloon above the ground.

21. One diagonal of a parallelogram is 20 centimeters long and at one end forms angles of 20° and 40° with the sides of the parallelogram. Find the lengths of the sides.

22. A ship at sea is sighted from two points, A and B, on the shore. A and B are 5 kilometers apart, and the line of sight from A to the ship forms an angle of 30° 18′ with $\overline{AB}$, while the line of sight from B to the ship forms an angle of 41° 12′ with $\overline{AB}$. How far is the ship from point A?

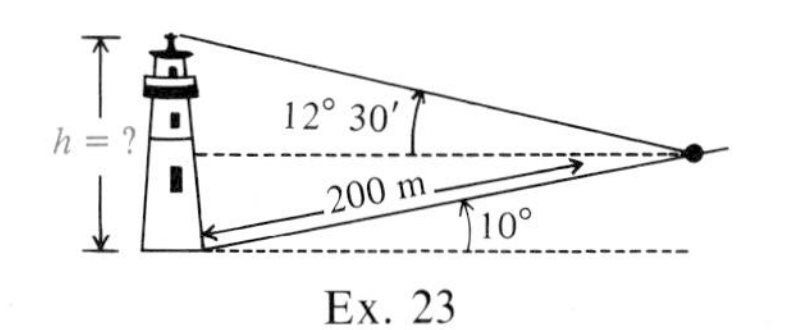

Ex. 23

23. A tower is built vertically on a hillside that slants upward 10° from the horizontal. From a point 200 meters uphill from the foot of the tower, the angle of elevation of the top of the tower is 12° 30′ (see figure). How high is the tower?

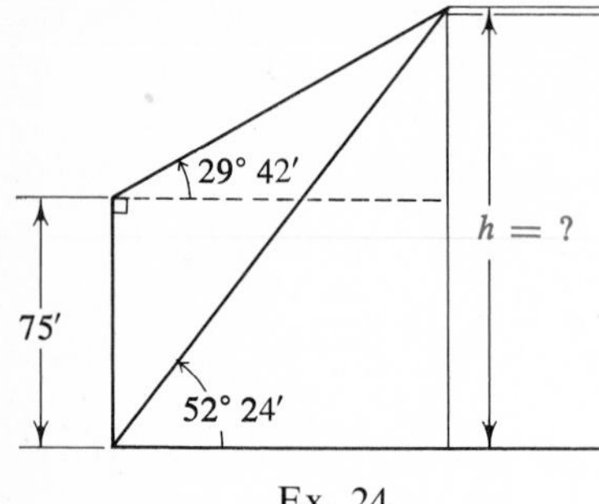

Ex. 24

24. From a window 75 meters above the ground, the measure of the angle of elevation to the top of a nearby building is 29° 42′ (see figure). From a point on the ground directly below the window, the measure of the angle of elevation of the top of the same building is 52° 24′. Find the height of the building.

B

Show that, in any triangle ABC, each of the following statements is true. *Hint:* In Exercises 25 and 26, start with the Law of Sines.

25. $\dfrac{a+b}{b} = \dfrac{\sin\alpha + \sin\beta}{\sin\beta}$

26. $\dfrac{a-b}{b} = \dfrac{\sin\alpha - \sin\beta}{\sin\beta}$

27. $\dfrac{a-b}{a+b} = \dfrac{\sin\alpha - \sin\beta}{\sin\alpha + \sin\beta}$

28. $\dfrac{a+b+c}{b} = \dfrac{\sin\alpha + \sin\beta + \sin\gamma}{\sin\beta}$

2.3 The Law of Cosines

The Law of Sines considered in Section 2.2 cannot be used directly to solve triangles in which the lengths of two sides and the measure of the angle formed by their respective rays are specified or in which only the lengths of the three sides are specified. Equations (1), (2), and (3) below can be used directly in such cases.

If α, β, and γ are the angles of any triangle, and a, b, and c are the lengths of the sides opposite α, β, and γ, respectively, then

$$c^2 = a^2 + b^2 - 2ab\cos\gamma, \tag{1}$$

$$b^2 = a^2 + c^2 - 2ac\cos\beta, \tag{2}$$

$$a^2 = b^2 + c^2 - 2bc\cos\alpha. \tag{3}$$

These statements are called the **Law of Cosines** and follow from a consideration of the distance formula (see Appendix A.4). Orienting triangle ABC with γ at the origin, as shown in Figure 2.7, we find that

$$\cos\gamma = \frac{x}{b} \quad \text{and} \quad \sin\gamma = \frac{y}{b},$$

from which the coordinates of A are

$$x = b\cos\gamma \quad \text{and} \quad y = b\sin\gamma.$$

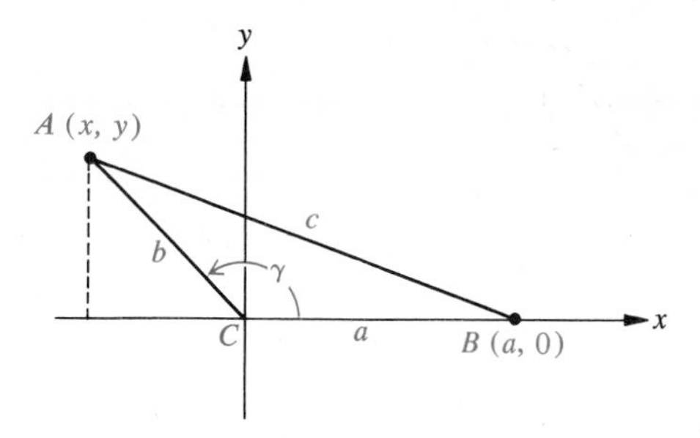

Figure 2.7

Of course, statements (1), (2), and (3) on page 62 would also be true if γ were an acute angle. From the distance formula,

$$\begin{aligned} c^2 &= (y - 0)^2 + (x - a)^2 \\ &= (b \sin \gamma - 0)^2 + (b \cos \gamma - a)^2 \\ &= b^2 \sin^2 \gamma + b^2 \cos^2 \gamma - 2ab \cos \gamma + a^2 \\ &= b^2(\sin^2 \gamma + \cos^2 \gamma) - 2ab \cos \gamma + a^2. \end{aligned}$$

Since $\sin^2 \gamma + \cos^2 \gamma = 1$ (see page 17), we have

$$c^2 = a^2 + b^2 - 2ab \cos \gamma.$$

If the triangle is oriented with A or B at the origin, Equations (2) and (3) in the Law of Cosines can be obtained in a similar way. The particular form of the equation that should be used will, of course, depend on the given information.

Notice that in Equation (1) if $\gamma = 90°$, then $\cos \gamma = \cos 90° = 0$ and the Law of Cosines reduces to

$$c^2 = a^2 + b^2 - 2ab(0),$$
$$c^2 = a^2 + b^2,$$

the familiar formula applicable to right triangles (the Pythagorean theorem).

Two Sides and the Included Angle

Now consider how the Law of Cosines can be used to solve a triangle when the measure of an angle and the lengths of the adjacent sides are known.

Example Given the measures $\beta = 75°$, $a = 7$, and $c = 5$ of an oblique triangle, find the length b.

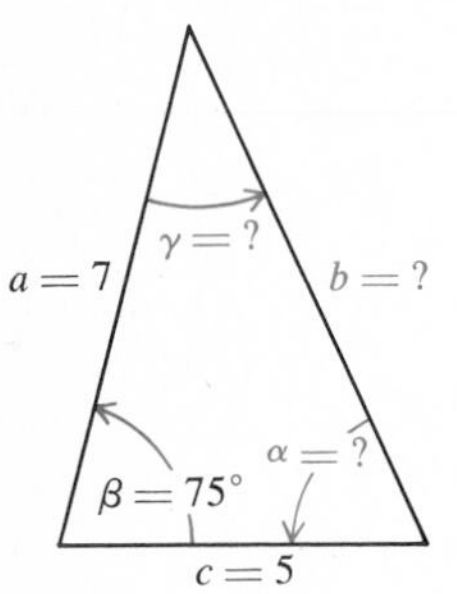

Solution Sketch a triangle showing the given information. The length b can be found by using the Law of Cosines expressed in the form $b^2 = a^2 + c^2 - 2ac \cos \beta$. Substituting the known values gives

$$\begin{aligned} b^2 &= 7^2 + 5^2 - 2(7)(5) \cos 75° \\ &\approx 49 + 25 - 70(0.2588) \approx 55.9; \\ b &\approx \sqrt{55.9}. \end{aligned}$$

Using a calculator,

$$b \approx \sqrt{55.9} \approx 7.48.$$

Alternatively, from Table I, Appendix C, we have $\sqrt{55} \approx 7.416$ and $\sqrt{56} \approx 7.483$. The difference $7.483 - 7.416 = 0.067$. Hence, using linear interpolation,

$$b \approx 7.416 + 0.9(0.067) \approx 7.48.$$

Three Sides

The Law of Sines could be used to complete the solution of the triangle in the above example since the length b, opposite the given angle β, is known. However, we shall use the Law of Cosines in the following example in order to show how it can be used to determine the measure of an angle if the lengths of three sides of a triangle are known.

Example To find the measure of α in the foregoing example, the equation

$$a^2 = b^2 + c^2 - 2bc \cos \alpha$$

is first written in the form

$$\cos \alpha = \frac{b^2 + c^2 - a^2}{2bc}.$$

From the previous calculations, $b^2 \approx 55.9$ and $b \approx 7.48$. Thus,

$$\cos \alpha \approx \frac{55.9 + 25 - 49}{2(7.48)(5)} \approx 0.4265.$$

Because α, an angle in a triangle, must have a measure less than 180°, and $\cos\alpha > 0$, then

$$\alpha \approx 64.8°.$$

Because $\alpha + \beta + \gamma = 180°$,

$$\gamma \approx 180° - 64.8° - 75° = 40.2°.$$

Thus, the solution to the triangle is

$$b \approx 7.48, \quad \alpha \approx 64.8°, \quad \text{and} \quad \gamma \approx 40.2°.$$

If, after first finding the length of the third side, $b = 7.48$ in the above example, the Law of Sines is used to find the measure of a second angle, the angle selected should be the smaller of the two remaining angles (the one opposite the shorter of the two given sides) and hence *an acute angle*. Thus, if the Law of Sines had been used to find the second angle in the above example, we would have found $\sin\gamma$ rather than $\sin\alpha$ because, of the two given sides, $c < a$ and γ is opposite c.

EXERCISE SET 2.3

A

Solve each triangle.

1. $b = 4$, $c = 3.5$, and $\alpha = 71°$

2. $c = 0.3$, $a = 0.1$, and $\beta = 70°$

3. $a = 3.2$, $b = 2.2$, and $\gamma = 75.3°$

4. $b = 6.0$, $a = 5.1$, and $\gamma = 83.5°$

5. $a = 0.7$, $c = 0.8$, and $\beta = 141°\,30'$

6. $b = 3.4$, $a = 2.1$, and $\gamma = 122°\,12'$

7. $b = 1.6$, $c = 3.2$, and $\alpha = 100°\,24'$

8. $c = 2.1$, $b = 4.3$, and $\alpha = 130°\,36'$

Solve the triangles for which the lengths of the three sides are given.

Example $a = 3.4$, $b = 2.7$, and $c = 1.3$. *(continued)*

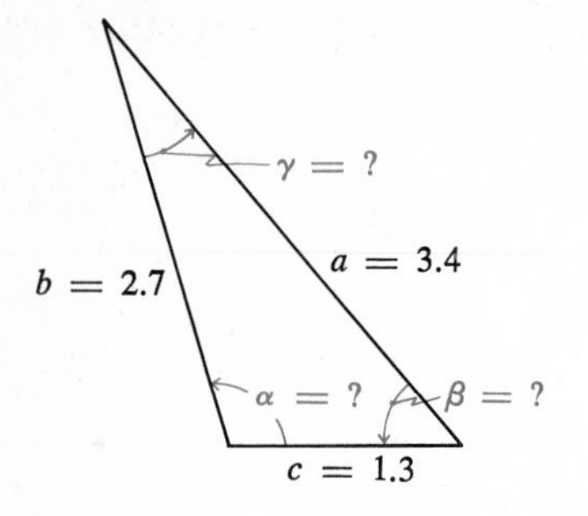

Solution Draw a sketch showing the given information. From the Law of Cosines,

$$\cos\alpha = \frac{b^2 + c^2 - a^2}{2bc} = \frac{(2.7)^2 + (1.3)^2 - (3.4)^2}{2(2.7)(1.3)}$$

$$= \frac{7.29 + 1.69 - 11.56}{7.02} \approx -0.3675.$$

Hence, $\tilde{\alpha} \approx 68.4°$. Then, because $\cos\alpha < 0$, α is an angle such that $90° < \alpha < 180°$. Thus,

$$\alpha \approx 180° - 68.4° = 111.6°.$$

From the Law of Sines,

$$\frac{\sin\alpha}{a} = \frac{\sin\beta}{b},$$

$$\frac{\sin 111.6°}{3.4} = \frac{\sin\beta}{2.7},$$

from which

$$\sin\beta = \frac{2.7(\sin 111.6°)}{3.4} \approx \frac{2.7(0.9298)}{3.4} \approx 0.7384.$$

Thus,

$$\beta \approx 47.6°.$$

From the fact that $\gamma = 180° - \alpha - \beta$,

$$\gamma \approx 180° - 111.6° - 47.6°$$

$$\approx 20.8°.$$

9. $a = 9$, $b = 7$, and $c = 5$

10. $a = 2.7$, $b = 5.1$, and $c = 4.4$

11. $a = 1.2$, $b = 9$, and $c = 10$

12. $a = 4.5$, $b = 7.5$, and $c = 5.8$

13. $a = 6$, $b = 8$, and $c = 12$

14. $a = 6$, $b = 12$, and $c = 13$

15. $a = 4$, $b = 5$, and $c = 6$

16. $a = 5$, $b = 7$, and $c = 8$

17. Find the measure of the smallest angle of the triangle whose sides have lengths 4.3, 5.1, and 6.3.

18. Find the measure of the smallest angle of the triangle whose sides have lengths 3.0, 4.2, and 3.8.

19. Find the measure of the largest angle of the triangle whose sides have lengths 2.9, 3.3, and 4.1.

20. Find the measure of the largest angle of the triangle whose sides have lengths 6.0, 8.2, and 9.4.

21. Find the area of the triangle in Exercise 17.

22. Find the area of the triangle in Exercise 18.

23. A ship is supposed to travel directly from port A to port B, a distance of 10 kilometers. After traveling a distance of 5 kilometers, the captain discovers he has been traveling 15° off course. At this point, how far is the ship from port B?

24. Two ships leave the same port on courses that form an angle of 37°12′. When one ship has traveled 12 kilometers, the other has traveled 18 kilometers. How far apart are the ships at that time?

25. Two sides of a parallelogram that is 6 centimeters by 8 centimeters include an angle of 67°. Find the length of the longer diagonal.

26. A tower 40 meters high is on top of a hill. A cable 50 meters long runs from the top of the tower to a point 20 meters down the hill. Find the angle the hillside forms with a horizontal line.

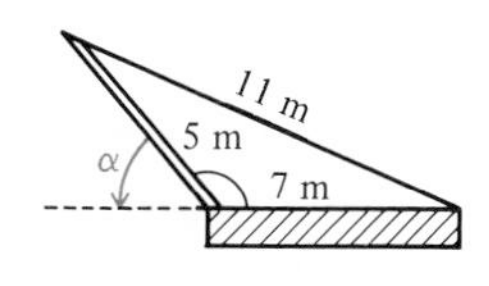

Ex. 27

27. A crane at the edge of a dock is supported by a cable 11 meters long, attached to the dock 7 meters from the base of the crane (see figure). If the crane arm is 5 meters long, find the acute angle that the crane arm forms with the horizontal.

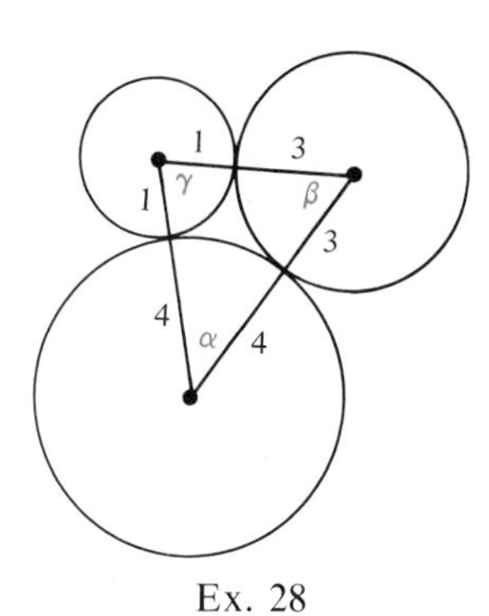

Ex. 28

28. Three circles with radii 1, 3, and 4 centimeters are tangent to each other (see figure). To the nearest 6′, find the three angles formed by the lines joining their centers.

B

The formulas in Exercises 29–32 where $s = \frac{1}{2}(a + b + c)$, *are sometimes used in the solution of triangles. Show that each statement is true for all* $\alpha \in A$. *Hint:* Use the fact that $\cos \alpha = \dfrac{b^2 + c^2 - a^2}{2bc}$.

29. $1 + \cos \alpha = \dfrac{(b + c + a)(b + c - a)}{2bc}$

30. $1 - \cos \alpha = \dfrac{(a - b + c)(a + b - c)}{2bc}$

31. $\dfrac{1 + \cos \alpha}{2} = \dfrac{s(s - a)}{bc}$.

32. $\dfrac{1 - \cos \alpha}{2} = \dfrac{(s - b)(s - c)}{bc}$.

33. Use the results of Exercises 31 and 32 to prove Hero's (or Heron's) formula, $\mathscr{A} = \sqrt{s(s - a)(s - b)(s - c)}$, which can be used to find the area of any triangle directly from the lengths of its three sides.

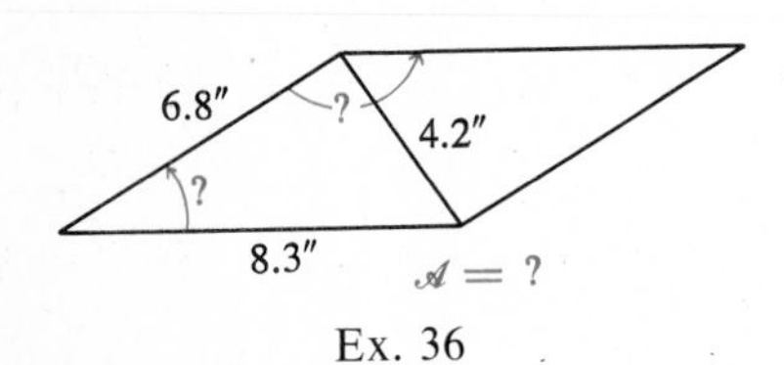

Ex. 36

34. Use Hero's formula to find the area of the triangle in Exercise 9 above.

35. Use Hero's formula to find the area of the triangle in Exercise 11 above.

36. Two sides of a parallelogram are of lengths 6.8 and 8.3 inches, and one of the diagonals is of length 4.2 inches (see figure). Find the area of the parallelogram.

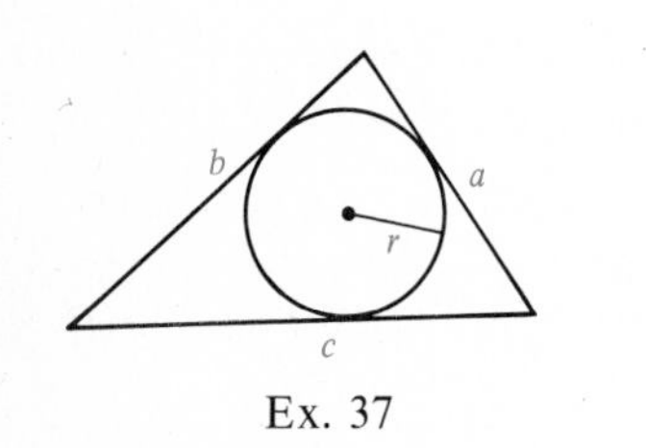

Ex. 37

37. Show that the radius r of an inscribed circle of a triangle (see figure) is given by

$$r = \sqrt{\frac{(s - a)(s - b)(s - c)}{s}}$$

where $s = \frac{1}{2}(a + b + c)$.

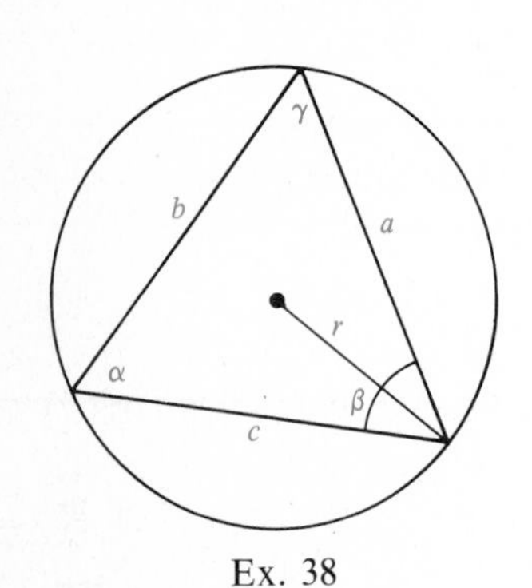

Ex. 38

38. Show that the radius r of a circumscribed circle of a triangle (see figure) is given by

$$r = \frac{a}{2 \sin \alpha} = \frac{b}{2 \sin \beta} = \frac{c}{2 \sin \gamma}.$$

2.4 Geometric Vectors

Some physical quantities, such as length and mass, are characterized by *magnitude alone*. If units of measure are specified, the quantities can be represented by real numbers and associated with points of a number line. In the description of other physical quantities such as force, velocity, and acceleration, *direction as well as magnitude* is required. Such a quantity can be represented by a pair of real numbers and associated with a line segment in the form of an arrow with length equal to the magnitude and pointing in an assigned direction.

Definition 2.1 *A two-dimensional* **geometric vector*** *designated by* $\vec{\mathbf{v}}$ *is a line segment with a specified direction.*

* The term *geometric vector* is used to differentiate this notion of a directed line segment from the notion of a vector as an ordered pair of real numbers.

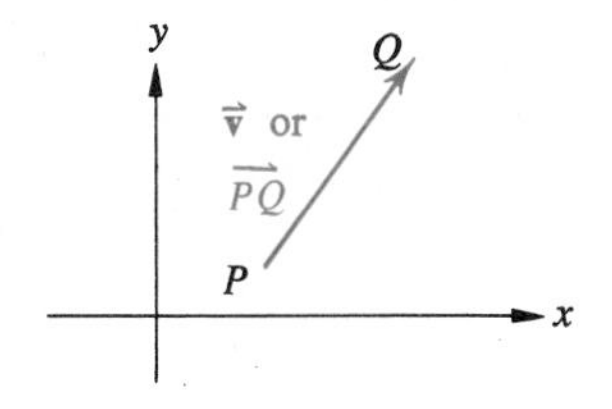

Figure 2.8

If the end points of a geometric vector are named, it is sometimes convenient to specify the vector by using these names. For example, in Figure 2.8 the geometric vector can also be designated by $\overrightarrow{\mathbf{PQ}}$, where the point named P is called the **initial point** and the point named Q is called the **terminal point.** For discussion purposes, the geometric vector in Figure 2.8 is shown in the framework of a rectangular coordinate system.

Sometimes a geometric vector is simply represented by $\mathbf{v}$ without using the arrow above the symbol. We are using the arrow to differentiate this symbol for a geometric vector from the symbol v as a variable for a real number. Also the arrow emphasizes the fact that a geometric vector has *magnitude* and *direction.*

Magnitude and Direction

We shall assume that each geometric vector in the plane has a length that is a real number that depends on the unit of measurement.

Definition 2.2 *For all geometric vectors* $\vec{v}$, *the* **norm** (*or* **magnitude**) *of* $\vec{v}$ *is the length of* $\vec{v}$ *and is designated by* $\|\vec{v}\|$.

When a geometric vector is considered in relation to a rectangular coordinate system, the following concept is useful.

Definition 2.3 *The* **direction angle** *of a geometric vector* $\vec{v}$ *is an angle* α, $-180° < \alpha \leq 180°$, *such that the initial side of* α *is the ray from the initial point of* $\vec{v}$ *parallel to the x-axis and directed in the positive x direction; the terminal side of* α *is the ray from the initial point of* $\vec{v}$ *and containing* $\vec{v}$.

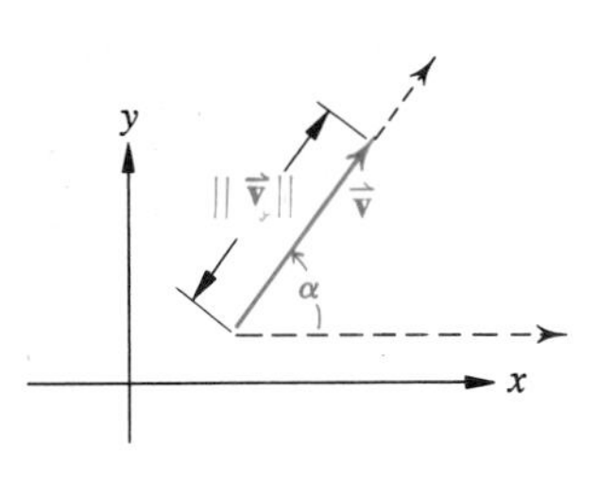

Figure 2.9

Figure 2.9 shows $\|\vec{v}\|$ and the direction angle α associated with a vector $\vec{v}$. From a vector viewpoint, the real number $\|\vec{v}\|$ is called a **scalar.**

Definition 2.4 *For all geometric vectors* $\vec{v}$, *the geometric vector* $-\vec{v}$ *has the same magnitude as* $\vec{v}$ *and has a direction angle whose measure differs from the measure of the direction angle of* $\vec{v}$ *by* 180°.

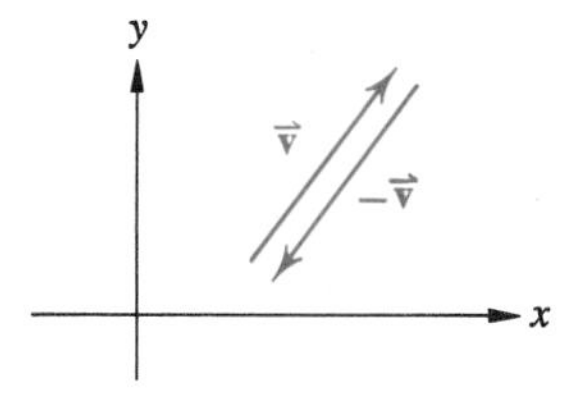

Figure 2.10

Figure 2.10 shows two vectors $\vec{v}$ and $-\vec{v}$.

Definition 2.5 *Two geometric vectors* $\vec{v}_1$ *and* $\vec{v}_2$ *are* **equivalent** *if they have the same magnitude and the same direction angle.*

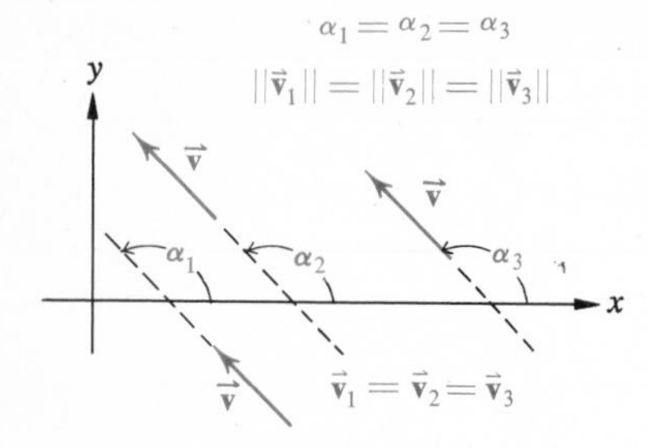

Figure 2.11

Figure 2.11 illustrates three equivalent vectors. We use the symbol = to indicate that two vectors are equivalent.

If we can imagine "sliding" a geometric vector so that it remains parallel to its original position in the plane, then any geometric vector can "slide" onto any vector equivalent to it.

Operations

Geometric vectors are very useful in setting up mathematical models of practical problems. However, before this can be done, certain operations must be defined.

Definition 2.6 *For all geometric vectors $\vec{v}_1$ and $\vec{v}_2$, such that the terminal point of $\vec{v}_1$ is the initial point of $\vec{v}_2$, $\vec{v}_1 + \vec{v}_2$ is the geometric vector having the same initial point as $\vec{v}_1$ and the same terminal point as $\vec{v}_2$. $\vec{v}_1 + \vec{v}_2$ is called the* **sum** *(or* **resultant***) of $\vec{v}_1$ and $\vec{v}_2$.*

As indicated in Figure 2.12a, if $\vec{v}_1$ and $\vec{v}_2$ are not in the relative locations required by Definition 2.6, we "slide" one of them, say $\vec{v}_2$, parallel to its original position into the correct location and then find the sum $\vec{v}_1 + \vec{v}_2$.

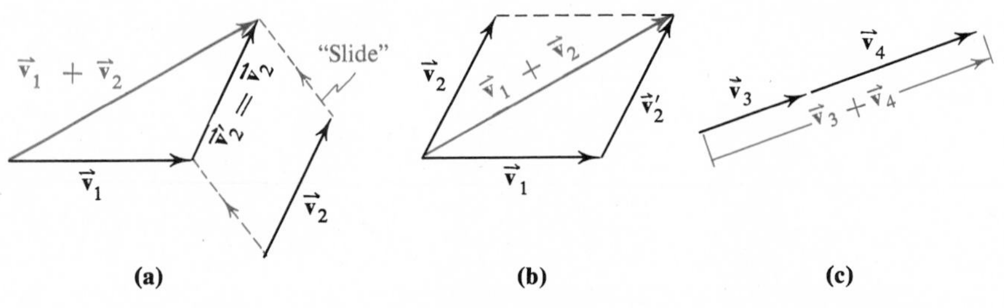

Figure 2.12

Because a triangle is formed by $\vec{v}_1$, $\vec{v}_2$, and $\vec{v}_1 + \vec{v}_2$, we sometimes say that geometric vectors are added according to "the triangle law." From the properties of a parallelogram, we have that, if $\vec{v}_1$ and $\vec{v}_2$ have the same initial point, as shown in Figure 2.12b, the sum $\vec{v}_1 + \vec{v}_2$ is a diagonal of the parallelogram with adjacent sides $\vec{v}_1$ and $\vec{v}_2$. Therefore, we sometimes also say that geometric vectors are added according to "the parallelogram law." Figure 2.12c shows the sum $\vec{v}_3 + \vec{v}_4$ if $\vec{v}_3$ and $\vec{v}_4$ are collinear (vectors lie on the same line).

Note that the symbol "+" in a vector sum is being used in a different sense from that of its ordinary usage to pair two real numbers in a sum, although the operation is commonly referred to as the "addition" of geometric vectors.

Example Construct the sum $\vec{v}_1 + \vec{v}_2$.

$\vec{v}_2$

$\vec{v}_1$

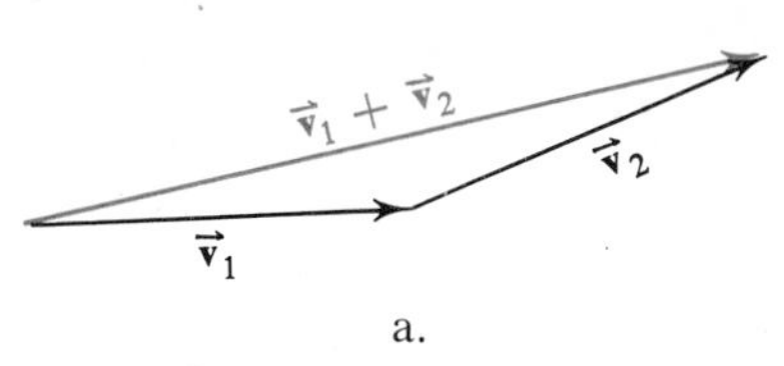

a.

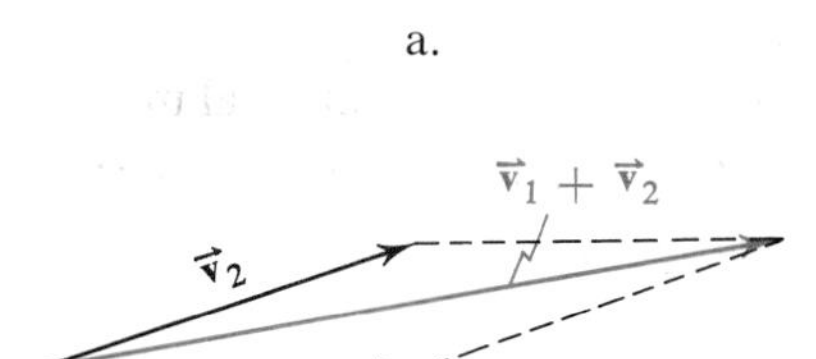

b.

Solution "Slide" the initial point of $\vec{v}_2$ onto the terminal point of $\vec{v}_1$. The sum $\vec{v}_1 + \vec{v}_2$ is the geometric vector from the initial point of $\vec{v}_1$ to the terminal point of $\vec{v}_2$ (figure a).

Alternate Solution "Slide" the initial point of $\vec{v}_2$ onto the initial point of $\vec{v}_1$ and form a parallelogram. The sum $\vec{v}_1 + \vec{v}_2$ is the diagonal of the parallelogram (figure b).

Definition 2.7 *For all geometric vectors $\vec{v}$ and any real number c, $c\vec{v}$ is a geometric vector with magnitude $\|c\vec{v}\| = |c| \cdot \|\vec{v}\|$, having the same direction angle as $\vec{v}$ if $c > 0$ and the same direction angle as $-\vec{v}$ if $c < 0$. The geometric vector $c\vec{v}$ is called a* **scalar multiple** *of $\vec{v}$.*

Example As shown in the figure, $2\vec{v}$ and $\frac{1}{2}\vec{v}$ are geometric vectors in the same direction as $\vec{v}$ such that $\|2\vec{v}\| = 2\|\vec{v}\|$ and $\|\frac{1}{2}\vec{v}\| = \frac{1}{2}\|\vec{v}\|$, while $-2\vec{v}$ is in the opposite direction such that $\|-2\vec{v}\| = |-2|\ \|\vec{v}\| = 2\|\vec{v}\|$.

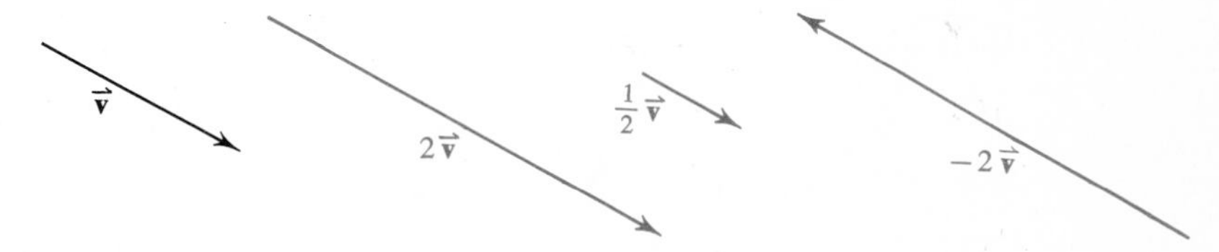

The symbol $\frac{\vec{v}}{c}$ is sometimes used to designate $\frac{1}{c}\vec{v}$. Thus, in the previous example, $\frac{1}{2}\vec{v}$ can be written $\frac{\vec{v}}{2}$.

EXERCISE SET 2.4

A

In Exercises 1–40, use the geometric vectors on page 72.

If $\|\vec{v}_1\| = 1$, estimate the magnitude and the measure of the direction angle for each of the following geometric vectors, where $\vec{v}_1$ is in the direction of the positive x-axis.

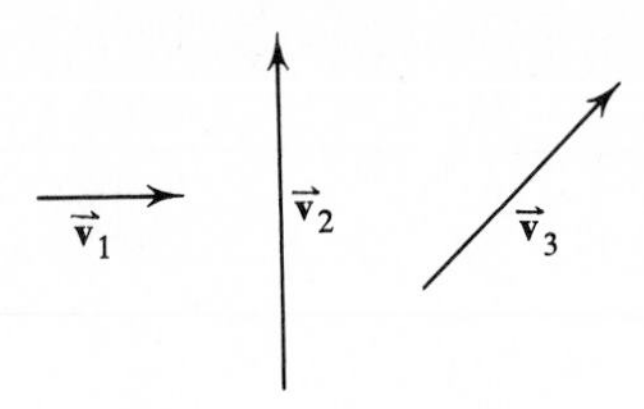

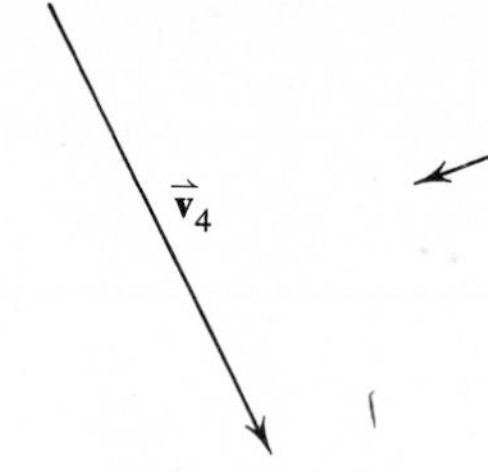

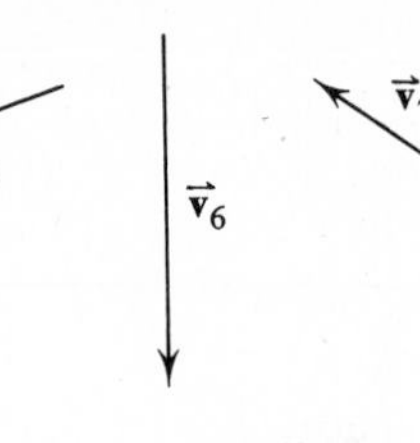

1. $\vec{v}_2$
2. $\vec{v}_3$
3. $\vec{v}_4$
4. $\vec{v}_5$
5. $\vec{v}_6$
6. $\vec{v}_7$

Using freehand methods, draw a single vector representing each of the following sums.

Example $(\vec{v}_1 + \vec{v}_3) + \vec{v}_7$

Solution See figure.

7. $\vec{v}_1 + \vec{v}_2$
8. $\vec{v}_1 + \vec{v}_3$
9. $\vec{v}_1 + \vec{v}_4$
10. $\vec{v}_1 + \vec{v}_5$
11. $\vec{v}_1 + \vec{v}_6$
12. $\vec{v}_1 + \vec{v}_7$
13. $\vec{v}_2 + \vec{v}_3$
14. $\vec{v}_2 + \vec{v}_4$
15. $(\vec{v}_2 + \vec{v}_5) + \vec{v}_6$
16. $(\vec{v}_2 + \vec{v}_6) + \vec{v}_7$
17. $(\vec{v}_2 + \vec{v}_3) + \vec{v}_5$
18. $\vec{v}_3 + (\vec{v}_4 + \vec{v}_7)$

Using freehand methods, draw each of the following.

19. $-\vec{v}_1$
20. $-\vec{v}_3$
21. $2\vec{v}_7$
22. $4\vec{v}_1$
23. $-2\vec{v}_5$
24. $-\vec{v}_6$
25. $\frac{1}{2}\vec{v}_3$
26. $-\frac{1}{3}\vec{v}_4$
27. $2\vec{v}_1 + 3\vec{v}_5$
28. $4\vec{v}_7 + 2\vec{v}_6$
29. $\frac{1}{2}\vec{v}_4 + \vec{v}_7$
30. $\vec{v}_2 + \frac{3}{4}\vec{v}_3$
31. $2\vec{v}_3 + (-3\vec{v}_1)$
32. $\vec{v}_7 + (-2\vec{v}_2)$

B

By appropriate freehand sketches show that each statement is true.

33. $\vec{v}_2 + \vec{v}_3 = \vec{v}_3 + \vec{v}_2$
34. $\vec{v}_5 + \vec{v}_4 = \vec{v}_4 + \vec{v}_5$
35. $(\vec{v}_6 + \vec{v}_7) + \vec{v}_2 = \vec{v}_6 + (\vec{v}_7 + \vec{v}_2)$
36. $\vec{v}_3 + (\vec{v}_7 + \vec{v}_5) = (\vec{v}_3 + \vec{v}_7) + \vec{v}_5$
37. $2(\vec{v}_3 + \vec{v}_5) = 2\vec{v}_3 + 2\vec{v}_5$
38. $(2 + 3)\vec{v}_7 = 2\vec{v}_7 + 3\vec{v}_7$
39. $\frac{1}{3}\vec{v}_1 + 2\vec{v}_1 = \frac{7}{3}\vec{v}_1$
40. $\vec{v}_2 + (-2\vec{v}_2) = -\vec{v}_2$

2.5 Applications of Geometric Vectors

Many kinds of practical problems involving such quantities as velocities, accelerations, forces, and directed line segments can be solved by the use of geometric vectors.

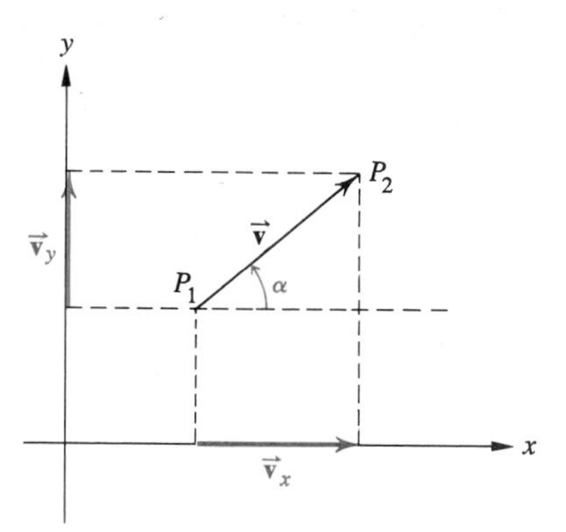

Figure 2.13

Projections

First, consider the *projections* of a geometric vector on the x- and y-axes of a Cartesian coordinate system. Figure 2.13 depicts a geometric vector $\vec{\mathbf{v}}$ and perpendicular lines drawn to the x- and y-axes from the endpoints P_1 and P_2 of $\vec{\mathbf{v}}$. The geometric vector projections $\vec{\mathbf{v}}_x$ and $\vec{\mathbf{v}}_y$ of $\vec{\mathbf{v}}$ on the x-axis and y-axis, respectively, are called the **rectangular components** of $\vec{\mathbf{v}}$; $\vec{\mathbf{v}}_x$ is called the **horizontal component** of $\vec{\mathbf{v}}$, and $\vec{\mathbf{v}}_y$ is called the **vertical component.** These components are determined by the magnitude of $\vec{\mathbf{v}}$ and its direction. Furthermore,

$$\vec{\mathbf{v}} = \vec{\mathbf{v}}_x + \vec{\mathbf{v}}_y.$$

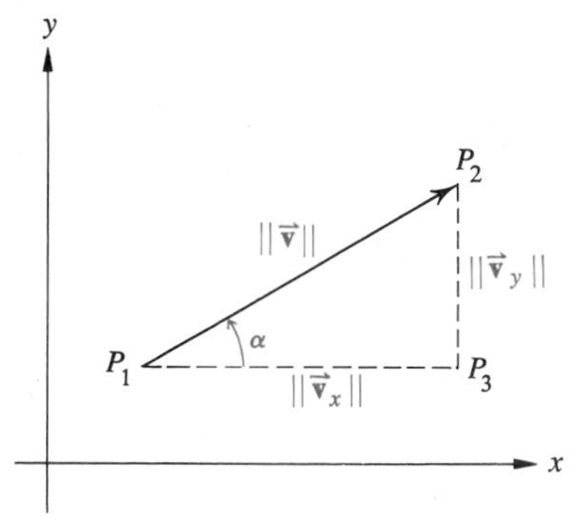

Figure 2.14

In Figure 2.14, $\overline{P_1P_3}$ is drawn parallel to the x-axis, and $\overline{P_2P_3}$ is drawn parallel to the y-axis. Since

$$|\sin \alpha| = \frac{\|\vec{\mathbf{v}}_y\|}{\|\vec{\mathbf{v}}\|},$$

$$\|\vec{\mathbf{v}}_y\| = \|\vec{\mathbf{v}}\| \cdot |\sin \alpha|. \tag{1}$$

Also, since

$$|\cos \alpha| = \frac{\|\vec{\mathbf{v}}_x\|}{\|\vec{\mathbf{v}}\|},$$

$$\|\vec{\mathbf{v}}_x\| = \|\vec{\mathbf{v}}\| \cdot |\cos \alpha|. \tag{2}$$

In Figure 2.14, α is shown to be a positive acute angle. From Definition 2.3, we have that $-180° < \alpha \leq 180°$ and hence $\cos \alpha$ or $\sin \alpha$ may be negative. Therefore because $\|\vec{\mathbf{v}}_x\|$, $\|\vec{\mathbf{v}}_y\|$, and $\|\vec{\mathbf{v}}\|$ are all positive, we use $|\sin \alpha|$ and $|\cos \alpha|$ in Equations (1) and (2).

If the geometric vectors $\vec{\mathbf{v}}_x$ and $\vec{\mathbf{v}}_y$ are known, the magnitude $\|\vec{\mathbf{v}}\|$ and direction angle α of $\vec{\mathbf{v}}$ can be determined from the relationships in a right triangle, as shown in Figure 2.14:

$$\|\vec{\mathbf{v}}\| = \sqrt{\|\vec{\mathbf{v}}_x\|^2 + \|\vec{\mathbf{v}}_y\|^2},$$

and

$$|\sin \alpha| = \frac{\|\vec{\mathbf{v}}_y\|}{\|\vec{\mathbf{v}}\|} \quad \text{or} \quad |\cos \alpha| = \frac{\|\vec{\mathbf{v}}_x\|}{\|\vec{\mathbf{v}}\|}.$$

Example The horizontal component, $\vec{\mathbf{v}}_x$, has its initial point at (4, 0) with $\|\vec{\mathbf{v}}_x\| = 6$. The vertical component, $\vec{\mathbf{v}}_y$, has its initial point at (0, 3) with $\|\vec{\mathbf{v}}_y\| = 8$. Find α and $\|\vec{\mathbf{v}}\|$ and sketch $\vec{\mathbf{v}}$. Assume $\vec{\mathbf{v}}_x$ and $\vec{\mathbf{v}}_y$ are directed in a positive direction.

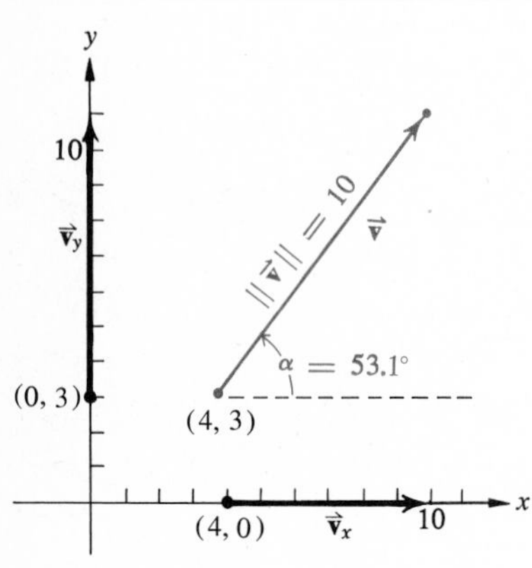

Solution The magnitude

$$\|\vec{\mathbf{v}}\| = \sqrt{\|\vec{\mathbf{v}}_x\|^2 + \|\vec{\mathbf{v}}_y\|^2}$$
$$= \sqrt{6^2 + 8^2} = \sqrt{100} = 10.$$

Because

$$\cos \alpha = \frac{6}{10} = 0.6000,$$
$$\alpha \approx 53.1°.$$

Thus, the geometric vector $\vec{\mathbf{v}}$ has its initial point at (4, 3), has a direction angle whose measure is approximately 53.1°, and has a magnitude of 10 units.

Force

Consider two forces, F_1 and F_2, acting on a particle at the point O, as in Figure 2.15. Since a force has magnitude and direction, we can represent it with a geometric vector, with magnitude given in terms of the units of force and with direction angle given by the direction of the force. It is shown in physics that the two forces represented by $\vec{\mathbf{v}}_1$ and $\vec{\mathbf{v}}_2$ can be replaced by a single equivalent force, F_3, associated with the vector $\vec{\mathbf{v}}_1 + \vec{\mathbf{v}}_2$, the resultant of the forces represented by $\vec{\mathbf{v}}_1$ and $\vec{\mathbf{v}}_2$. The resultant $\vec{\mathbf{v}}_1 + \vec{\mathbf{v}}_2$ can be obtained, as discussed in Section 2.4, by constructing the parallelogram with adjacent sides $\vec{\mathbf{v}}_1$ and $\vec{\mathbf{v}}_2$ and drawing the diagonal $\vec{\mathbf{v}}_1 + \vec{\mathbf{v}}_2$, which is then associated with the resultant force F_3.

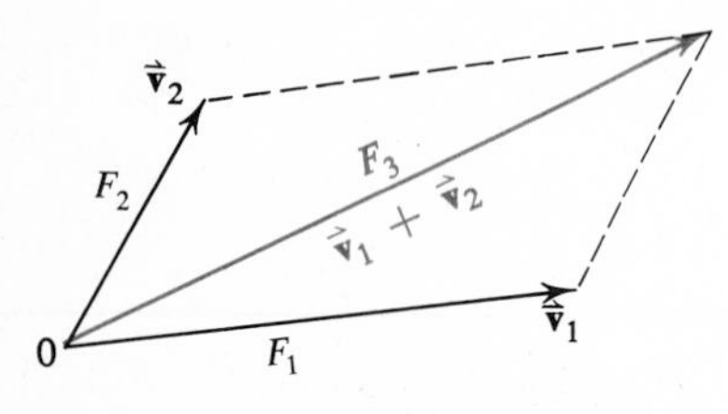

Figure 2.15

Example Two forces, the first one of 5 pounds and the second of 12 pounds, act on a body at right angles to each other. Find the magnitude of the resultant force and the measure of the angle that the resultant force makes with the 5-pound force.

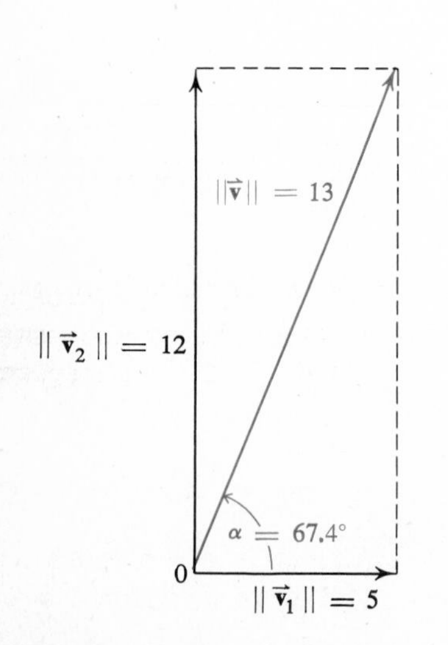

Solution Because $\|\vec{\mathbf{v}}_1\| = 5$ and $\|\vec{\mathbf{v}}_2\| = 12$,

$$\|\vec{\mathbf{v}}\| = \sqrt{5^2 + 12^2} = \sqrt{169} = 13,$$

and

$$\cos \alpha = \frac{5}{13} \approx 0.3846.$$

Hence, $\alpha \approx 67.4°$. Thus, the magnitude of the resultant force is 13 pounds, and the angle it makes with the 5-pound force has a measure of approximately 67.4°.

In cases where the measure of an angle between two forces is not 90°, it is sometimes convenient to use the Law of Cosines and the Law of Sines in order to find the resultant force.

Example A force of 5 pounds and another force of 8 pounds act on an object at an angle of 60° with respect to each other. Find the magnitude of the resultant force and the angle it forms with respect to the 8-pound force.

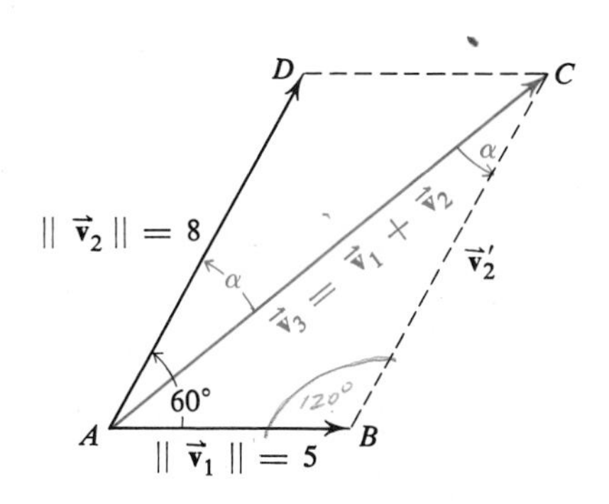

Solution The figure shows the forces of 5 pounds and 8 pounds represented by the geometric vectors $\vec{v}_1$ and $\vec{v}_2$, respectively, and the resultant $\vec{v}_3 = \vec{v}_1 + \vec{v}_2$. Since $\vec{v}_2'$ is equivalent to $\vec{v}_2$, and since the consecutive angles of a parallelogram are supplementary,

$$\angle ABC = 180° - 60° = 120°,$$

and it follows from the Law of Cosines that

$$\begin{aligned}\|\vec{v}_3\|^2 &= \|\vec{v}_1\|^2 + \|\vec{v}_2\|^2 - 2\|\vec{v}_1\| \cdot \|\vec{v}_2\| \cos 120° \\ &= 5^2 + 8^2 - 2(5)(8)(-\cos 60°) \\ &= 25 + 64 + 80(0.5000) = 129.\end{aligned}$$

Therefore, $\|\vec{v}_3\| = \sqrt{129} \approx 11.4$. From the Law of Sines,

$$\frac{\|\vec{v}_1\|}{\sin \alpha} = \frac{\|\vec{v}_3\|}{\sin 120°},$$

or, equivalently,

$$\sin \alpha = \frac{\|\vec{v}_1\| \sin 120°}{\|\vec{v}_3\|} \approx \frac{5(0.8660)}{11.4} \approx 0.3798.$$

Hence, $\alpha \approx 22.3°$. Thus, the forces of 5 and 8 pounds, acting at an angle of 60° with respect to each other, can be represented by the single resultant force of 11.4 pounds. The angle it makes with the 8-pound force has a measure of approximately 22.3°.

Velocity

Geometric vectors also can facilitate the solution of problems involving velocities and accelerations. As examples, we shall look at sev-

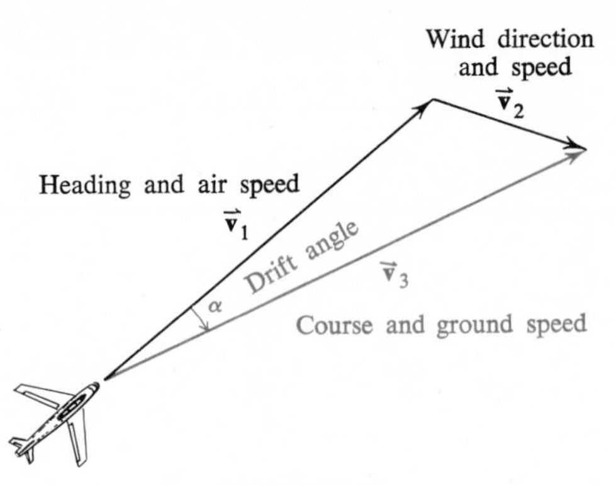

Figure 2.16

eral air navigation problems. First, however, we consider some terminology that is commonly used. The **heading** of an airplane is the direction in which it is pointed, and the **course** is the direction in which it is actually flying over the ground. Its **air speed** is the speed relative to the air, and its **ground speed** is the speed relative to the ground. The two directions and two speeds may differ because of wind effect.

If we use one geometric vector, say $\vec{v}_1$, to represent the heading and air speed of an airplane (see Figure 2.16) and another, say $\vec{v}_2$, to represent the wind direction and speed, then $\vec{v}_3 = \vec{v}_1 + \vec{v}_2$ represents the course and ground speed. The angle α between the geometric vectors $\vec{v}_1$ and $\vec{v}_3$ is called the **drift angle.** The measures of the angles for the heading, wind, and course directions in air navigation are generally given in relation to the Earth's meridian measured clockwise from the ray directed toward true north. This angle is called a **bearing.** In this case, a positive number is assigned as the measure of an angle formed by a rotation in a clockwise direction. In graphical representations, we shall use the positive y-axis as the true north direction, and the positive x-axis as the true east direction. For example, in Figure 2.17, the bearing of the geometric vector $\vec{v}_1$ is 120°, that of $\vec{v}_2$ is 285°, and that of $\vec{v}_3$ is 354°.*

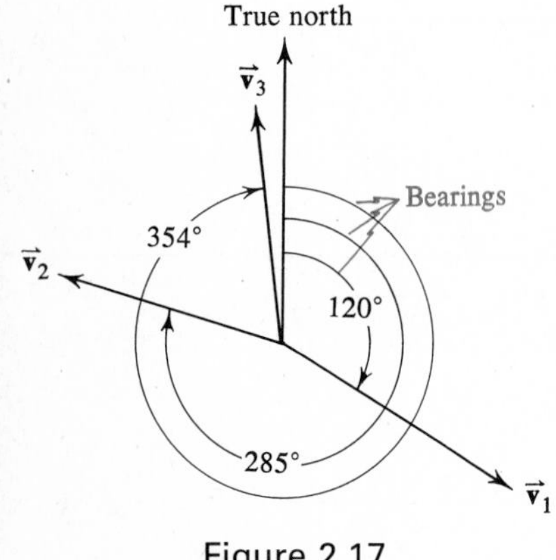

Figure 2.17

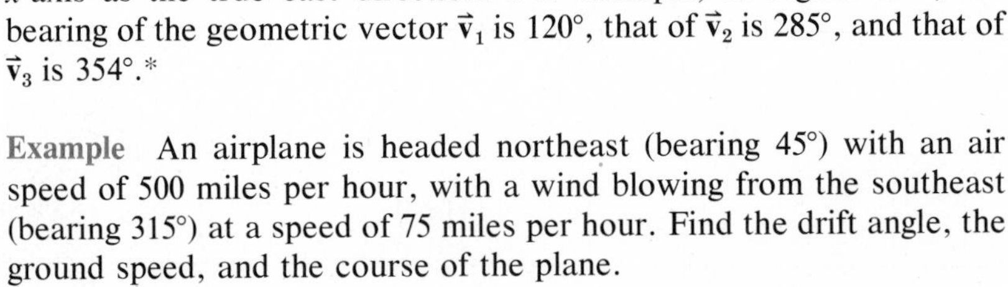

Example An airplane is headed northeast (bearing 45°) with an air speed of 500 miles per hour, with a wind blowing from the southeast (bearing 315°) at a speed of 75 miles per hour. Find the drift angle, the ground speed, and the course of the plane.

* Other types of bearings, each of which has its own reference, or initial, ray from which the angle is measured, are also defined. A bearing used in surveying is designated by an acute angle measured to the east or west of a ray pointing toward true north or true south, for example, N 35° E, N 38° W, S 22° E, and S 47° W.

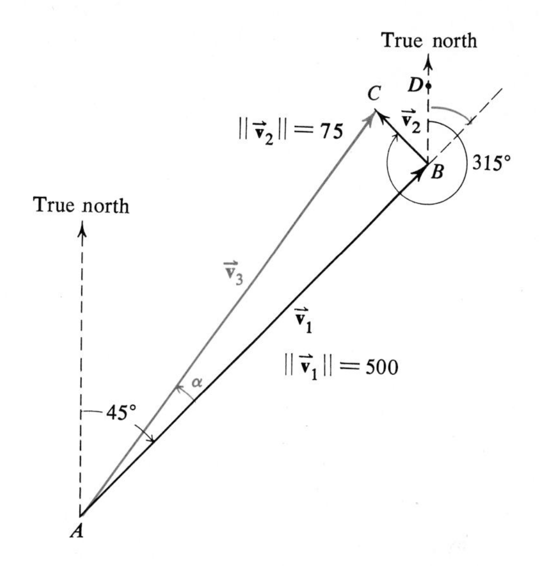

Solution The figure shows three vectors: $\vec{\mathbf{v}}_1$ representing the heading and air speed, $\vec{\mathbf{v}}_2$ representing the wind direction and speed, and $\vec{\mathbf{v}}_3$ representing the course and ground speed. From geometry we have that

$$\angle ABC = 315° - 45° - 180° = 90°.$$

Because $\angle ABC$ is a right angle,

$$\tan \alpha = \frac{75}{500} = 0.1500,$$

and the drift angle $\alpha \approx 8.5°$. The bearing, or direction, of the course is

$$45° - \alpha = 45° - 8.5° = 36.5°.$$

Now, because $\cos \alpha = 500/\|\vec{\mathbf{v}}_3\|$,

$$\|\vec{\mathbf{v}}_3\| = \frac{500}{\cos 8.5°} \approx \frac{500}{0.9890} \approx 506.$$

Thus, the ground speed (magnitude of the course vector) is approximately 506 miles per hour.

EXERCISE SET 2.5

A

The direction angle α *and* $\|\vec{\mathbf{v}}\|$ *are given. Find* $\|\vec{\mathbf{v}}_x\|$ *and* $\|\vec{\mathbf{v}}_y\|$.

Example $\alpha = 60°$ and $\|\vec{\mathbf{v}}\| = 8$

Solution The figure shows the geometric vector $\vec{\mathbf{v}}$ and its rectangular components $\vec{\mathbf{v}}_x$ and $\vec{\mathbf{v}}_y$. From Equations (1) and (2) on page 73,

$$\|\vec{\mathbf{v}}_x\| = \|\vec{\mathbf{v}}\|\,|\cos 60°| = 8\left(\frac{1}{2}\right) = 4,$$

$$\|\vec{\mathbf{v}}_y\| = \|\vec{\mathbf{v}}\|\,|\sin 60°| = 8\left(\frac{\sqrt{3}}{2}\right) = 4\sqrt{3}.$$

1. $\alpha = 45°$ and $\|\vec{\mathbf{v}}\| = 5\sqrt{2}$
2. $\alpha = 30°$ and $\|\vec{\mathbf{v}}\| = 6$
3. $\alpha = 90°$ and $\|\vec{\mathbf{v}}\| = 15$
4. $\alpha = 120°$ and $\|\vec{\mathbf{v}}\| = 2$
5. $\alpha = -30°$ and $\|\vec{\mathbf{v}}\| = 12$
6. $\alpha = -45°$ and $\|\vec{\mathbf{v}}\| = 2\sqrt{2}$
7. $\alpha = -120°$ and $\|\vec{\mathbf{v}}\| = 8$
8. $\alpha = -150°$ and $\|\vec{\mathbf{v}}\| = 4$

In Exercises 9–14, the two given forces act on a point in a plane at an angle with the specified measure. Find the magnitude of the resultant force and the angles the resultant force makes with the given forces.

Example Forces of 10 and 20 kilograms act on a point at an angle of 45° with respect to each other.

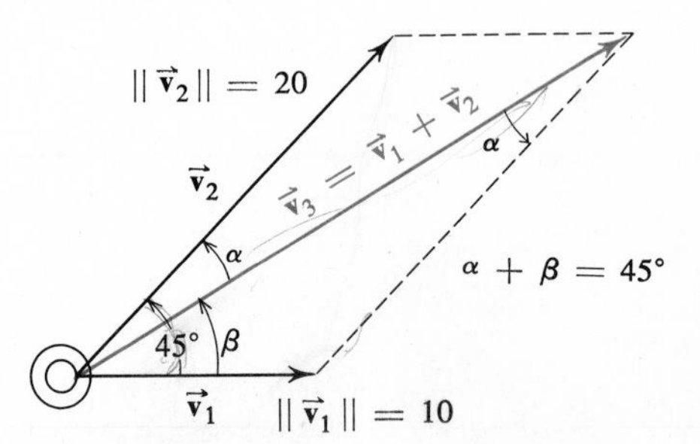

Solution The figure shows the given forces $\vec{\mathbf{v}}_1$ and $\vec{\mathbf{v}}_2$ and the resultant $\vec{\mathbf{v}}_3 = \vec{\mathbf{v}}_1 + \vec{\mathbf{v}}_2$. Since $\alpha + \beta = 45°$, the measure of the angle opposite $\vec{\mathbf{v}}_3$ is 135°. From the fact that $\cos 135° = -\cos 45°$ and from the Law of Cosines,

$$\|\vec{\mathbf{v}}_3\|^2 = 10^2 + 20^2 - 2(10)(20)(-\cos 45°)$$

$$\approx 100 + 400 + 283 \approx 783,$$

and by means of a calculator or by using interpolation from Table I, $\|\vec{\mathbf{v}}_3\| \approx 28.0$. From the Law of Sines,

$$\frac{\|\vec{v}_1\|}{\sin \alpha} = \frac{\|\vec{v}_3\|}{\sin 135^\circ};$$

so

$$\sin \alpha = \frac{\|\vec{v}_1\| \sin 135^\circ}{\|\vec{v}_3\|} \approx \frac{10(0.7071)}{28.0} \approx 0.2525.$$

Hence, $\alpha \approx 14.6^\circ$. If $\alpha + \beta = 45^\circ$, then

$$\beta \approx 45^\circ - 14.6^\circ = 30.4^\circ.$$

Thus, the resultant force is approximately 28.0 kilograms. The angles this force makes with the force of 10 kilograms and with the force of 20 kilograms have measures of approximately 30.4° and 14.6°, respectively.

9. Forces of 3 and 4 pounds, acting at an angle of 90° with respect to each other.

10. Forces of 5 and 7 pounds, acting at an angle of 30° with respect to each other.

11. Forces of 15 and 20 kilograms, acting at an angle of 75° with respect to each other.

12. Forces of 8 and 11 kilograms, acting at an angle of 132° 48′ with respect to each other.

13. Forces of 6 and 10 kilograms acting at an angle of 120° with respect to each other.

14. Forces of 20 and 30 kilograms, acting at an angle of 100° with respect to each other.

15. The resultant of two forces acting at an angle of 90° with respect to each other is 50 kilograms. If one of the forces is 30 kilograms, find the other force and the angle it makes with the resultant.

16. Two equal forces acting at an angle of 60° with respect to each other have a resultant of 80 kilograms. Find the magnitude of each force.

17. Two forces acting at an angle of 60° with respect to each other have a resultant of 50 pounds. If one force acts at an angle of 20° with respect to the resultant, find the magnitude of each force.

18. Two forces, one of 100 kilograms and the other 200 kilograms, have a resultant of 250 kilograms. Find the angle formed by the two forces.

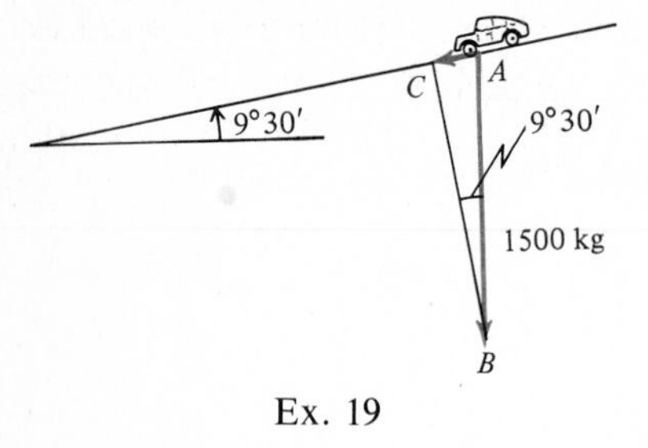

Ex. 19

19. An auto, weighing 1500 kilograms, is on a slope that makes an angle of 9° 30′ with the horizontal. Find the force ($\|\mathbf{AC}\|$ in figure) that pulls the auto down the hill.

20. A steel ball, weighing 50 kilograms, is on an inclined surface. What angle does the plane make with the horizontal if a force of 10 kilograms is pulling the ball down the surface?

21. A force of 110 kilograms is needed to keep a weight of 200 kilograms from sliding down an inclined plane. What angle does the plane make with the horizontal? *Note:* The force that keeps it from sliding is opposite to the force pulling it down the plane, and has the same magnitude.

22. A force of 200 kilograms is needed to keep a weight of 320 kilograms from sliding down an inclined ramp. What angle does the plane make with the horizontal?

23. A bullet has an acceleration of 2 meters/second² in a direction making an angle of 35° with the horizontal. Find the horizontal and vertical components ($\|\vec{\mathbf{v}}_x\|$ and $\|\vec{\mathbf{v}}_y\|$) of its acceleration.

24. A space rocket has an acceleration of 4.2 meters per second² in a direction making an angle of 73° with the horizontal. Find the horizontal and vertical components of the rocket's acceleration.

25. A crew can row a boat at a speed of 9.2 kilometers per hour in still water. If the crew heads across a river at right angles to the current and finds its "drift angle" to be 6°, find the speed of the current.

26. A man can swim at a speed of 2 kilometers per hour in still water. If he heads across a river at right angles to a current of 5 kilometers per hour, find his speed in relation to the land and the direction in which he actually moves.

27. An airplane is headed southeast (bearing 135°) with an air speed of 600 miles per hour, with the wind blowing from the northeast (bearing 225°) at a speed of 120 miles per hour. Find the drift angle, the ground speed, and the course of the airplane.

28. Solve Exercise 27 if the wind is from the southwest (bearing 45°) at a speed of 60 miles per hour.

29. An airplane is headed on a bearing of 250° with an air speed of 425 miles per hour. The course has a bearing of 262°. The ground speed is 475 miles per hour. Find the drift angle, the wind direction, and the wind speed.

30. Solve Exercise 29 if the course has a bearing of 241°.

31. A pilot wishes to fly on a course bearing 90° and with a ground speed of 600 kilometers per hour. If a wind is blowing from the north (bearing 180°) with a speed of 50 kilometers an hour, what must be the heading and air speed of the aircraft?

32. Solve Exercise 31 if the wind is from the northwest (bearing 135°).

B

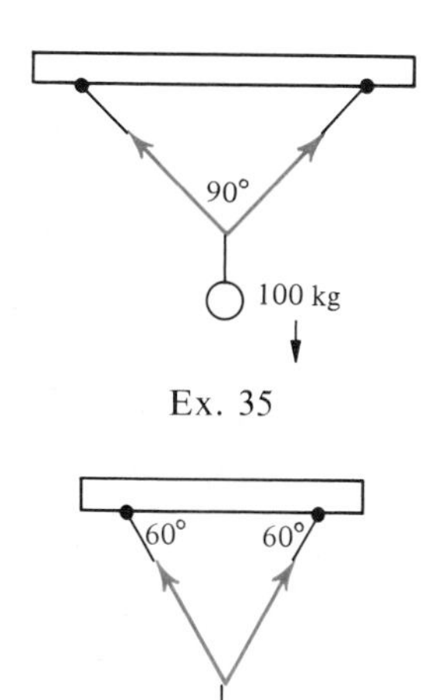

33. Two ships leave a harbor at the same time, the first traveling at 20 knots (nautical miles per hour) on a course bearing 40°, and the second at 25 knots on a course bearing 300°. After one hour, how far apart are they and what is the bearing of a course from the first ship to the second?

34. Three forces of 25, 35, and 45 kilograms act in directions having bearings of 240°, 290°, and 340°, respectively. Find a force, acting due east, and a force, acting due south, that would counterbalance these three forces.

35. A 100-kilogram weight hangs from a rope that makes an angle of 90° where it is attached to the weight (see figure). What is the tension (force) on each half of the rope?

36. A 50-kilogram weight hangs from a rope whose two ends are tied at 60° angles to a support (see figure). What is the tension on each half of the rope?

2.6 Uniform Circular Motion

If the measure of a central angle is in radians, then the length of the intercepted arc of a circle with a given radius can be found directly.

Arc Length

Recall from Section 1.1 that α^{R} indicates how many times the length of the radius r of the circle has been used as a unit length along the arc of a circle. Hence, the length s of an intercepted arc is given by

$$s = r \cdot \alpha^{R}. \tag{1}$$

The length s is a directed length; it is a positive or negative real number as α^{R} is a positive or negative real number, respectively. If the measure of an angle is given in degrees, the measure must first be changed to radian measure before Equation (1) can be applied.

Examples On each circle with the given radius, find the length (to the nearest tenth) of the arc intercepted by the given angle. Use $\pi \approx 3.14$.

a. $r = 4$ cm; $\alpha = 2^{R}$

b. $r = 6$ cm; $\alpha = -150^\circ$

Solutions

a. From Equation (1) on page 81, $s = 4 \cdot 2 = 8.0$. Therefore the arc is 8 centimeters in length.

b. From Equation II (page 3), the related radian measure of -150° is

$$\alpha = \left[\frac{\pi}{180}(-150)\right]^{R} = -\frac{5\pi}{6}^{R}.$$

From Equation (1) above,

$$s = 6\left(-\frac{5\pi}{6}\right) = -5\pi.$$

Since $\pi \approx 3.14$, the arc length to the nearest tenth is -15.7 centimeters.

a.

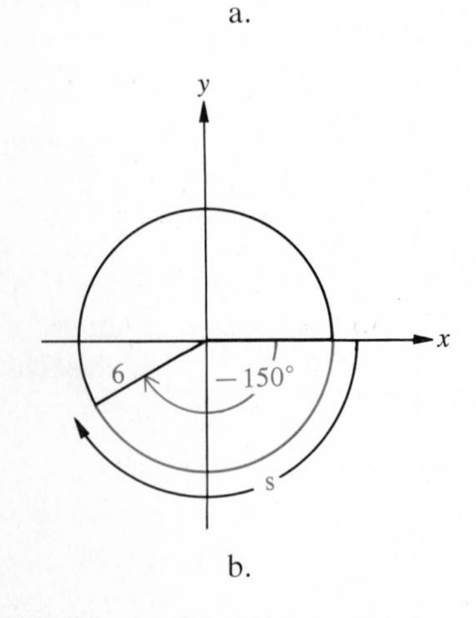

b.

Linear and Angular Velocity

Now consider a point moving along an arc of a circle at a constant (uniform) speed. If the point moves from point P_1 to P_2 in time t, the distance traveled in time t equals the length of the arc $s = r \cdot \alpha^{R}$, as shown in Figure 2.18. The angle α is called the **angle of rotation.**

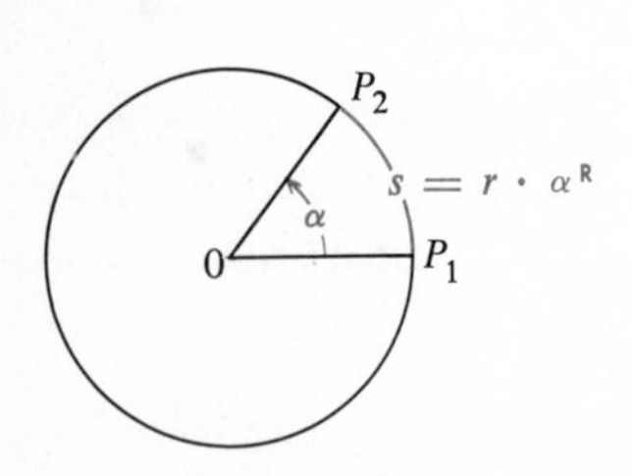

Figure 2.18

Definition 2.8 *The* **linear velocity** v *of a point moving along an arc of a circle at a constant speed is given by*

$$v = \frac{\textbf{directed length of the arc}}{\textbf{time}} = \frac{r \cdot \alpha^{R}}{t}. \tag{2}$$

Example The earth rotates $2\pi^{R}$ in 24 hours. Assuming the radius of the earth to be 4000 miles, the linear velocity of an object on the equator is given by

$$v = \frac{4000 \cdot 2\pi}{24}$$

$$\approx 1047 \text{ miles per hour.}$$

Definition 2.9 *The* **angular velocity ω** *(the Greek omega) of a point moving along an arc of a circle at a constant speed is given by*

$$\omega = \frac{\textbf{directed measure of the angle of rotation}}{\textbf{time}}.$$

If the measure of the angle is in radians,

$$\omega = \frac{\alpha^{\mathrm{R}}}{t}, \tag{3}$$

where ω is given in radians per unit time. If the measure of the angle is in degrees,

$$\omega = \frac{\alpha^{\circ}}{t}, \tag{3a}$$

where ω is given in degrees per unit time.

Notice that the linear velocity and angular velocity are both positive if the motion is in a counterclockwise direction because the length of the arc and the measure of the angle of rotation are both positive. The linear velocity and angular velocity are both negative if the motion is in a clockwise direction.

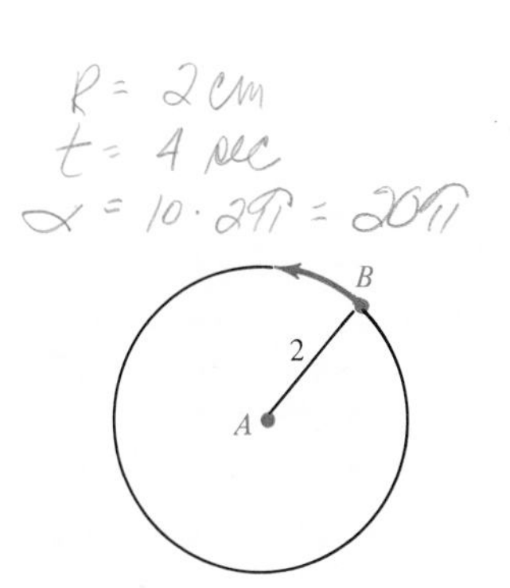

Example In the figure, point B revolves in a circular path about A making 10 complete revolutions in 4 seconds. Find the linear and the angular velocities of point B if the radius of the circle is 2 centimeters. Use $\pi \approx 3.14$.

Solution One revolution corresponds to an angle of rotation of 2π radians. From Definition 2.8, with $r = 2$,

$$v = \frac{2 \cdot 10(2\pi)}{4} = 10\pi \approx 31.4.$$

Hence the linear velocity is approximately 31.4 centimeters per second.

From Definition 2.9,

$$\omega = \frac{10 \cdot 2\pi}{4} = 5\pi \approx 15.7.$$

Therefore, the angular velocity is approximately 15.7 radians per second.

Substituting ω, the left-hand member of (3), for α^R/t in the right-hand member of (2), we have the linear velocity v of a point moving along the arc of a circle at a constant speed given by

$$v = r\omega, \tag{4}$$

where r is the length of the radius of the circle and ω is the angular velocity of the moving point in *radians per unit time*.

Example A turbine with a radius of length 0.74 meter turns at the rate of 10,000 revolutions per minute. What is the linear velocity of a point on the rim of the turbine? Use $\pi \approx 3.14$.

Solution One revolution corresponds to 2π radians. Therefore, 10,000 revolutions per minute corresponds to an angular velocity of $20{,}000\pi$ radians per minute. Using Equation (4), with $r = 0.74$,

$$v = r\omega = 0.74 \cdot 20{,}000\pi \approx 46{,}472.$$

Hence, the linear velocity is approximately 46,500 meters per minute.

EXERCISE SET 2.6

A

On each circle with the given radius, find the length of the arc (to the nearest tenth of a unit) intercepted by each angle with given measure. Use $\pi \approx 3.14$.

Example $r = 9$ centimeters; $240°$

Solution From Equation II on page 3, the related radian measure of $240°$ is

$$\left(\frac{\pi}{180} \cdot 240\right)^R = \frac{4\pi^R}{3}.$$

Thus,

$$s = r \cdot \alpha^R = 9 \cdot \frac{4\pi}{3} = 12\pi \approx 37.7,$$

and the length of the intercepted arc is approximately 37.7 centimeters.

1. $r = 2$ meters; 3.1^R

2. $r = 30$ centimeters; 1.5^R

3. $r = \pi$ decimeters; 150°

4. $r = 2\pi$ meters; 210°

5. $r = 3$ feet; $-135°$

6. $r = 20$ inches; $-240°$

Find the measure (to the nearest tenth of a degree) of the central angle that intercepts each arc on a circle with given radius.

Example $s = 15$ centimeters; $r = 5$ centimeters

Solution Substituting 5 for r and 15 for s in $s = r \cdot \alpha^{\mathrm{R}}$ yields

$$15 = 5 \cdot \alpha^{\mathrm{R}}.$$

Thus

$$\alpha^{\mathrm{R}} = \frac{15^{\mathrm{R}}}{5} = 3^{\mathrm{R}}.$$

From Equation I on page 3,

$$\alpha^{\circ} = \left(\frac{180}{\pi} \cdot 3\right)^{\circ} \approx 171.9°.$$

7. $s = 31$ decimeters; $r = 14$ decimeters

8. $s = 2.5$ meters; $r = 2.5$ meters

9. $s = 4.9$ centimeters; $r = 1$ centimeter

10. $s = 14.20$ inches; $r = 1.27$ inches

In Exercises 11–16, consider a flywheel with uniform circular motion, where v is the linear velocity of a point on the rim and ω is the angular velocity. Find v, r, or ω to the nearest tenth of a unit. Use $\pi \approx 3.14$.

Example Find ω in degrees per hour if $v = 0.5$ meter per minute and $r = 0.63$ meter.

Solution 0.5 meter per minute is 60(0.5) or 30 meters per hour. From Equation (4) above, we have that

$$30 = 0.63\omega,$$

from which ω equals approximately 47.6 radians per hour. Therefore, in degrees per hour,

$$\omega = 47.6\left(\frac{180}{\pi}\right) \approx 2728.7.$$

11. Find v if $r = 0.3$ meter and $\omega = 12$ radians per second.

12. Find r if $v = 10$ centimeters per second and $\omega = 10$ radians per second.

13. Find ω in radians per second if $v = 180$ centimeters per second and $r = 6$ centimeters.

14. Find ω in radians per second if $v = 4$ meters per second and $r = 0.2$ meter.

15. Find v if $r = 9$ decimeters and $\omega = -3$ revolutions per second.

16. Find v if $r = 4$ decimeters and $\omega = -120$ degrees per second.

17. The center line of a freeway curve is laid out as an arc of a circle of radius 300 meters. What is the length of the center line of the curve if it subtends an angle of 20° at the center of the circle?

18. A pendulum 24 decimeters long oscillates 4.2° on each side of its vertical position. Find the length of arc through which the end of the pendulum swings.

19. Two points on the earth's surface (assuming the earth to be a perfect sphere) are located so that the radii from the center of the earth to each point form an angle of 25°. Assuming the earth's radius to be 4000 miles, find the distance between the points measured along the earth's surface.

20. A wheel 10 inches in diameter is turning at a rate of 200 revolutions per minute. If it is rolling on the ground without slipping, how far *in feet* will it travel in 5 minutes?

21. A belt moving at a rate of 100 centimeters per minute is turning a pulley at an angular velocity of 8 revolutions per minute. Find the pulley diameter.

22. The rear gear of a bicycle has a radius of 10 centimeters and the front gear has a radius of 25 centimeters. Through what angle (in radians) does the front gear turn if the back gear turns through 5 radians?

23. A propeller, 2 meters from tip to tip, is rotating at a rate of 800 revolutions per minute. Find the linear velocity of a tip of the propeller in meters per minute.

24. Find the linear velocity, in centimeters per minute, of the tip of an hour hand of a clock, if the hour hand has a length of 1.5 decimeters.

25. A racing driver moves around a curved track at a speed of 160 miles per hour. If the radius of the circle, of which the driver's path is an arc, is 1500 feet, what is the driver's angular speed in radians per minute around the curve?

26. Using 8000 miles for the diameter of the earth, find the linear velocity, in miles per hour, of a point on the equator.

27. Assuming that the earth's orbit about the sun is circular, what is the angular velocity of the earth in radians per hour (assuming 365 days per year and 24 hours per day)?

28. Using the information in Exercise 27, what is the linear velocity in miles per hour of the earth in its orbit around the sun if the circular path of the earth has a radius of 9.3×10^7 miles?

29. Assuming that the moon's orbit about the earth is circular with a radius of 2.4×10^5 miles and that it takes 28 days for one orbit, what is the linear velocity of the moon in miles per hour?

30. The planet Mars moves around the sun in 686 earth-days. If the radius of Mars' orbit (assumed to be circular) around the sun is approximately 227,000,000 kilometers, find the linear velocity of Mars in its orbit in kilometers per hour. Assume 1 earth-day equals 24 hours.

B

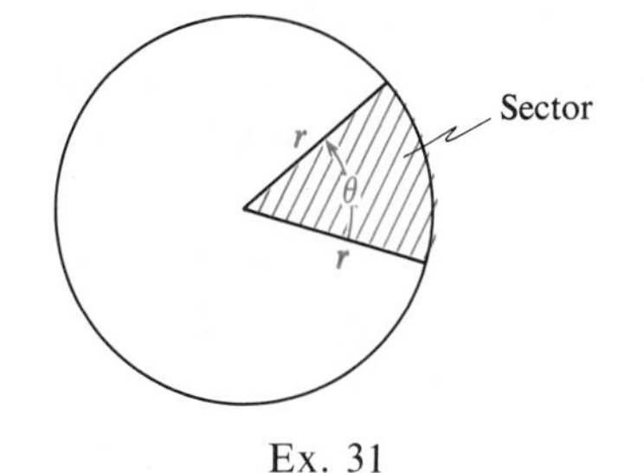

Ex. 31

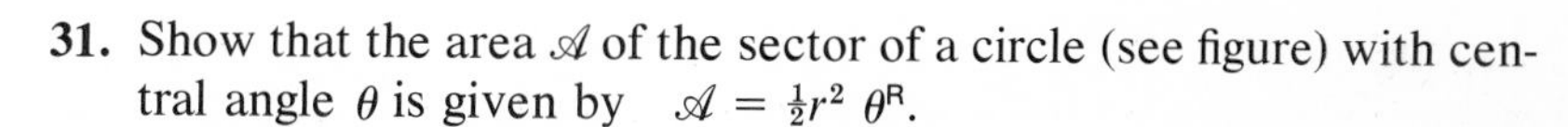

31. Show that the area $\mathscr{A}$ of the sector of a circle (see figure) with central angle θ is given by $\mathscr{A} = \frac{1}{2}r^2\,\theta^{\mathsf{R}}$.

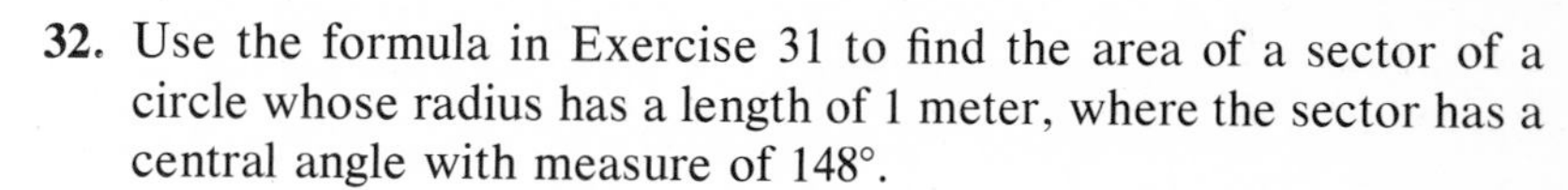

32. Use the formula in Exercise 31 to find the area of a sector of a circle whose radius has a length of 1 meter, where the sector has a central angle with measure of 148°.

2.7 Simple Harmonic Motion

In Section 1.7 we noted that the sine function has an infinite number of elements in the domain (set of all angles) that are paired with a given element in the range. For example,

$$\sin\frac{\pi}{6}^{\mathsf{R}} = \frac{1}{2}, \qquad \sin\frac{13\pi}{6}^{\mathsf{R}} = \frac{1}{2}, \qquad \sin\left(-\frac{11\pi}{6}\right)^{\mathsf{R}} = \frac{1}{2}.$$

In fact, for all $\alpha \in A$ and $k \in J$,

$$\sin(\alpha + k \cdot 2\pi)^{\mathsf{R}} = \sin\alpha^{\mathsf{R}}. \tag{1}$$

The property of being *periodic*, illustrated by the example above, is possessed by many functions, including each of the trigonometric functions. It is defined as follows.

Definition 2.10 *A function f is* **periodic** *if there exists some number a ($a \neq 0$) such that, for all elements x in the domain, $x + a$ is in the domain and*

$$f(x + a) = f(x).$$

The smallest positive value of a for which this is true is the **period** *of the function.*

In Equation (1), if $k = 1$, we have

$$\sin(\alpha + 2\pi)^{\mathrm{R}} = \sin \alpha^{\mathrm{R}}.$$

Hence, the period of the sine function is $2\pi^{\mathrm{R}}$.
The cosine function is also periodic because

$$\cos(\alpha + k \cdot 2\pi)^{\mathrm{R}} = \cos \alpha^{\mathrm{R}}, \quad k \in J.$$

Since for $k = 1$,

$$\cos(\alpha + 2\pi)^{\mathrm{R}} = \cos \alpha^{\mathrm{R}},$$

it follows that the period of the cosine function is also $2\pi^{\mathrm{R}}$.

Circular Functions

We shall now introduce some functions that are closely related to the trigonometric functions and can be used as models for periodic phenomena. First, however, we note that it is common practice to omit the letter ^{R} in specifying the radian measure of an angle. Thus, we write expressions such as $\sin \dfrac{\pi}{6}^{\mathrm{R}}$ and $\cos 2^{\mathrm{R}}$ as $\sin \dfrac{\pi}{6}$ and $\cos 2$, respectively.

The trigonometric functions have been defined so that the *domains are sets of angles* and the ranges are sets of real numbers. Let us now consider functions in which the *domains are sets of numbers* that are measures, in radian units, of the respective angles and whose ranges are the same as the respective ranges of the analogous trigonometric functions. Because the radian measures of angles are determined by the lengths of arcs on a circle (see Section 1.1), these functions are called **circular functions.** Furthermore, each of the six circular functions is given the same name as the corresponding trigonometric function.

Thus, we can view an expression such as $\sin(\pi/6)$ as an element in the range of the *trigonometric* sine function, where the corresponding element in the domain is an *angle* with a measure of $\pi/6$, or as an element in the range of the *circular* sine function, where the corresponding element in the domain is the *real number* $\pi/6$. Table 1.2 (page 16), Table III in Appendix C, or a calculator set on the RAD setting, which we have been using to obtain values of the trigonometric functions, can now also be used to obtain values of the circular functions. However, in this case the elements in the domain are viewed as real numbers rather than as angles.

Examples Find numerical values for each circular function value.

a. $\sin \frac{\pi}{3}$ b. $\cos 4$

Solutions

a. Using Table 1.2, we have that

$$\sin \frac{\pi}{3} = \frac{\sqrt{3}}{2}.$$

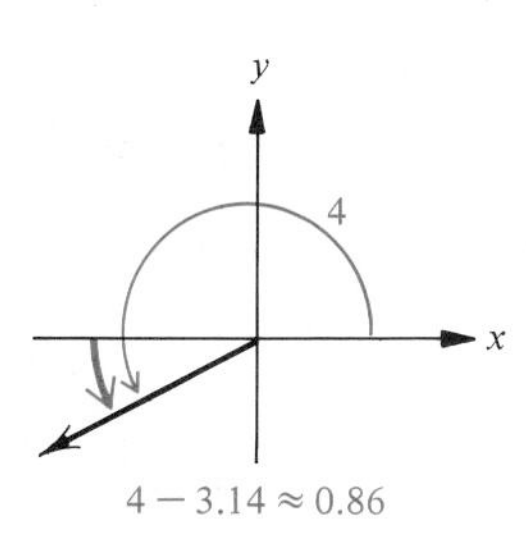

$4 - 3.14 \approx 0.86$

b. Using Table III in Appendix C, or a calculator set on the RAD setting, and $\pi \approx 3.14$, we have that

$$\begin{aligned} \cos 4 &\approx -\cos(0.86) \\ &\approx -0.6524. \end{aligned}$$

The usefulness of the circular functions arises from the fact that, like the trigonometric functions, they are periodic and that the real numbers that are elements in the domain can be viewed as measures of things other than angles. For the latter reason, the variables frequently used to represent unspecified elements in the set R of real numbers (r, s, t, x, y, etc.) are used to represent elements in the domain of the circular functions. For example, in the field of physics, any oscillatory (periodic) motion satisfying an equation of the form

$$d = A \sin Bt \qquad \text{or} \qquad d = A \cos Bt,$$

where d is a measure of displacement, A and B are constants for the particular motion, and t is a measure of time, is called **simple harmonic motion.** Using tables for function values and the fact that sine and cosine functions are periodic, we can now solve some simple problems pertaining to such motion.

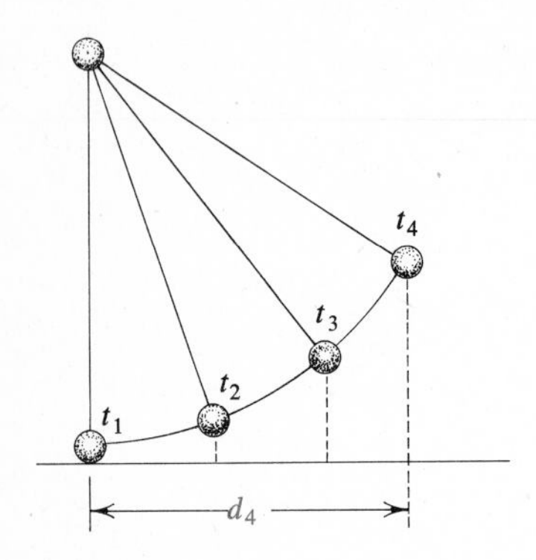

Example The approximate horizontal displacement of the bob on an oscillating pendulum (see figure) is described by $d = K \sin 2\pi t$, where K and d are expressed in centimeters and t is expressed in seconds. Using $\pi = 3.14$, find d if $K = 10$ and $t = 1.1$.

Solution Substituting 10 for K and 1.1 for t in the equation gives

$$\begin{aligned} d &= 10 \sin 2\pi(1.1) = 10 \sin 2.2\pi \\ &= 10 \sin(2.2\pi - 2\pi) = 10 \sin 0.2\pi \\ &\approx 10 \sin(0.2)(3.14) = 10 \sin 0.628. \end{aligned}$$

Thus, from Table III, or by using a calculator,

$$d \approx 10 \sin 0.628 \approx 10(0.5875) \approx 5.88.$$

The displacement is 5.88 centimeters.

As we have noted, the trigonometric functions with domains that are sets of angles and the circular functions with domains that are sets of real numbers are closely related. Hence, we will simply refer to both functions as trigonometric functions.

EXERCISE SET 2.7

A

A spring is displaced from its resting position as shown in the figure. The displacement d is given by $d = A \cos \omega t$, *where A, the initial displacement, and d are expressed in centimeters, t is expressed in seconds, and* $\omega = 6$.

Example If the initial displacement A is 12, find d when $t = 2\pi$.

Solution Since $A = 12$, $\omega = 6$, and $t = 2\pi$,

$$d = A \cos \omega t = 12 \cos 6(2\pi) = 12 \cos 12\pi.$$

Since $\cos 12\pi = \cos(0 + 6 \cdot 2\pi) = \cos 0 = 1$, we have

$$d = 12(1) = 12.$$

The displacement is 12 centimeters.

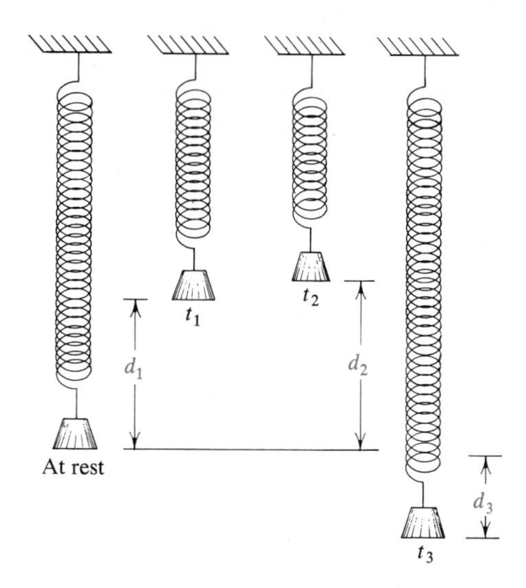

1. If the initial displacement is 8, find d when t equals 2π, 4π, and 6π.

2. If the initial displacement is -8, find d when t equals 2π, 4π, and 6π.

3. If the initial displacement is 10, find d when $t = 4$ and when $t = 6$.

4. If the initial displacement is -10, find d when $t = 4$ and when $t = 6$.

As in the example on page 90, the horizontal displacement d of the bob on an oscillating pendulum is described by the equation $d = K \sin 2\pi t$, *where K and d are expressed in centimeters and t is expressed in seconds. Use* $\pi \approx 3.14$.

5. Find K if $d = 20$ when $t = \frac{1}{6}$.

6. Find K if $d = 10$ when $t = \frac{1}{12}$.

7. Find d if $t = 1.2$ and $K = 22$.

8. Find d if $t = 2.3$ and $K = 20$.

The instantaneous voltage E in a current at time t is given by the equation $E = 200 \sin 3700t$, *where E is expressed in volts and t is expressed in seconds. Find the voltage at each of the following times. Use* $\pi \approx 3.14$.

Example $t = 0.04$

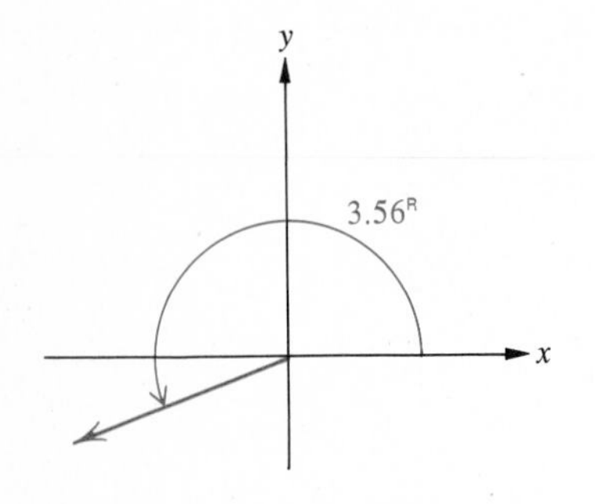

Solution

$$E = 200 \sin 3700t = 200 \sin 3700(0.04)$$
$$= 200 \sin 148 = 200 \sin(148 - 23 \cdot 2\pi)$$
$$\approx 200 \sin 3.56.$$

Since 3.56^R is in Quadrant III,

$$\sin 3.56 \approx -\sin(3.56 - 3.14) = -\sin 0.42.$$

Therefore, from Table III or by using a calculator,

$$E \approx 200 \sin 0.42 \approx -200(0.4078) \approx -81.6.\text{*}$$

The voltage is -81.6 volts.

9. $t = 0.001$ **10.** $t = 0.002$ **11.** $t = 0.005$

12. $t = 0.008$ **13.** $t = 0.012$ **14.** $t = 0.020$

The instantaneous current i in an alternating system is given by the equation $i = I_{max} \sin \omega t$, *where* I_{max} *and* ω *are constants, i is expressed in amperes, and t is expressed in seconds. Find the current at each of the following times if* $I_{max} = 0.05$ *and* $\omega = 377$. *Use* $\pi \approx 3.14$.

15. $t = 0.01$ **16.** $t = 0.03$ **17.** $t = 0.2$

18. $t = 0.4$ **19.** $t = 1.0$ **20.** $t = 3.0$

The instantaneous voltage E in a circuit at time t is given by

$$E = 1.2 \sin 997\,\pi t, \tag{1}$$

where E is expressed in volts and t is expressed in seconds. Find the voltage at each of the following times, where in each case time is given in seconds.

* If the function value sin 148 were obtained *directly* on a calculator instead of using $-\sin 0.42$, the value obtained for E would be approximately -67.7. The large difference from the answer -81.6 in the example is the result of the more precise value for π that is programmed into the calculator. To ten places, $\pi = 3.1415926536$.

Example $t = 4.5$

Solution Substituting 4.5 for t in Equation (1) we obtain

$$E = 1.2 \sin 997\pi(4.5) = 1.2 \sin 4486.5\pi$$
$$= 1.2 \sin\left(\frac{1}{2} + 4486\right)\pi = 1.2 \sin\left(\frac{\pi}{2} + 2243 \cdot 2\pi\right)$$
$$= 1.2 \sin\frac{\pi}{2} = 1.2(1) = 1.2.$$

The voltage is 1.2 volts.

21. $t = 0.5$ **22.** $t = 1.5$ **23.** $t = 2.5$

24. $t = 3.5$ **25.** $t = 6$ **26.** $t = 10$

The pressure P in a traveling sound wave is described by the equation $P = 10 \sin 200\pi[t - (x/1000)]$, *where x is the distance from the source of the sound in centimeters, t is the time expressed in seconds, and P is expressed in dynes per square centimeter. Find P for each of the following conditions. Use* $\pi \approx 3.14$.

27. $x = 200$ and $t = 0.006$ **28.** $x = 330$ and $t = 0.01$

29. $x = 660$ and $t = 0.02$ **30.** $x = 1650$ and $t = 0.05$

B

31. The displacement d of a point on a vibrating spring under certain conditions is given by $d = \frac{1}{50} \sin 50t - t \cos 50t$, where d is in feet and t is in seconds. Find the displacement when $t = 0.3$ second.

32. It is shown in calculus that an object having simple harmonic motion as described on page 89 has velocity v given by $v = AB \cos Bt$ and an acceleration given by $a = -AB^2 \sin Bt$. The motion of a certain spring is expressed by $d = 3 \sin(\pi t/2)$, where d is a displacement in centimeters and t is in seconds.

a. Express the velocity and the acceleration of a point on the free end of the spring in terms of t.

b. Find the velocity and acceleration when t equals 2, 3, and 4 seconds.

Chapter Summary

2.1 The six trigonometric ratios of an acute angle α may be expressed in terms of the sides of a right triangle, where a is the length of the side opposite α, b is the length of the side adjacent to α, and c is the length of the hypotenuse:

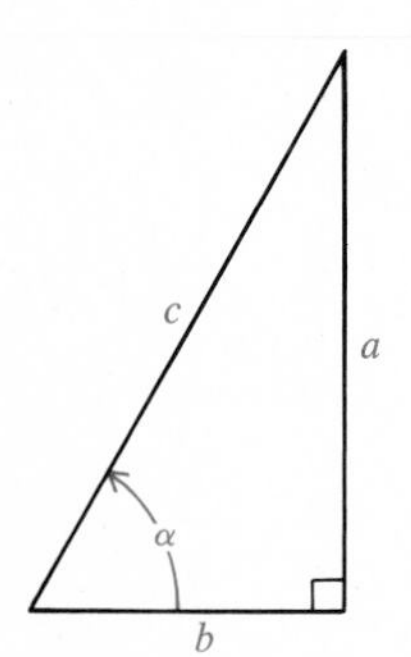

$$\sin \alpha = \frac{a}{c}, \qquad \csc \alpha = \frac{c}{a},$$

$$\cos \alpha = \frac{b}{c}, \qquad \sec \alpha = \frac{c}{b},$$

$$\tan \alpha = \frac{a}{b}, \qquad \cot \alpha = \frac{b}{a}.$$

2.2 The **Law of Sines:**

If α, β, and γ are the angles of any triangle, and if a, b, and c are the lengths of the sides opposite α, β, and γ, respectively, then

$$\frac{\sin \alpha}{a} = \frac{\sin \beta}{b} = \frac{\sin \gamma}{c}.$$

This law is used if the measures of two angles and the length of any side, or if the lengths of two sides and the measure of an angle opposite one of these sides, are given. In the latter case, certain ambiguities may arise. Two, one, or no triangles may exist.

2.3 The **Law of Cosines:**

If α, β, and γ are the angles of any triangle, and a, b, and c are the lengths of the sides opposite α, β, and γ, respectively, then

$$c^2 = a^2 + b^2 - 2ab \cos \gamma,$$

$$b^2 = a^2 + c^2 - 2ac \cos \beta,$$

$$a^2 = b^2 + c^2 - 2bc \cos \alpha.$$

This law is used if the lengths of two sides and the measure of the included angle, or if the lengths of three sides, are given.

2.4 A **geometric vector** $\vec{\mathbf{v}}$ is a line segment with a specified direction. Its **norm** or **magnitude** is designated by $\|\vec{\mathbf{v}}\|$. The **direction angle** is the angle α, where $-180° < \alpha \leq 180°$, such that:

The initial side of α is the ray from the initial point of $\vec{\mathbf{v}}$ parallel to the x-axis and directed in the positive x direction; the terminal side of α is the ray from the initial point of $\vec{\mathbf{v}}$ and containing $\vec{\mathbf{v}}$.

The **sum** $\vec{v}_1 + \vec{v}_2$ is defined as the geometric vector having as its initial point the initial point of $\vec{v}_1$ and as its terminal point the terminal point of $\vec{v}_2$, and where the terminal point of $\vec{v}_1$ is the initial point of $\vec{v}_2$. When viewed in this way, geometric vectors are said to be added according to "the triangle law." When viewed as the diagonal of a parallelogram with adjacent sides $\vec{v}_1$ and $\vec{v}_2$, geometric vectors are said to be added according to "the parallelogram law."

If $\vec{v}$ is any geometric vector and c is any real number (scalar), then $c\vec{v}$ is a geometric vector collinear to $\vec{v}$ with magnitude multiplied by $|c|$.

2.5 The magnitudes of the **geometric vector projections,** $\vec{v}_x$ and $\vec{v}_y$, of $\vec{v}$ on the x-axis and y-axis, respectively, are given by

$$\|\vec{v}_x\| = \|\vec{v}\| \cdot |\cos \alpha| \qquad \text{and} \qquad \|\vec{v}_y\| = \|\vec{v}\| \cdot |\sin \alpha|.$$

Geometric vectors can help to visualize and thus facilitate the solution of problems involving such quantities as force, velocity, and acceleration, since these quantities have both magnitude and direction.

2.6 The length s of the arc of a circle of radius of length r intercepted by an angle α whose vertex is at the center of the circle is given by

$$s = r \cdot \alpha^{R}.$$

The **linear velocity** v of a point moving along an arc of a circle at a constant speed is given by

$$v = \frac{s}{t} = \frac{r \cdot \alpha^{R}}{t},$$

where t is the time required to move through the distance s. The **angular velocity** ω of the moving point is given by

$$\omega = \frac{\alpha}{t}.$$

The linear velocity v and the angular velocity ω of a point on a circle with radius of length r are related by the equation

$$v = r\omega,$$

where ω is given in radians per unit time.

2.7 A function f is **periodic** if there exists a number $a (a \neq 0)$ such that for each x in the domain, $x + a$ is in the domain and $f(x + a) = f(x)$.

The smallest positive value of a for which this is true is the **period** of the function.

Functions in which the elements in the domains are numbers that are measures (in radian units) of angles and whose ranges are the same as the respective ranges of the analogous trigonometric functions are called **circular functions.** Tables in radian measure (or a calculator) that are used to obtain elements in the ranges of the trigonometric functions can also be used to obtain elements in the ranges of the circular functions.

Review Exercises

A

2.1 *Solve each right triangle. In each case,* $\gamma = 90°$.

1. $c = 14$, $\beta = 51°$ **2.** $c = 6.7$, $a = 5.4$

3. $a = 12.9$, $\alpha = 32° \, 36'$ **4.** $b = 9.72$, $\alpha = 71° \, 18'$

2.2 *Solve each triangle.*

5. $c = 13.6$, $\alpha = 30.3°$, $\beta = 72.2°$

6. $c = 1.4$, $\alpha = 135° \, 6'$, $\gamma = 34° \, 54'$

Determine the number of triangles that satisfy the conditions in each exercise and solve each triangle.

7. $b = 3.9$, $a = 5$, $\beta = 42.7°$ **8.** $b = 37$, $a = 51$, $\alpha = 135° \, 30'$

2.3 *Solve each triangle.*

9. $a = 1.9$, $b = 2.3$, $\gamma = 58° \, 12'$ **10.** $a = 3.6$, $b = 4.2$, $c = 6.1$

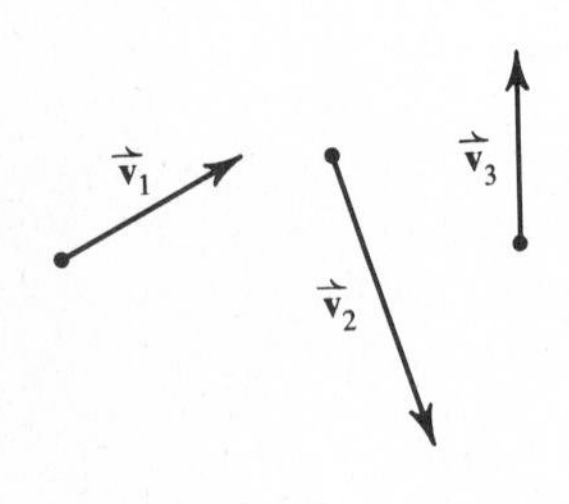

2.4 *Use freehand methods and the geometric vectors shown at the left to draw a single vector to represent each of the following.*

11. $(\vec{v}_1 + \vec{v}_2) + \vec{v}_3$ **12.** $-3\vec{v}_1$

13. $2\vec{v}_2 + \vec{v}_3$ **14.** $2\vec{v}_1 - 3\vec{v}_3$

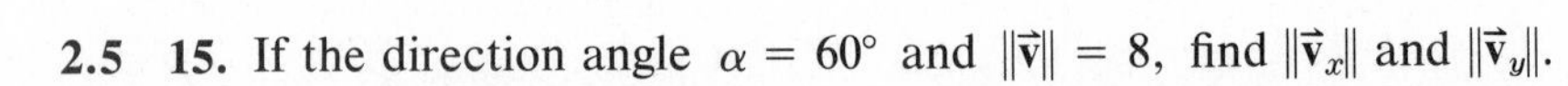

2.5 **15.** If the direction angle $\alpha = 60°$ and $\|\vec{v}\| = 8$, find $\|\vec{v}_x\|$ and $\|\vec{v}_y\|$.

16. If the direction angle $\alpha = -135°$ and $\|\vec{v}\| = 6$, find $\|\vec{v}_x\|$ and $\|\vec{v}_y\|$.

17. The resultant of two forces acting at an angle of 90° with respect to each other is 130 kilograms. If one of the forces is 50 kilograms, find the other force and the angle it makes with the resultant.

18. A man can swim at a speed of 3 miles per hour in still water. If he heads across a river at right angles to a current of 6 miles per hour, find his speed in relation to the land and the direction in which he actually moves.

19. An airplane is headed on a bearing of 280° with an air speed of 800 kilometers per hour. The course has a bearing of 292°. The ground speed is 860 kilometers per hour. Find the wind direction and wind speed.

20. A force of 150 kilograms is needed to keep a steel ball weighing 250 kilograms from rolling down an inclined plane. What angle does the plane make with the horizontal?

2.6 **21.** Find the length of the arc intercepted on a circle by a central angle α with measure 2.2^R if the radius of the circle is 3.1 centimeters.

22. Find the measure of the central angle which intercepts an arc with length 2.1 meters on a circle whose radius is 9.1 meters in length.

23. Find the linear velocity v of a point on a circle moving with constant speed if $r = 2$ inches and $\omega = 6$ radians per minute.

24. Find the angular velocity ω of a point on a circle moving with constant speed if $r = 1.5$ feet and $v = 19.7$ feet per second.

2.7 **25.** The instantaneous current i in an alternating system is given by the equation $i = I_{max} \sin \omega t$, where I_{max} and ω are constants, i is expressed in amperes, and t is expressed in seconds. Find the current at $t = 0.02$ if $I_{max} = 0.04$ and $\omega = 352$.

26. The horizontal displacement d of the bob on an oscillating pendulum is described by the equation $d = K \sin 2\pi t$, where K and d are expressed in centimeters and t is expressed in seconds. Find K if $d = 15$ when $t = 3/4$.

27. The instantaneous voltage E in a circuit at time t is given by the equation $E = 1.3 \sin 980\pi t$, where E is expressed in volts and t is expressed in seconds. Find the voltage when $t = 3$.

28. The pressure P in a traveling sound wave is described by the equation $P = 10 \sin 200\pi[t - (x/1000)]$, where x is the distance from the source of the sound in centimeters, t is the time expressed in seconds, and P is expressed in dynes per square centimeter. Find P if $x = 100$ and $t = 0.003$.

3 Graphical Representations

The periodic characteristics of the trigonometric and circular functions can be exhibited very effectively by graphing these functions in a Cartesian coordinate system. You may want to review the material on graphing in Appendix A.4 before starting this chapter.

3.1 Sine and Cosine

In graphing the trigonometric and circular functions, it is convenient to view line segments on the x-axis of a Cartesian coordinate system as obtained by unwinding the arc on a unit circle with the same length, as shown in Figure 3.1. A similar association could be made for the lengths of arcs in a clockwise direction and line segments on the negative axis.

The graphs of the sine and cosine functions, defined by the two equations $y = \sin x$ and $y = \cos x$, consist of the set of points corresponding to all ordered pairs (x, y) in the respective functions. The replacement set of x is considered to be *either* the set of all angles *or* the set of real numbers. For this reason, it is often convenient to designate replacements for x in terms of rational multiples of π and to designate units on the x-axis in the same way without using the symbol for radian units.

Although Table III is available to find values for $\sin x$ and $\cos x$ for many values of x, function values from Table 1.2 (page 16) with which you are probably now familiar, and the notion of a reference angle, are

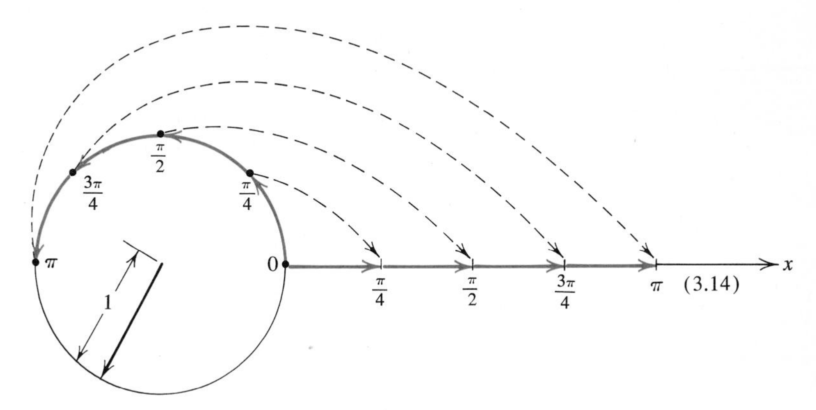

Figure 3.1

sufficient to obtain good approximations to the graphs of these functions for $0 \leq x \leq 2\pi$.

We first consider the graph of

$$y = \sin x. \tag{1}$$

In order not to distort the form of the graph of (1) and the following graphs, we shall use the same units of measurement on the x-axis and y-axis. To obtain selected ordered pairs in the function shown in Figure 3.2a on page 100, we first note from Table 1.2 (or construction of an appropriate right triangle or from memory) that

$$\sin \frac{\pi}{4} = \frac{1}{\sqrt{2}} = \frac{\sqrt{2}}{2} \approx 0.7,$$

and then we obtain

$$\sin \frac{3\pi}{4} = \sin \frac{\pi}{4} \approx 0.7,$$

$$\sin \frac{5\pi}{4} = -\sin \frac{\pi}{4} \approx -0.7,$$

$$\sin \frac{7\pi}{4} = -\sin \frac{\pi}{4} \approx -0.7.$$

The fact that the sine function is continuous implies that its graph contains no "breaks." Furthermore, since sine function values increase as x increases from 0 to $\pi/2$ and then decrease as x increases from $\pi/2$ to $3\pi/2$, etc., we connect the points in Figure 3.2a with a

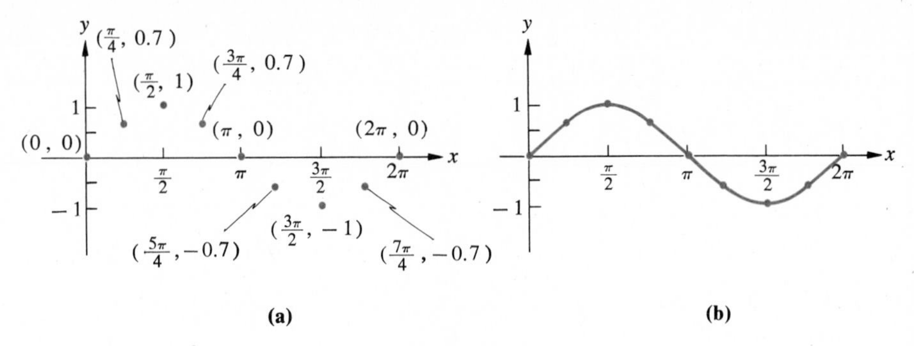

Figure 3.2

smooth curve to produce Figure 3.2b. Now, because the sine function is periodic with period 2π, that is,

$$\sin(x + k \cdot 2\pi) = \sin x, \quad k \in J,$$

we repeat the graph in Figure 3.2b in both directions and obtain Figure 3.3, which is part of the graph of $y = \sin x$ for all angles x or real numbers x. The graph over one period is called a **cycle** of the wave.

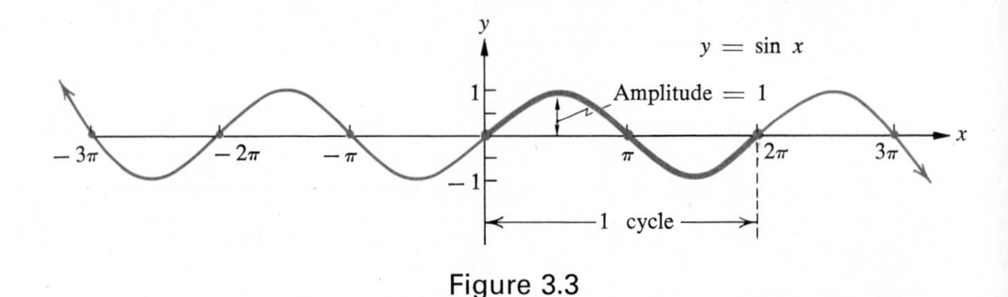

Figure 3.3

Definition 3.1 *The* **amplitude** *of a sine wave is the absolute value of half the difference of the maximum and minimum ordinates of the wave. This number is also called the amplitude of the function.*

Thus, the amplitude of the wave in Figure 3.3 is

$$\left|\frac{1}{2}[1 - (-1)]\right| = 1.$$

Several important notions of the sine function are evident from the graph. Observe that the range of the function is

$$\{y \mid -1 \leq y \leq 1\}.$$

Also observe that the zeros of the function, values of x for which $\sin x = 0$ and which are the x-coordinates of the points where the curve intersects the x-axis, are $k\pi$, $k \in J$.

We obtain the graph of

$$y = \cos x \tag{2}$$

in the same way that we obtained the graph of $y = \sin x$. Some ordered pairs in the function for $0 \leq x \leq 2\pi$ are obtained using Table 1.2. First noting that $\cos \pi/4 \approx 0.7$, we then obtain

$$\cos \frac{3\pi}{4} = -\cos \frac{\pi}{4} \approx -0.7,$$

$$\cos \frac{5\pi}{4} = -\cos \frac{\pi}{4} \approx -0.7,$$

$$\cos \frac{7\pi}{4} = \cos \frac{\pi}{4} \approx 0.7.$$

The graphs of the ordered pairs are shown on the plane in Figure 3.4a. In this case function values decrease as x increases from zero to π and they increase as x increases from π to 2π. We connect the points in Figure 3.4a with a smooth curve to produce Figure 3.4b.

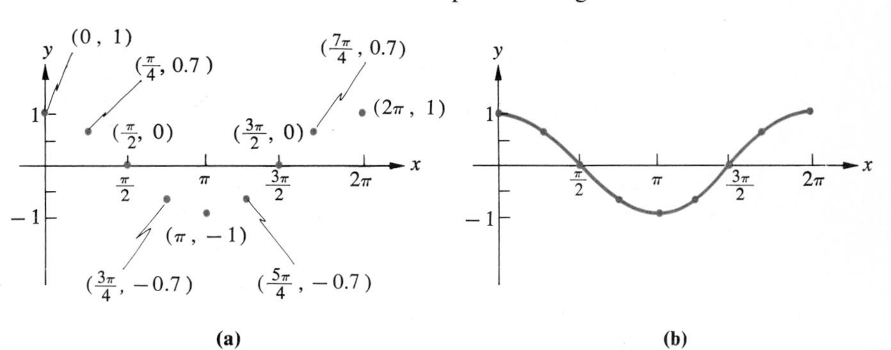

Figure 3.4

Since the cosine function is also periodic with period 2π,

$$\cos(x + k \cdot 2\pi) = \cos x, \quad k \in J,$$

we repeat the pattern in Figure 3.4b and obtain Figure 3.5, which is part of the graph of $y = \cos x$ for all angles x or real numbers x. We observe that the range of the function is

$$-1 \leq y \leq 1$$

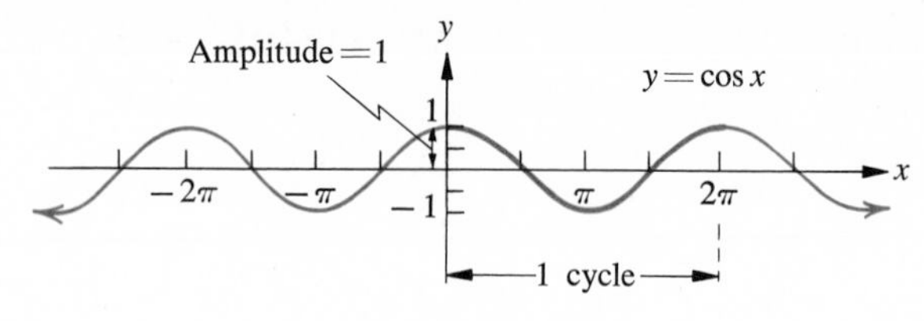

Figure 3.5

and therefore the amplitude of the wave is 1. Furthermore, we note that the zeros of the function are $\pi/2 + k\pi$, $k \in J$.

Observe that the sine function and the cosine function have the same "form." For this reason both curves are called **sine waves** or **sinusoidal waves.**

EXERCISE SET 3.1

A

1. Graph the elements of

$$\left\{-2\pi, \frac{-3\pi}{2}, -\pi, \frac{-\pi}{2}, 0, \frac{\pi}{2}, \pi, \frac{3\pi}{2}, 2\pi\right\}$$

on the number line below that is scaled in integral units. Use $\pi \approx 3.14$.

−5 0 5

2. Graph the elements of

$$\left\{0, \frac{\pi}{4}, \frac{\pi}{2}, \frac{3\pi}{4}, \pi, \frac{5\pi}{4}, \frac{3\pi}{2}, \frac{7\pi}{4}, 2\pi\right\}$$

on a number line scaled in integral units as shown in Problem 1. Use $\pi \approx 3.14$.

3. Graph the elements of

$$\left\{0, \frac{\pi}{4}, \frac{\pi}{2}, \frac{3\pi}{4}, \pi, \frac{5\pi}{4}, \frac{3\pi}{2}, \frac{7\pi}{4}, 2\pi\right\}$$

on the number line below that is scaled in multiples of π.

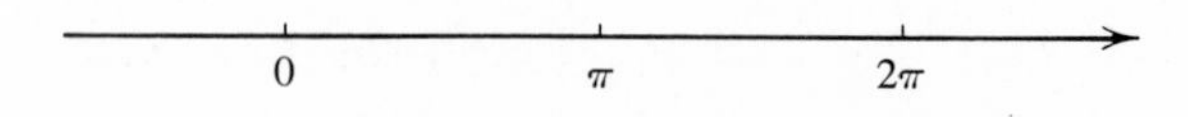

4. Graph the elements of

$$\left\{-2\pi, \frac{-7\pi}{4}, \frac{-3\pi}{2}, \frac{-5\pi}{4}, -\pi, \frac{-3\pi}{4}, \frac{-\pi}{2}, \frac{-\pi}{4}, 0\right\}$$

on a number line scaled in multiples of π as shown in Exercise 3.

5. Graph the elements of

$$\left\{-2\pi, \frac{-5\pi}{3}, \frac{-4\pi}{3}, -\pi, \frac{-2\pi}{3}, \frac{-\pi}{3}, 0, \frac{\pi}{3}, \frac{2\pi}{3}, \pi, \frac{4\pi}{3}, \frac{5\pi}{3}, 2\pi\right\}$$

on a number line scaled in multiples of π.

6. Graph the elements of

$$\left\{0, \frac{\pi}{6}, \frac{\pi}{3}, \frac{\pi}{2}, \frac{2\pi}{3}, \frac{5\pi}{6}, \pi, \frac{7\pi}{6}, \frac{4\pi}{3}, \frac{3\pi}{2}, \frac{5\pi}{3}, \frac{11\pi}{6}, 2\pi\right\}$$

on a number line scaled in multiples of π.

In Exercises 7–18, (**a**) *graph the function defined by each equation by graphing one cycle and then repeating the pattern and* (**b**) *specify the zeros and the amplitude of the function.*

Example $y = 2 \sin x$

Solution

a. Some ordered pairs in the function defined by the equation are obtained in order to graph one cycle. The systematic tabulation of the data shown is helpful. The ordered pairs (x, y) are then graphed as shown, connected with a smooth curve, and the pattern of the wave duplicated. Table III or a calculator can be used to obtain additional points to construct the graph.

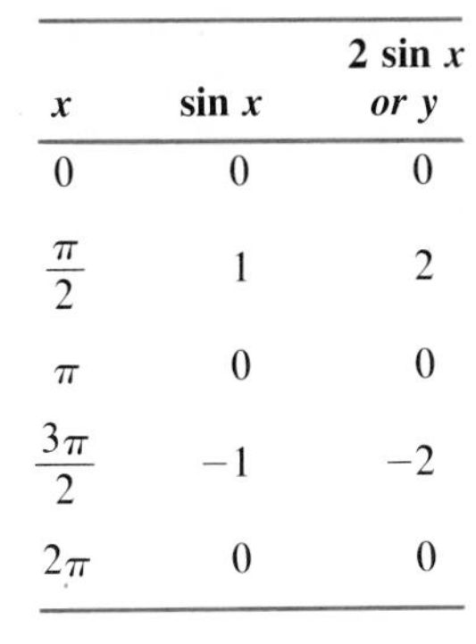

x	$\sin x$	$2 \sin x$ *or* y
0	0	0
$\frac{\pi}{2}$	1	2
π	0	0
$\frac{3\pi}{2}$	-1	-2
2π	0	0

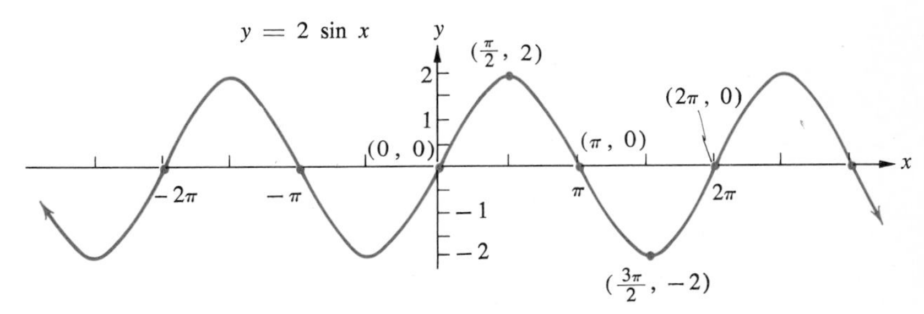

b. Zeros: $k\pi$, $k \in J$; amplitude: $\left|\frac{1}{2}[2 - (-2)]\right| = 2.$

7. $y = 3 \sin x$ **8.** $y = 2 \cos x$ **9.** $y = \frac{1}{2} \cos x$

10. $y = \frac{1}{3} \sin x$ **11.** $y = -\sin x$ **12.** $y = -\cos x$

Example $y = \cos 2x$

Solution
a. Some ordered pairs in the function defined by the equation are obtained in order to graph one cycle. The ordered pairs (x, y) obtained from the table are then graphed as shown, connected with a smooth curve, and the pattern of the wave duplicated.

x	$2x$	$\cos 2x$ *or* y
0	0	1
$\frac{\pi}{4}$	$\frac{\pi}{2}$	0
$\frac{\pi}{2}$	π	-1
$\frac{3\pi}{4}$	$\frac{3\pi}{2}$	0
π	2π	1

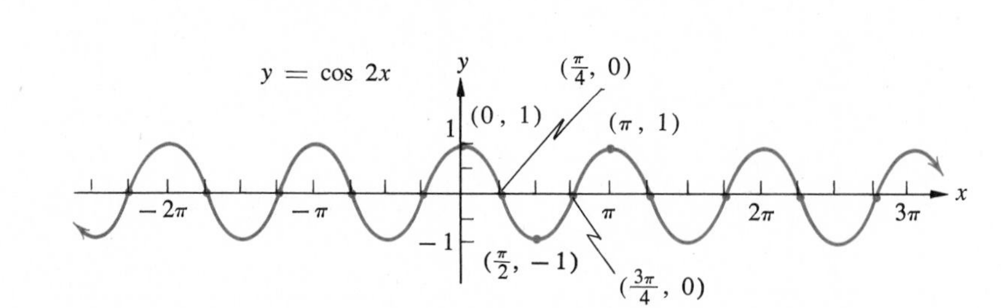

b. Zeros: $x = \frac{\pi}{4} + k \cdot \frac{\pi}{2}, k \in J$; amplitude: $\left|\frac{1}{2}[1 - (-1)]\right| = 1.$

13. $y = \sin 2x$ **14.** $y = \sin \frac{1}{2} x$ **15.** $y = \cos \frac{1}{2} x$

16. $y = \cos \frac{1}{3} x$ **17.** $y = \sin \left(x - \frac{\pi}{2}\right)$ **18.** $y = \cos \left(x + \frac{\pi}{4}\right)$

B

Graph two cycles of each of the following.

Example $y = \cos \frac{\pi}{3} x$

Solution In such cases where the coefficient of x is a multiple of π, it is helpful to label the x-axis in rational units and to select rational numbers for x, as shown in the table. Some ordered pairs in the function defined by the equation are graphed as shown and connected with a smooth curve. Then the pattern of the wave is duplicated.

x	$\frac{\pi}{3}x$	$\cos\frac{\pi}{3}x$ or y
0	0	1
1	$\frac{\pi}{3}$	$\frac{1}{2}$
2	$\frac{2\pi}{3}$	$-\frac{1}{2}$
3	π	-1
4	$\frac{4\pi}{3}$	$-\frac{1}{2}$
5	$\frac{5\pi}{3}$	$\frac{1}{2}$
6	2π	1

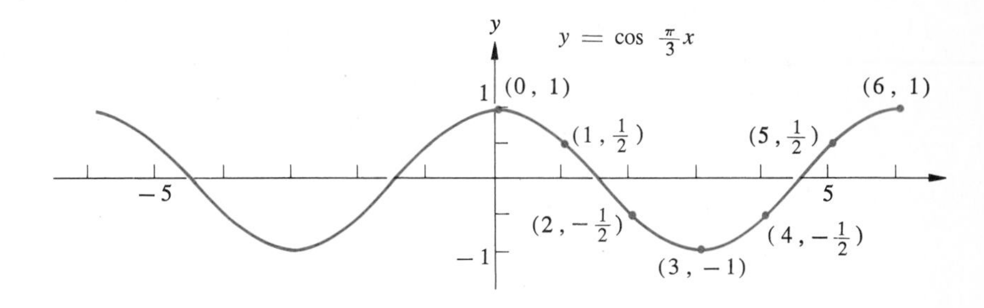

19. $y = \cos\frac{\pi}{2}x$

20. $y = \sin\frac{\pi}{2}x$

21. $y = \sin \pi x$

22. $y = \cos \pi x$

23. Graph $y = \dfrac{\sin x}{x}$, $-\pi < x < \pi$, and describe the graph as x approaches zero. Is the graph periodic? Why or why not?

24. Graph $y = x \sin x$, $-4\pi \leq x \leq 4\pi$.

25. Graph $y = \dfrac{\cos x}{x}$, $-\pi < x < \pi$, and describe the graph as x approaches zero. Is the graph periodic? Why or why not?

26. Graph $y = x \cos x$, $-4\pi \leq x \leq 4\pi$.

3.2 General Sine Waves

Functions defined by equations of the form

$$y = A \sin B(x + C) \qquad \text{or} \qquad y = A \cos B(x + C),$$

where $A, B, C, x, y \in R$ and $A, B \neq 0$, always have sine waves for their graphs. Depending upon the values for A, B, and C, these sine waves have different amplitudes, different periods, and different horizontal positions with respect to the origin. To facilitate graphing such functions, we now consider these factors in more detail.

Amplitude

First consider how the value of A in the defining equation affects the graph of the sine function or the cosine function. For each value of x, each ordinate of the graph of

$$y = A \sin x$$

is A times the ordinate of the graph of $y = \sin x$. The effect of A, then, is to multiply the amplitude of the graph of $y = \sin x$ by $|A|$. The graph of $y = A \cos x$ is related to the graph of $y = \cos x$ in a similar way.

Example Graph $y = 2 \cos x$ over the interval $-2\pi \leq x \leq 4\pi$.

Solution

1. Sketch the graph of $y = \cos x$, over the interval $0 \leq x \leq 2\pi$, as a reference.
2. The amplitude of the graph of $y = 2 \cos x$ is 2. Therefore, sketch the desired graph on the same coordinate system over the same interval by making each ordinate 2 times the ordinate of the graph of $y = \cos x$.
3. Repeat the cycle over the interval $-2\pi \leq x \leq 4\pi$.

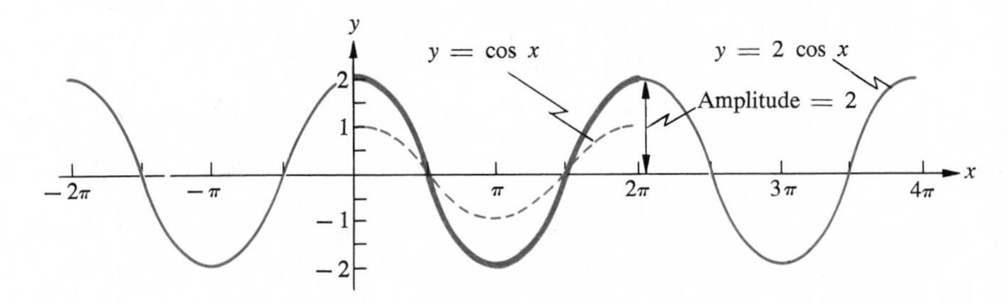

The first cycle is shown sketched with a heavier line for emphasis.

Period

Now consider how different values for B affect the graph of

$$y = \sin Bx \qquad \text{or} \qquad y = \cos Bx.$$

A complete cycle of the graph of $y = \sin x$ (or $y = \cos x$) occurs over the period 2π. Hence, a cycle of the graph of $y = \sin Bx$ (or $y = \cos Bx$) is obtained as Bx increases from 0 to 2π. Therefore, the period of either function is the solution of the equation

$$Bx = 2\pi \qquad \text{or} \qquad x = \frac{2\pi}{B}.$$

Hence, the period P of a function defined by an equation of the form $y = \sin Bx$ or $y = \cos Bx$ is

$$\mathbf{P = \frac{2\pi}{|B|}}.$$

The absolute value of B is used so that the period is a positive number.

Example Graph $y = \sin 3x$, over the interval $0 \le x \le 2\pi$.

Solution

1. Sketch the graph of $y = \sin x$, over the interval $0 \le x \le 2\pi$, as a reference.
2. The period $P = 2\pi/|B| = 2\pi/3$. Therefore, sketch a complete cycle of the desired graph on the same set of axes over the interval $0 \le x \le 2\pi/3$.
3. Repeat the cycle obtained in step 2 over the interval $0 \le x \le 2\pi$.

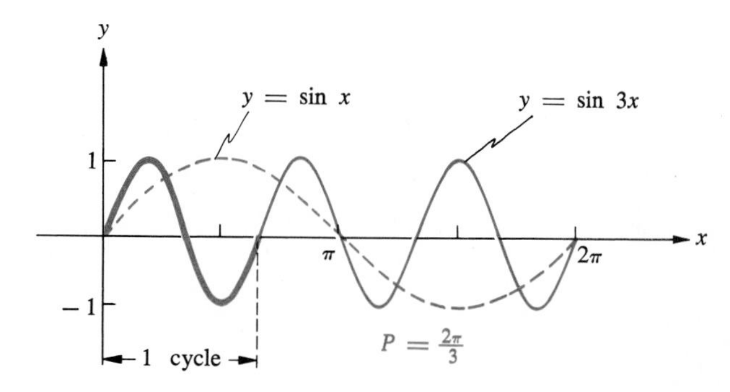

Phase Shift

Finally, consider how different values for C affect the graph of

$$y = \sin(x + C) \qquad \text{or} \qquad y = \cos(x + C).$$

For $x = -C$,

$$\sin(x + C) = \sin(-C + C) = \sin 0 = 0,$$

and, for any x_1, the value of $\sin(x + C)$ at $x_1 - C$ is the same as the value of $\sin x$ at x_1. That is,

$$\sin[(x_1 - C) + C] = \sin x_1.$$

For this reason, the graph of $y = \sin(x + C)$ is shifted C units to the left of the graph of $y = \sin x$ if $C > 0$ and shifted $|C|$ units to the right of the graph of $y = \sin x$ if $C < 0$.

Definition 3.2 *The number $|C|$ is the* **phase shift** *of the graph of periodic functions defined by equations of the form*

$$y = \sin(x + C) \qquad \textit{or} \qquad y = \cos(x + C).$$

Note that the phase shift $|C|$ is a positive number whether the shift of the graph of $y = \text{trig}(x + C)$ is to the right or to the left of the graph of $y = \text{trig } x$.

Example Graph $y = \sin\left(x - \frac{\pi}{3}\right)$ over the interval $-2\pi \leq x \leq 4\pi$.

Solution

1. Sketch the graph of $y = \sin x$, over the interval $0 \leq x \leq 2\pi$, as a reference.
2. Since $C = -\pi/3$, the graph of $y = \sin x$ is shifted $|-\pi/3|$ units to the right to obtain the desired graph.
3. Sketch the graph over the interval $-2\pi \leq x \leq 4\pi$.

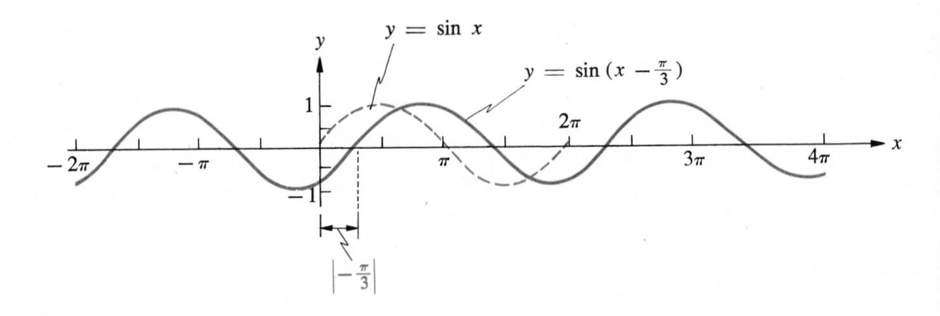

We can now use all information about the effects of A, B, and C on the graphs of functions defined by the equation $y = A \sin B(x + C)$ or by the equation $y = A \cos B(x + C)$ to facilitate making such graphs. An example is shown in Exercise Set 3.2.

The Period P, a Positive Rational Number

If the period P of a graph is a positive rational number, instead of a multiple of π such as we encountered in the examples above, it is helpful to use selected rational values as elements of the domain and to label the x-axis in rational units to facilitate sketching the graph.

Example Graph $y = \frac{1}{2} \cos \pi x$ over the interval $-5 \leq x < 5$.

Solution

1. The phase shift $|C|$ is zero. Therefore compare the equation with

$$y = A \cos Bx.$$

It is evident that the graph of $y = \frac{1}{2} \cos \pi x$ has

$$\text{amplitude: } |A| = \frac{1}{2}; \quad \text{period: } P = \frac{2\pi}{|B|} = \frac{2\pi}{\pi} = 2.$$

2. Sketch the graph, marking the x-axis in rational units.

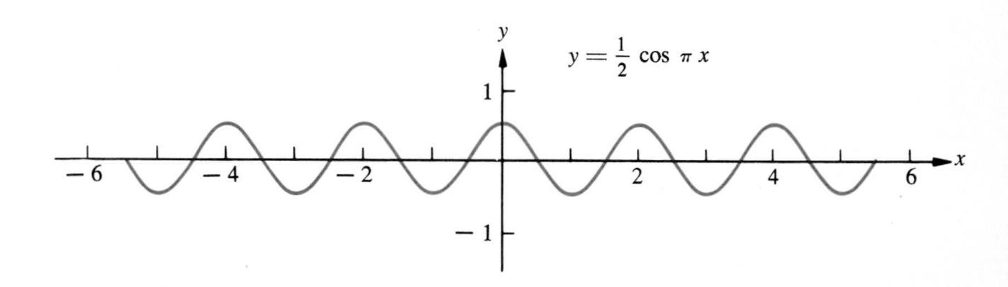

EXERCISE SET 3.2

A

Sketch the graph of each equation over the interval $-2\pi \leq x \leq 2\pi$.

Example $y = \dfrac{1}{2} \cos 2x$

Solution

1. Compare the equation with $y = A \cos Bx$. It is evident that the graph of $y = \frac{1}{2} \cos 2x$ has

$$\text{amplitude: } |A| = \frac{1}{2}; \quad \text{period: } P = \frac{2\pi}{|B|} = \frac{2\pi}{2} = \pi.$$

2. Sketch the graph.

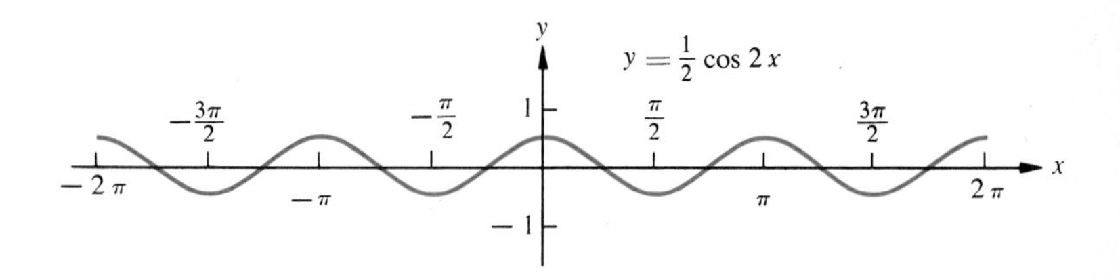

1. $y = 3 \sin x$ **2.** $y = 3 \cos x$ **3.** $y = -2 \cos x$

4. $y = -5 \sin x$ **5.** $y = \dfrac{1}{2} \sin x$ **6.** $y = \dfrac{1}{3} \cos x$

7. $y = \cos 2x$

8. $y = \sin 3x$

9. $y = \sin \frac{1}{2} x$

10. $y = \cos \frac{1}{3} x$

11. $y = 2 \sin 2x$

12. $y = \frac{1}{3} \cos \frac{1}{2} x$

Example $y = 2 \cos \left(3x + \frac{\pi}{2}\right)$

Solution

1. Rewrite the equation by factoring 3 from the expression in parentheses.

$$y = 2 \cos 3 \left(x + \frac{\pi}{6}\right)$$

2. Compare this equation with $y = A \cos B(x + C)$. It is evident that the graph of $y = 2 \cos 3[x + (\pi/6)]$ has

$$\text{amplitude: } |A| = 2; \quad \text{period: } P = \frac{2\pi}{|B|} = \frac{2\pi}{3};$$

phase shift: $\frac{\pi}{6}$ unit to the left of the graph of $y = 2 \cos 3x$.

3. Sketch the graph.

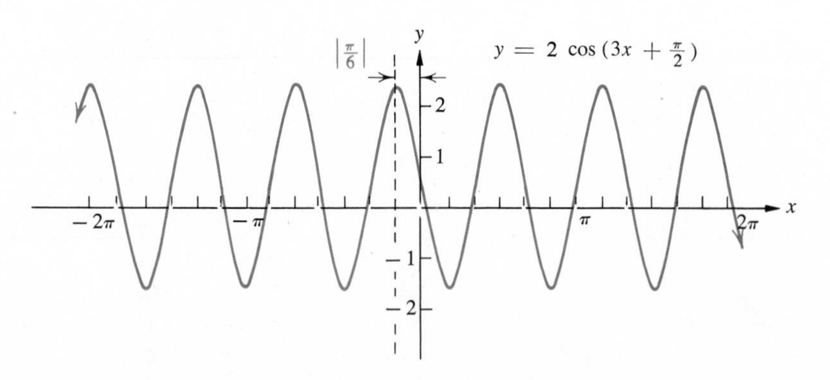

13. $y = \cos(x + \pi)$

14. $y = \sin \left(x - \frac{\pi}{2}\right)$

15. $y = 2 \sin \left(x - \frac{\pi}{4}\right)$

16. $y = \frac{1}{2} \cos \left(x + \frac{\pi}{6}\right)$

17. $y = 2 \cos \left(2x + \frac{\pi}{2}\right)$

18. $y = \frac{1}{2} \sin \left(\frac{1}{2} x - \frac{\pi}{4}\right)$

19. $y = \frac{1}{2} \sin \left(\frac{1}{2}x + \frac{\pi}{2}\right)$

20. $y = 2 \cos(2x - \pi)$

B

Graph two cycles of each of the following with positive elements in the domain. *Hint:* Scale the x-axis in rational units 1, 2, 3, ...

Example $y = \sin \pi x$

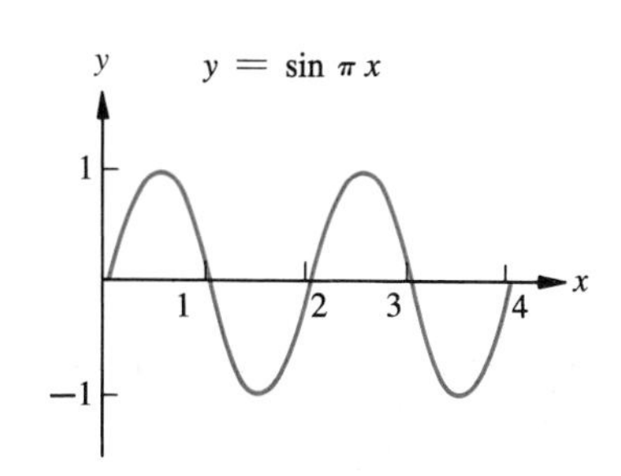

Solution

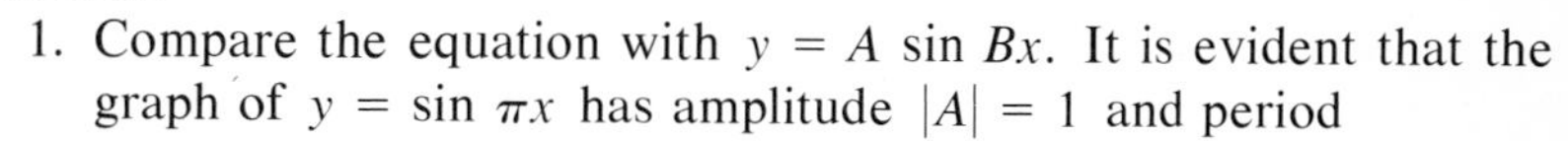

1. Compare the equation with $y = A \sin Bx$. It is evident that the graph of $y = \sin \pi x$ has amplitude $|A| = 1$ and period

$$P = \frac{2\pi}{|B|} = \frac{2\pi}{\pi} = 2.$$

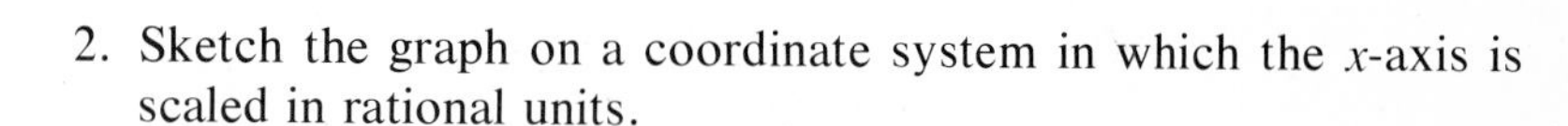

2. Sketch the graph on a coordinate system in which the x-axis is scaled in rational units.

21. $y = \cos \pi x$

22. $y = \sin \frac{\pi}{2} x$

23. $y = \frac{1}{2} \sin \frac{\pi}{3} x$

24. $y = 2 \cos \frac{\pi}{4} x$

3.3 Graphical Addition

Functions defined by equations such as

$$y = 2 \sin x + \cos x \qquad \text{and} \qquad y = x + \cos x$$

can be graphed by methods similar to those we used in the preceding section. We can find some elements in the range for several elements in the domain by first using available tables to find values $\cos x$ and $\sin x$ and then performing the indicated operations. Then we can graph the set of ordered pairs obtained and complete the graph by connecting these points with a smooth curve.

An alternative, and somewhat simpler, method involves the **graphical addition of ordinates.** Two examples illustrate this method.

Example Graph $y = 2 \sin x + \cos x$ over the interval $0 \le x \le 2\pi$.

Solution Sketch the graphs of $y_1 = 2 \sin x$ and $y_2 = \cos x$ on the same set of axes over the given interval. The ordinate of the graph of $y = 2 \sin x + \cos x$ for x is the algebraic sum of the ordinates of the graph of $y_1 = 2 \sin x$ and the graph of $y_2 = \cos x$. For example, for $x = \pi/3$,

$$2 \sin x = 2\left(\frac{\sqrt{3}}{2}\right) \approx 1.7 \qquad \text{and} \qquad \cos x = 0.5,$$

from which

$$y = 2 \sin x + \cos x \approx 1.7 + 0.5 = 2.2.$$

The ordinate 2.2 can be approximated graphically by "adding" vertical directed segments, $\overrightarrow{AC}$ and $\overrightarrow{AB}$, from the x-axis to the respective curves at $x = \pi/3$, as shown in the figure. Other ordinates can be found graphically in a similar way. The points in the graph of

$$y = 2 \sin x + \cos x$$

for which $2 \sin x$ or $\cos x$ equals zero are particularly easy to locate. Connect the points with a smooth curve to obtain the graph shown.

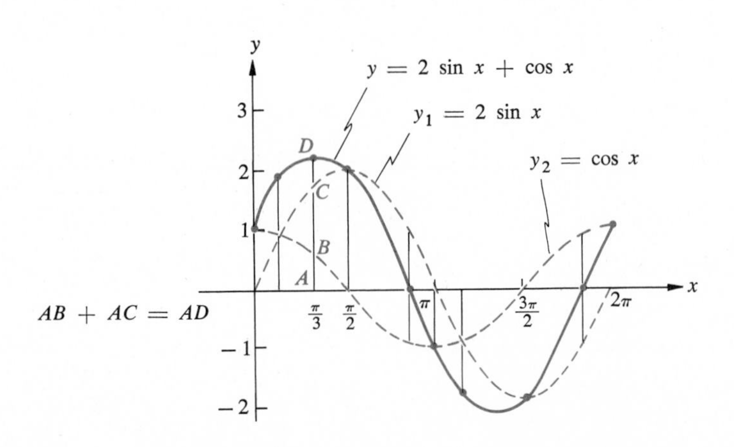

Example Graph $y = x + \cos x$ over the interval $-\pi \leq x \leq 2\pi$.

Solution Sketch the graphs of $y_1 = x$ and $y_2 = \cos x$ on the same set of axes over the given interval. Then "add" directed segments for arbitrary values of x. The points in the graph of $y = x + \cos x$ for which $\cos x = 0$ are easy to locate. Connect the points with a smooth curve to obtain the graph shown.

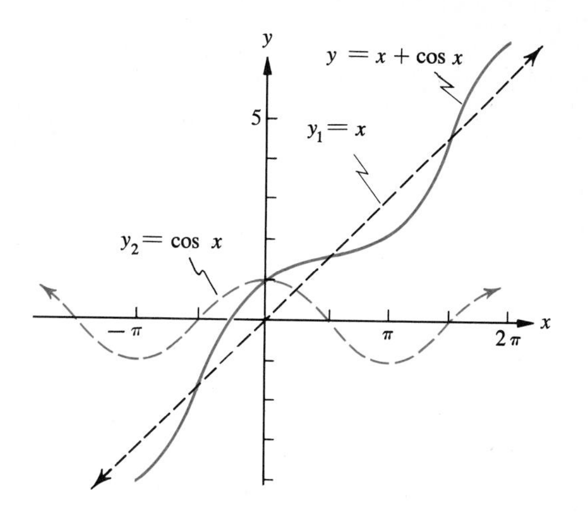

EXERCISE SET 3.3

A

Sketch the graph of each equation over the interval $-\pi \leq x \leq 2\pi$.

Example $y = \sin x + \cos x$

Solution

1. Graph $y_1 = \sin x$ and $y_2 = \cos x$ on the same set of axes.
2. Add the ordinates of several pairs of points for several values of x and draw a smooth curve.

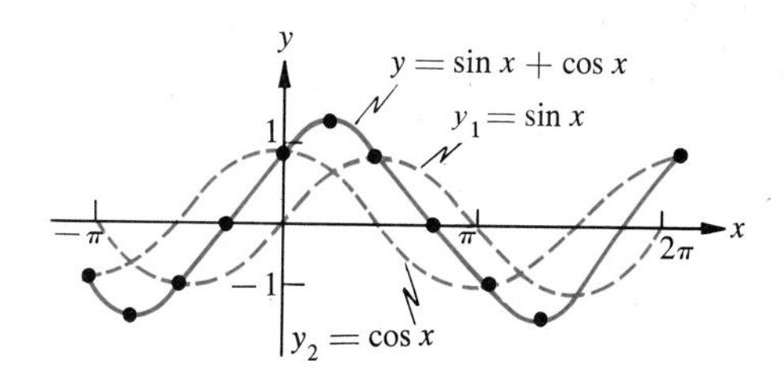

1. $y = 2 \cos x + \sin x$

2. $y = \cos 2x + \frac{1}{2} \sin x$

3. $y = \frac{1}{2} \cos x + \sin \frac{1}{2} x$

4. $y = \cos \frac{1}{2} x + \frac{1}{3} \sin x$

5. $y = \sin 2x + \frac{1}{2} \cos x$

6. $y = \sin \frac{1}{2} x + \cos x$

7. $y = 2 \sin x + \cos \frac{1}{2} x$

8. $y = \sin \frac{1}{2} x + 2 \cos \frac{1}{2} x$

9. $y = 2 \sin x - \cos x$

10. $y = \cos x - \frac{1}{2} \sin x$

11. $y = \cos \frac{1}{2} x - 2 \sin x$

12. $y = 2 \cos x - \frac{1}{3} \sin x$

13. $y = 2 + 2 \cos x$

14. $y = 1 - 2 \sin x$

15. $y = x + \sin x$

16. $y = \frac{x}{2} - \cos x$

B

In Exercises 17–22, sketch the graph of each equation over the specified interval by using "addition of ordinates."

17. $y = x - 2 + \cos x,\ 0 \le x \le 2\pi.$
Hint: First graph $y_1 = x - 2$ and $y_2 = \cos x$.

18. $y = 3 - x + \sin x,\ 0 \le x \le 2\pi.$
Hint: First graph $y_1 = 3 - x$ and $y_2 = \sin x$.

19. $y = 2 \cos x + \frac{x^2}{12},\ 0 \le x \le 2\pi$

20. $y = 2 \sin x + \frac{1}{x},\ 0 \le x \le 2\pi$

21. $y = \sin \pi x - x + 4,\ 0 \le x \le 4$

22. $y = \cos \frac{\pi}{2} x - \frac{x^2}{4},\ 0 \le x \le 2\pi$

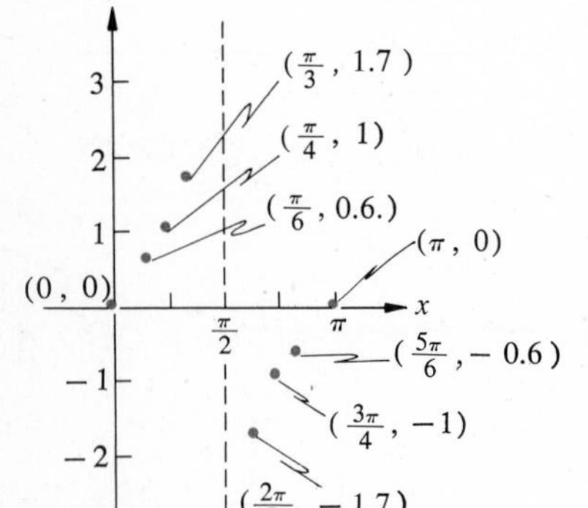

Figure 3.6

3.4 Graphs of Other Trigonometric Functions

We can graph the tangent, cotangent, secant, and cosecant functions on a Cartesian coordinate system in the same way that we graphed the sine and cosine functions in Section 3.2. Although the graphs of these functions are not sine waves, the graphs do exhibit the periodic nature and other important features of the functions.

Tangent Function

The period of the tangent function is π. Therefore we need consider some special values for x only in the interval 0 to π. Using Table 1.2, we obtain some ordered pairs in the function and graph these ordered pairs as shown in Figure 3.6. Observe that $\sqrt{3} \approx 1.7$ and that

$$\frac{1}{\sqrt{3}} = \frac{\sqrt{3}}{3} \approx \frac{1.7}{3} \approx 0.6.$$

Since $\tan x = \sin x/\cos x$, the function defined by $y = \tan x$ is continuous in the domain

$$0 \leq x < \frac{\pi}{2} \quad \text{and} \quad \frac{\pi}{2} < x \leq \pi$$

($\cos x \neq 0$ in these intervals), and we connect the points in Figure 3.6 with a smooth curve to produce one cycle (heavy line). Then we repeat this pattern in both directions to obtain Figure 3.7. Now Table III or a calculator can be used to obtain additional points to assist in graphing this function.

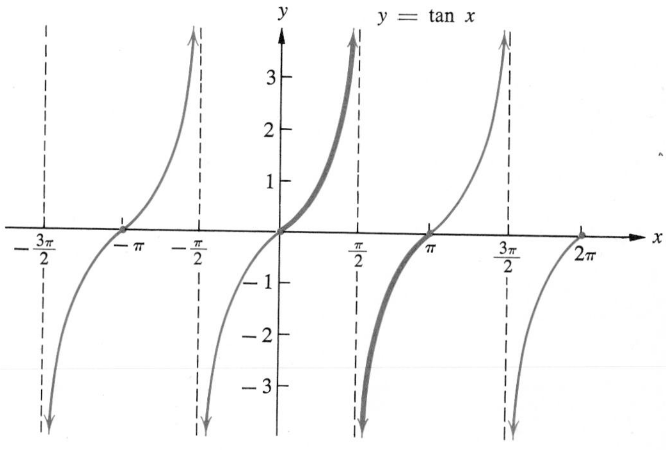

Figure 3.7

Because $\tan x$ is undefined for $x = \pi/2$, there is no point in the graph of the function when $x = \pi/2$. Notice, however, that $|\tan x|$ increases indefinitely as x is taken closer and closer to $\pi/2$. The vertical dashed lines in Figures 3.6 and 3.7 are called **asymptotes.** The entire set of asymptotes to the curve is the set of lines which are the graphs of

$$x = \frac{\pi}{2} + k\pi, \quad k \in J.$$

Furthermore, observe that the zeros of the function that are associated with the points where the curve intersects the x-axis are $k\pi$, $k \in J$.

Cotangent Function

The graph of the function defined by $y = \cot x$ can also be obtained by plotting points in the interval $0 \leq x < \pi$. For $x \neq k(\pi/2)$, $k \in J$, however,

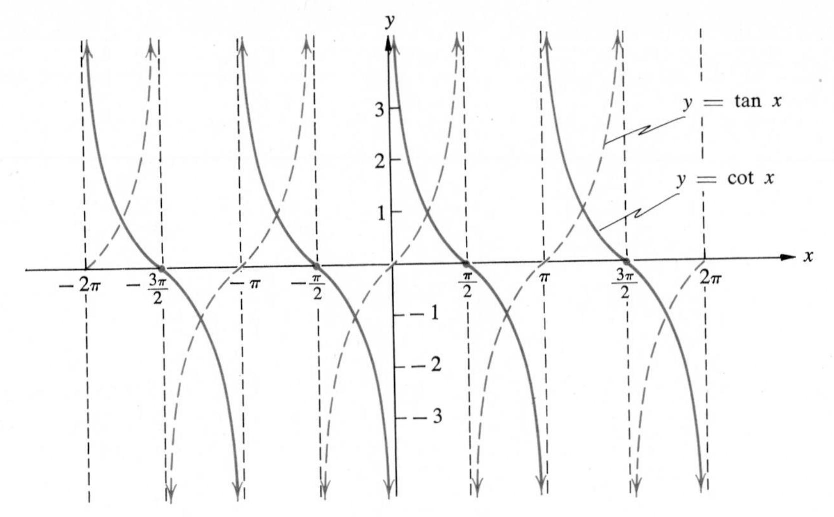

Figure 3.8

$$\cot x = \frac{1}{\tan x}.$$

Thus, each ordinate of the graph of $y = \cot x$, $x \neq k(\pi/2)$, is the reciprocal of the corresponding ordinate of the graph of $y = \tan x$. With this information we can quickly sketch the graph of $y = \cot x$ if we first sketch of $y = \tan x$ on the same set of axes. The graphs appear in Figure 3.8. Note that the zeros of the function defined by $y = \cot x$ are

$$\frac{\pi}{2} + k\pi, \quad k \in J,$$

and that the asymptotes to the curve are given by

$$x = k\pi, \quad k \in J.$$

Also, the range of the tangent or cotangent function is $\{y \mid y \in R\}$.

Secant and Cosecant Functions

The graphs of the functions defined by $y = \sec x$ and $y = \csc x$ can be obtained by using the fact that, since

$$\sec x = \frac{1}{\cos x}, \quad x \neq \frac{\pi}{2} + k\pi, \quad k \in J, \tag{1}$$

and

$$\csc x = \frac{1}{\sin x}, \quad x \neq k\pi, \quad k \in J, \tag{2}$$

each ordinate of the graph of $y = \sec x$ is the reciprocal of the corresponding ordinate of the graph of $y = \cos x$, and each ordinate of the graph of $y = \csc x$ is the reciprocal of the corresponding ordinate of the graph of $y = \sin x$, except for the noted restrictions. The construction of the graphs of $y = \sec x$ and $y = \csc x$ are left as exercises.

Because the ranges of the cosine and sine functions are in the interval between -1 and 1 and include these values, we have from Equations (1) and (2) that the range of either the secant or the cosecant function is $\{y \mid y \leq -1 \text{ or } y \geq 1\}$. Thus, there are no zeros of these functions.

Ranges and Periods

TABLE 3.1

Function	Range
$y = \sin x$	$-1 \leq y \leq 1$
$y = \cos x$	$-1 \leq y \leq 1$
$y = \tan x$	$y \in R$
$y = \cot x$	$y \in R$
$y = \sec x$	$y \leq -1$ or $y \geq 1$
$y = \csc x$	$y \leq -1$ or $y \geq 1$

The ranges of the six circular functions have been briefly considered in Chapter 1 and in this chapter. For convenience we summarize this data in Table 3.1.

Variations in the constants A, B, and C in equations such as

$$y = A \tan B(x + C) \tag{3}$$

affect their graphs in much the same way that these constants affect the graphs of $y = A \sin B(x + C)$ and $y = A \cos B(x + C)$. However, in the graphs of equations of the form of Equation (3), the coefficient A is not referred to as the amplitude. Furthermore, because the tangent function and the cotangent function have period π, the period P of functions defined by equations of the form $y = \tan Bx$ or $y = \cot Bx$ is given by

$$P = \frac{\pi}{|B|}.$$

EXERCISE SET 3.4

A

Sketch the graph of the function defined by each equation over the interval $0 \leq x \leq 2\pi$ *and specify* **(a)** *the range,* **(b)** *the equations of the asymptotes, and* **(c)** *the zeros of the function (if they exist).*

Example $y = \tan 2x$

Solution Since the period of $y = \tan x$ is π, the period of $y = \tan 2x$ is $\pi/2$. The graph is sketched as shown.

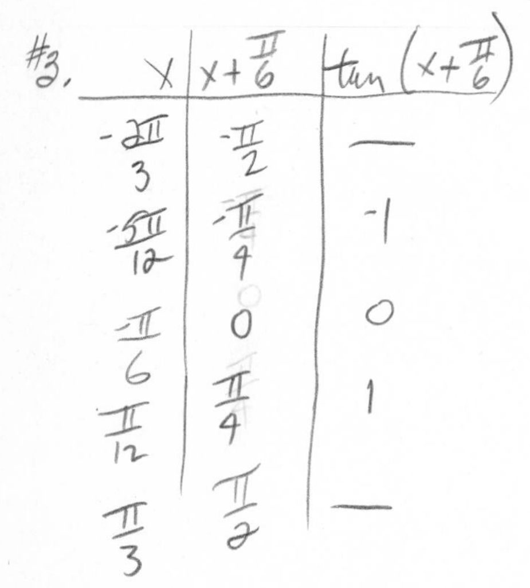

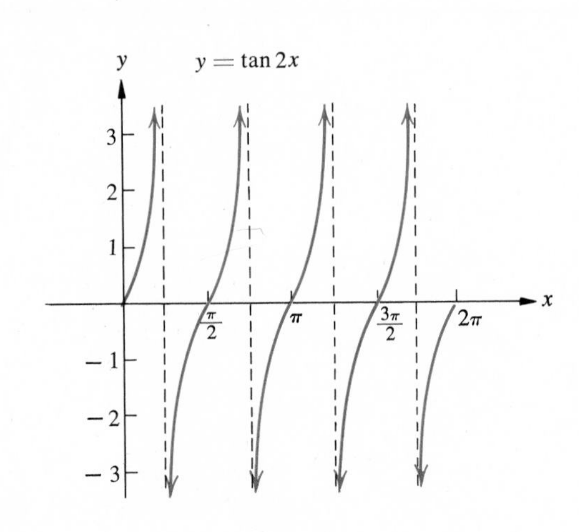

a. Range: $\{y \mid y \in R\}$

b. Since the equations of the asymptotes of $y = \tan x$ are

$$x = \frac{\pi}{2} + k\pi, \quad k \in J,$$

the equations of the asymptotes of $y = \tan 2x$ are

$$x = \frac{1}{2}\left(\frac{\pi}{2} + k\pi\right) = \frac{\pi}{4} + k \cdot \frac{\pi}{2}, \quad k \in J.$$

c. Zeros: $\left\{x \mid x = k \cdot \frac{\pi}{2},\ k \in J\right\}$

1. $y = \tan 3x$ **2.** $y = \tan \frac{1}{2} x$ **3.** $y = \tan\left(x + \frac{\pi}{6}\right)$

4. $y = \tan\left(x - \frac{\pi}{4}\right)$ **5.** $y = \cot \frac{1}{2} x$ **6.** $y = \cot 2x$

7. Graph $y = \sin x$, $-2\pi \le x \le 2\pi$ and then use the fact that $\csc x = \dfrac{1}{\sin x}$, $x \ne k\pi$, $k \in J$, to graph $y = \csc x$ on the same set of axes over the same interval.

8. Graph $y = \cos x$, $-2\pi \leq x \leq 2\pi$ and then use the fact that $\sec x = \dfrac{1}{\cos x}$, $x \neq \dfrac{\pi}{2} + k \in J$, to graph $y = \sec x$ on the same set of axes over the same interval.

9. Use the graph obtained in Exercise 7 to write the equations of the asymptotes of the graph of $y = \csc x$.

10. Use the graph obtained in Exercise 8 to write the equations of the asymptotes of the graph of $y = \sec x$.

11. Use the fact that $\frac{1}{2} \sec x = \dfrac{1}{2 \cos x}$ to graph $y = \frac{1}{2} \sec x$ over the interval $-\pi \leq x \leq 2\pi$.

12. Use the fact that $2 \csc x = \dfrac{1}{\frac{1}{2} \sin x}$ to graph $y = 2 \csc x$ over the interval $-\pi \leq x \leq 2\pi$.

13. Use the fact that $\csc 2x = \dfrac{1}{\sin 2x}$ to graph $y = \csc 2x$ over the interval $-\pi \leq x \leq 2\pi$.

14. Use the fact that $\sec \frac{1}{2}x = \dfrac{1}{\cos \frac{1}{2}x}$ to graph $y = \sec \frac{1}{2}x$ over the interval $-\pi \leq x \leq 2\pi$.

15. Graph $y = \sec \left(x + \dfrac{\pi}{4}\right)$ over the interval $-\pi \leq x \leq 2\pi$.

16. Graph $y = \csc \left(x - \dfrac{\pi}{6}\right)$ over the interval $-\pi \leq x \leq 2\pi$.

3.5 Graphing Systems of Equations

It is sometimes convenient to use systems of equations involving trigonometric function values to specify conditions on variables. As in algebraic equations, the solution of a system of equations is the intersection of the solution set of each equation in the system.

In this section we will consider graphical methods of approximating solutions of such systems. In Chapter 4 we will consider algebraic methods to find such solutions.

Example Find all elements (approximate values) in the intersection $\{(x, y) \mid y = 2 \sin x\} \cap \{(x, y) \mid y = \frac{3}{2}\}$ in the interval $0 \leq x \leq 2\pi$ by graphical methods.

(continued)

Solution Graph each set and approximate the coordinates of the points of intersection.

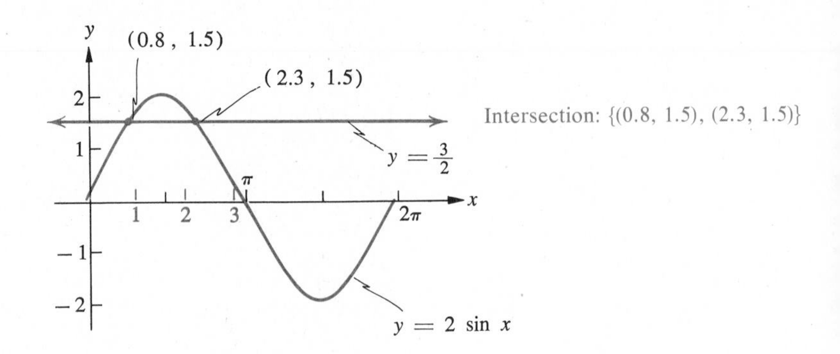

Parametric Equations

Sometimes it is convenient to define a relation (or function) in which the variable representing an element in the domain and the variable representing an element in the range are each related to a third variable, frequently a variable representing time or an angle. For example, the pair of equations

$$x = t - 3 \qquad \text{and} \qquad y = t + 1$$

express x in terms of t and y in terms of t. Such equations are called **parametric equations;** t is the **parameter.** To graph the relation between x and y as shown in the example below, we can choose arbitrary real number replacement values for t, and then obtain the corresponding values for x and y. Alternatively, by solving the system of equations, we can eliminate the parameter t to first obtain an equation in the variables x and y. Then we can graph the equation in the usual way.

Example

a. Graph the function defined by the parametric equations

$$x = t - 3 \qquad \text{and} \qquad y = t + 1 \tag{1}$$

by assigning arbitrary values for t to obtain some ordered pairs (x, y) in the function.

b. Eliminate t and find an equation in x and y.

Solution

a. Choosing 0, 1, 2, and 3 as replacements for t, the corresponding ordered pairs (x, y) shown in the figure can be obtained. The construction of a table is sometimes helpful to obtain these ordered pairs.

t	x	y
0	−3	1
1	−2	2
2	−1	3
3	0	4

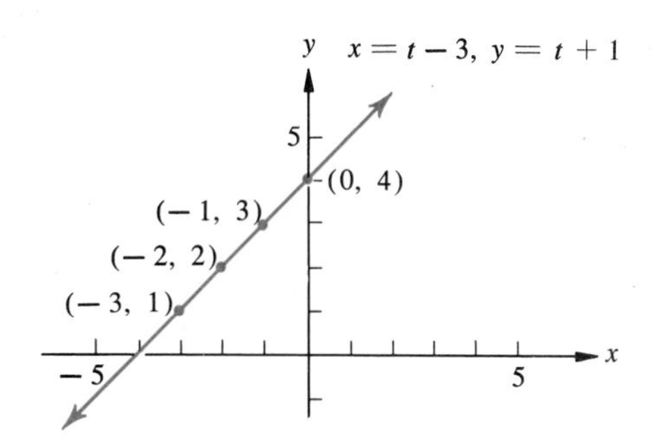

b. Equations (1) can be written equivalently as

$$t = x + 3 \qquad \text{and} \qquad t = y - 1,$$

from which

$$x + 3 = y - 1 \quad \leftrightarrow \quad x - y = -4,^{*}$$

the equation of a straight line in x and y.

The next example illustrates a case where the parameter is an angle.

Example

a. Graph the function defined by the parametric equations

$$x = 3 \cos \alpha \qquad \text{and} \qquad y = 4 \sin \alpha$$

by assigning arbitrary values for α to obtain some ordered pairs (x, y) in the function.

b. Eliminate α and find an equation in x and y.

Solution

a. Choosing selected elements of $\{\alpha \mid 0 \leq \alpha < 2\pi\}$ as replacements for α, we can obtain the corresponding ordered pairs (x, y) as shown in the figure on page 122. The tabulation of values is helpful.

(continued)

* The symbol $\leftrightarrow$ should be read "is equivalent to."

α	x $3 \cos \alpha$	y $4 \sin \alpha$
0	3	0
$\frac{\pi}{6}$	$\frac{3\sqrt{3}}{2} \approx 2.6$	2
$\frac{\pi}{4}$	$\frac{3}{\sqrt{2}} \approx 2.1$	$\frac{4}{\sqrt{2}} \approx 2.8$
$\frac{\pi}{2}$	0	4
$\frac{5\pi}{6}$	-2.6	2
.	.	.
.	.	.
.	.	.
2π	3	0

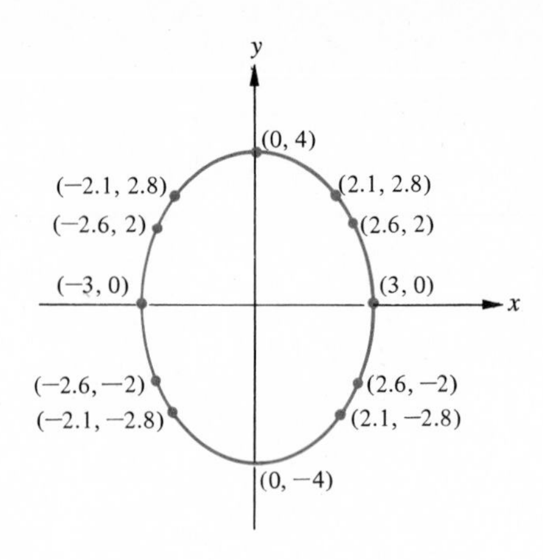

b. Motivated by the fact that $\cos^2 \alpha + \sin^2 \alpha = 1$, we first solve the given equations for $\cos \alpha$ and $\sin \alpha$.

$$x = 3 \cos \alpha \quad \leftrightarrow \quad \cos \alpha = \frac{x}{3};$$

$$y = 4 \sin \alpha \quad \leftrightarrow \quad \sin \alpha = \frac{y}{4}.$$

Hence,

$$\cos^2 \alpha + \sin^2 \alpha = \frac{x^2}{9} + \frac{y^2}{16}.$$

Because $\cos^2 \alpha + \sin^2 \alpha = 1$,

$$\frac{x^2}{9} + \frac{y^2}{16} = 1,$$

from which

$$16x^2 + 9y^2 = 144.$$

EXERCISE SET 3.5

By graphical methods, find elements (approximate values) in each set over the interval $0 \le x \le 2\pi$.

Example $\{(x, y) \mid y = \tan x\} \cap \{(x, y) \mid y = 1\}$

Solution Sketch the graph of $y = \tan x$ and $y = 1$ as shown. The coordinates of the points of intersection are estimated.

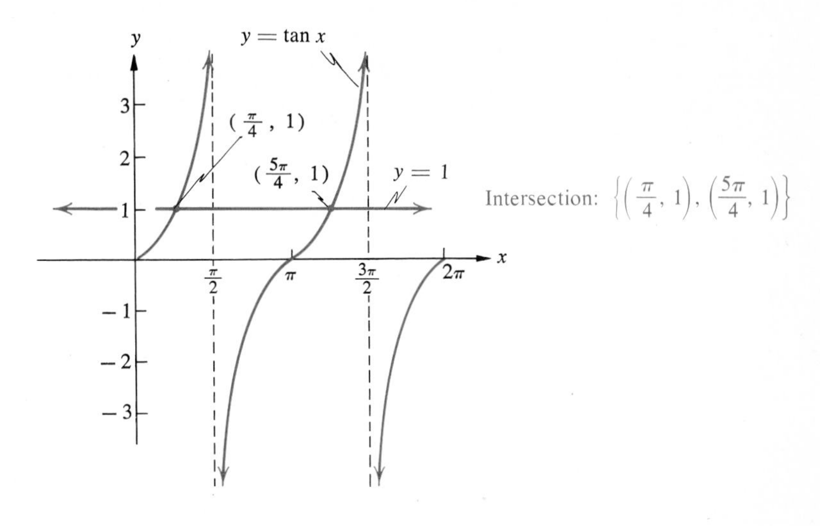

1. $\{(x, y) \mid y = \cos x\} \cap \{(x, y) \mid y = 1\}$
2. $\{(x, y) \mid y = \sin 2x\} \cap \{(x, y) \mid y = -1\}$
3. $\left\{(x, y) \mid y = \cos \frac{1}{2} x\right\} \cap \left\{(x, y) \mid y = \frac{1}{2}\right\}$
4. $\{(x, y) \mid y = 2 \sin x\} \cap \left\{(x, y) \mid y = -\frac{1}{2}\right\}$
5. $\{(x, y) \mid y = \tan x\} \cap \{(x, y) \mid y = -1\}$
6. $\{(x, y) \mid y = \cot x\} \cap \{(x, y) \mid y = \sqrt{3}\}$

Solve each system by graphical methods over the interval $0 \leq x \leq 2\pi$.

Example $y = \sin x$

$y = 2 \cos x$

Solution

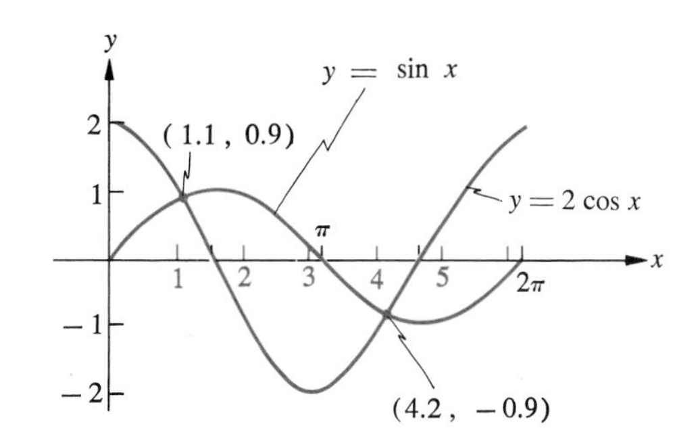

The solution set is $\{(1.1, 0.9), (4.2, -0.9)\}$

7. $y = \sin 2x$

$y = \frac{1}{2} \sin x$

8. $y = \cos 2x$

$y = 2 \sin x$

9. $y = 2 \cos 2x$

$y = \sin \frac{1}{2} x$

10. $y = \cos x$

$y = 2 \cos \frac{1}{2} x$

In Exercises 11–18, (**a**) *graph the curve represented by the parametric equations by assigning values to* α*, where* $0 \le \alpha < 2\pi$*, to obtain some ordered pairs* (x, y)*, and* (**b**) *eliminate the parameter to find an equation in* x *and* y.

Example $x = 2 \cos \alpha, \ y = \sin \alpha$

Solution

a.

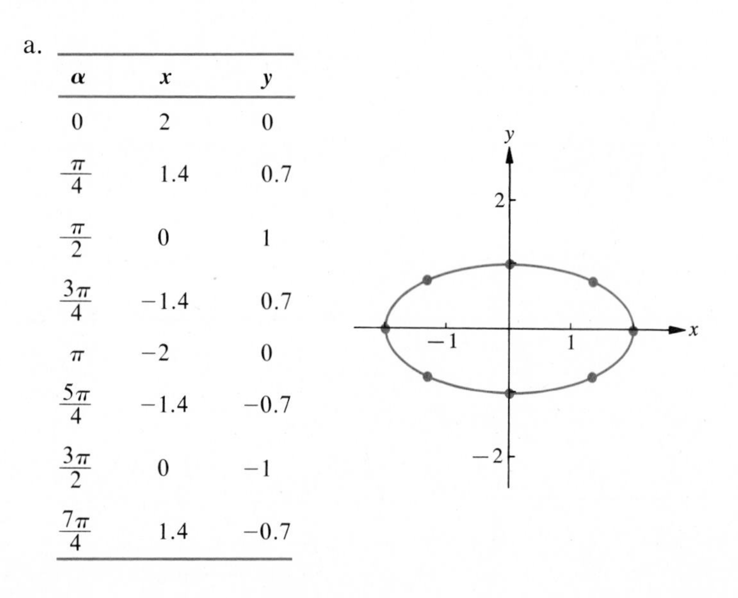

α	x	y
0	2	0
$\frac{\pi}{4}$	1.4	0.7
$\frac{\pi}{2}$	0	1
$\frac{3\pi}{4}$	−1.4	0.7
π	−2	0
$\frac{5\pi}{4}$	−1.4	−0.7
$\frac{3\pi}{2}$	0	−1
$\frac{7\pi}{4}$	1.4	−0.7

b. Solve for $\cos \alpha$ and $\sin \alpha$:

$$\cos \alpha = \frac{x}{2} \quad \text{and} \quad \sin \alpha = y,$$

from which

$$\cos^2 \alpha = \frac{x^2}{4} \quad \text{and} \quad \sin^2 \alpha = y^2.$$

Since $\cos^2 \alpha + \sin^2 \alpha = 1$, we have

$$\frac{x^2}{4} + y^2 = 1,$$

from which

$$x^2 + 4y^2 = 4.$$

11. $x = \cos \alpha,\ y = \sin \alpha$

12. $x = 2 \cos \alpha,\ y = 2 \sin \alpha$

13. $x = 4 \cos \alpha,\ y = 3 \sin \alpha$

14. $x = 2 \cos \alpha,\ y = 5 \sin \alpha$

15. $x = 5 \cos \alpha,\ y = 2 \sin \alpha$

16. $x = 3 \cos \alpha,\ y = 6 \sin \alpha$

17. $x = -\cos \alpha,\ y = 3 \sin \alpha$

18. $x = 2 \cos \alpha,\ y = -\sin \alpha$

B

In Exercises 19 and 20 obtain one equation in x and y.

19. $\begin{cases} x = 2 \cot t \\ y = 2 \sin^2 t \end{cases}$

20. $\begin{cases} x = \sin t \\ y = \cos^2 t(2 + \sin t) \end{cases}$

21. A curve called the **cycloid** is determined by the equations

$$x = a(\alpha - \sin \alpha),$$
$$y = a(1 - \cos \alpha).$$

Graph these equations in an x,y coordinate system for $a = 1$ and $0 \le \alpha \le 2\pi$.

22. Graph the equations in Exercise 21 for $a = 2$ and for $a = 3$.

3.6 Inverse Trigonometric Functions

In Section 1.4 we obtained values for x in the domains of the trigonometric functions

$$y = \text{trig } x$$

for given values of y in the range. Recall that we restricted the values of y to be greater than 0 and the values of x to be between 0 and $\pi/2$. For example, the least positive x for which $\cos x = 1/\sqrt{2}$ is $\pi/4$. However, note in Figure 3.9 on page 126 that each value of x such that

$$x = \frac{\pi}{4} + k \cdot 2\pi, \quad k \in J$$

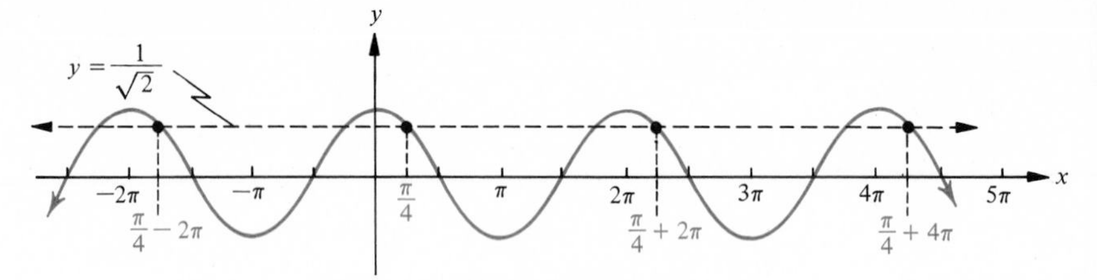

Figure 3.9

is a solution of $\cos x = 1/\sqrt{2}$ because the cosine function has a period 2π.

Similar considerations are also applicable to other circular functions.

In this section we are interested in obtaining elements in the domains of the trigonometric functions for all elements in their ranges. Furthermore, we will restrict the domains in such a way that there will be *one and only one* element in the domain for a given element in its range.

Inverse of a Function

The inverse of a function is the relation obtained by interchanging the components of every ordered pair in the function (see Appendix A.6). Hence, the defining equation of the inverse relation of a function can be obtained by interchanging the variables in the defining equation of the original function. Furthermore, the graph of the inverse of a function is the reflection of the graph of the function with respect to the graph of the equation $y = x$. Consider the graph of the function

$$y = \sin x,$$

with domain R and range $\{y \mid -1 \leq y \leq 1\}$, and its inverse

$$x = \sin y,$$

with domain $\{x \mid -1 \leq x \leq 1\}$ and range $y \in R$, in Figure 3.10. Notice that the graph of $x = \sin y$ is the reflection of the graph of $y = \sin x$ in the graph of the equation $y = x$. Further, notice that

$$x = \sin y$$

defines a relation that is not a function because for each element x in its domain, there is an unlimited number of elements y_1, y_2, etc., in its range as shown in Figure 3.11.

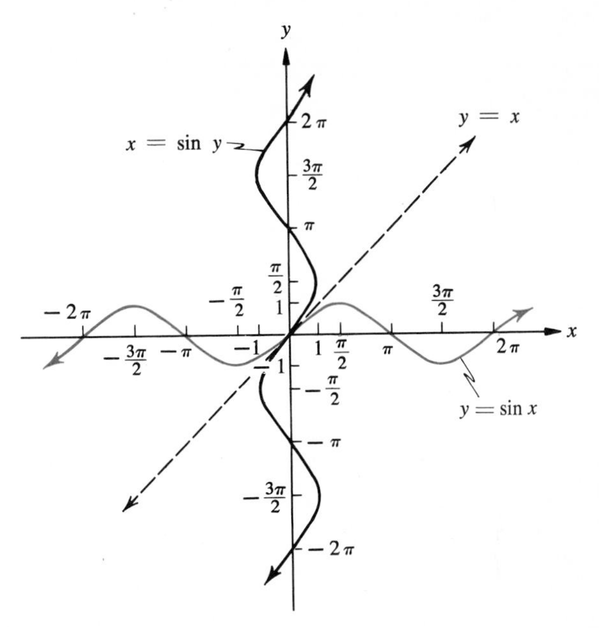

Figure 3.10

Inverse Functions

Each trigonometric function has an inverse relation. By suitably restricting the domain of the function or the range of the respective inverse relation, we can define an inverse function for each circular function. For example, if we restrict the domain of the sine function to $\{x \mid -\pi/2 \leq x \leq \pi/2\}$, then its inverse

$$x = \sin y, \quad -\frac{\pi}{2} \leq y \leq \frac{\pi}{2}, \tag{1}$$

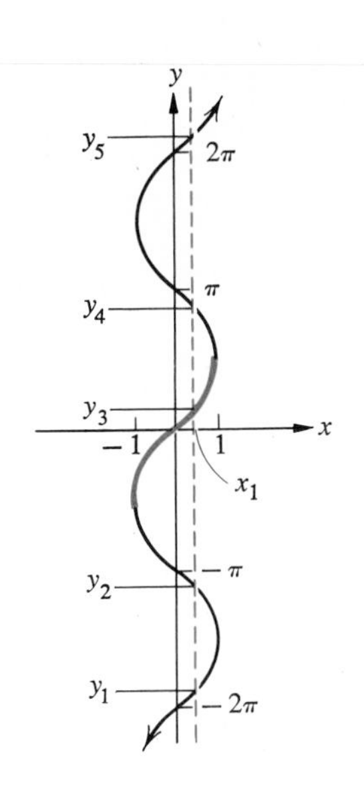

Figure 3.11

defines an inverse function (there is only one y for each x in the domain). Its graph is shown in color in Figure 3.11.

Generally, we wish to write the defining Equation (1) in a form in which y is expressed explicitly in terms of x. This is done by using the inverse notation introduced in Section 1.4. For example,

$$x = \sin y, \quad -\frac{\pi}{2} \leq y \leq \frac{\pi}{2} \quad \leftrightarrow \quad y = \text{Arcsin } x \quad \leftrightarrow \quad y = \text{Sin}^{-1} x,$$

where again we note that "A" and "S" are capitalized.

Similar notation is used for the other inverse circular functions, as shown in Definition 3.3.

Definition 3.3 *The inverse notation corresponding to the six inverse trigonometric functions is*

$$y = \operatorname{Sin}^{-1} x \quad \leftrightarrow \quad x = \sin y, \quad -\frac{\pi}{2} \le y \le \frac{\pi}{2},$$

$$y = \operatorname{Cos}^{-1} x \quad \leftrightarrow \quad x = \cos y, \quad 0 \le y \le \pi,$$

$$y = \operatorname{Tan}^{-1} x \quad \leftrightarrow \quad x = \tan y, \quad -\frac{\pi}{2} < y < \frac{\pi}{2},$$

$$y = \operatorname{Csc}^{-1} x \quad \leftrightarrow \quad x = \csc y, \quad -\frac{\pi}{2} \le y < 0 \quad \text{or} \quad 0 < y \le \frac{\pi}{2},$$

$$y = \operatorname{Sec}^{-1} x \quad \leftrightarrow \quad x = \sec y, \quad 0 \le y < \frac{\pi}{2} \quad \text{or} \quad \frac{\pi}{2} < y \le \pi,$$

$$y = \operatorname{Cot}^{-1} x \quad \leftrightarrow \quad x = \cot y, \quad -\frac{\pi}{2} \le y < 0 \quad \text{or} \quad 0 < y \le \frac{\pi}{2}.$$

The function values in $\operatorname{Sin}^{-1} x$, $\operatorname{Cos}^{-1} x$, etc., are sometimes called the **principal values** in the corresponding relation.

Notice that the ranges of the inverse functions are specified in the definition rather than following the usual procedure of stating the domain. Furthermore, the choices for the ranges are somewhat arbitrary. Those that have been chosen here involve small values of y and include values between 0 and $\pi/2$, have relatively simple graphs, and yield a one-to-one correspondence between elements in the domain and range. They are also the values that are obtained with most calculators.

As we noted in Section 1.4, it is sometimes easier to interpret a relationship expressed in inverse notation if it is rewritten as an equation without using such notation.

Example The equation $y = \operatorname{Arcsin}(-1/2)$ can be rewritten equivalently without inverse notation as

$$\sin y = -\frac{1}{2}, \quad -\frac{\pi}{2} \le y \le \frac{\pi}{2}.$$

This equation is satisfied for $y = -\pi/6$. Hence,

$$y = \operatorname{Arcsin}\left(-\frac{1}{2}\right) = -\frac{\pi}{6}.$$

With a little practice you will be able to find an inverse function value directly from the expression involving the inverse notation.

Examples Find the value of each of the following by using Table 1.2.

a. $\text{Arccos}\,\dfrac{1}{\sqrt{2}}$ b. $\text{Tan}^{-1}\dfrac{1}{\sqrt{3}}$ c. $\text{Csc}^{-1}\dfrac{2}{\sqrt{3}}$

Solutions

a. From Definition 3.3, $\pi/4$ is the number between 0 and $\pi/2$ such that $\cos(\pi/4) = 1/\sqrt{2}$. Hence,

$$\text{Arccos}\,\frac{1}{\sqrt{2}} = \frac{\pi}{4}.$$

b. $\pi/6$ is the number between 0 and $\pi/2$ such that tan $\pi/6 = 1/\sqrt{3}$. Hence,

$$\text{Tan}^{-1}\frac{1}{\sqrt{3}} = \frac{\pi}{6}.$$

c. Because $\csc x = 1/\sin x$, we want a number x between 0 and $\pi/2$ such that $\csc x = 2/\sqrt{3}$ or $\sin x = \sqrt{3}/2$. From Table 1.2, $\pi/3$ is that number. Hence,

$$\text{Csc}^{-1}\frac{2}{\sqrt{3}} = \frac{\pi}{3}.$$

Careful attention should be given to the ranges specified in Definition 3.3 when finding values for $\text{Trig}^{-1}\,x$ for negative values of x.

Examples Find the value of each of the following using Table 1.2 or an appropriate reduction formula.

a. $\text{Arccos}\left(-\dfrac{1}{\sqrt{2}}\right)$ b. $\text{Tan}^{-1}\left(-\dfrac{1}{\sqrt{3}}\right)$ c. $\text{Csc}^{-1}\left(-\dfrac{2}{\sqrt{3}}\right)$

Solutions

a. In the preceding example, note that Arccos $1/\sqrt{2} = \pi/4$. For $\text{Arccos}(-1/\sqrt{2})$, the number between $\pi/2$ and π is $\pi - (\pi/4)$, or $3\pi/4$, such that $\cos 3\pi/4 = -1/\sqrt{2}$. Hence,

$$\text{Arccos}\left(-\frac{1}{\sqrt{2}}\right) = \frac{3\pi}{4}.$$

(continued)

b. Since $-\pi/6$ is the number between 0 and $-\pi/2$ such that $\tan(-\pi/6) = -1/\sqrt{3}$,

$$\text{Tan}^{-1}\left(-\frac{1}{\sqrt{3}}\right) = -\frac{\pi}{6}.$$

c. Since $-\pi/3$ is the number between 0 and $-\pi/2$ such that $\csc(-\pi/3) = -2/\sqrt{3}$,

$$\text{Csc}^{-1}\left(-\frac{2}{\sqrt{3}}\right) = -\frac{\pi}{3}.$$

Graphs of the Inverse Trigonometric Functions

The graph of the Arcsine function (shown in color in Figure 3.11) and the graphs of the other inverse functions, which can be obtained by

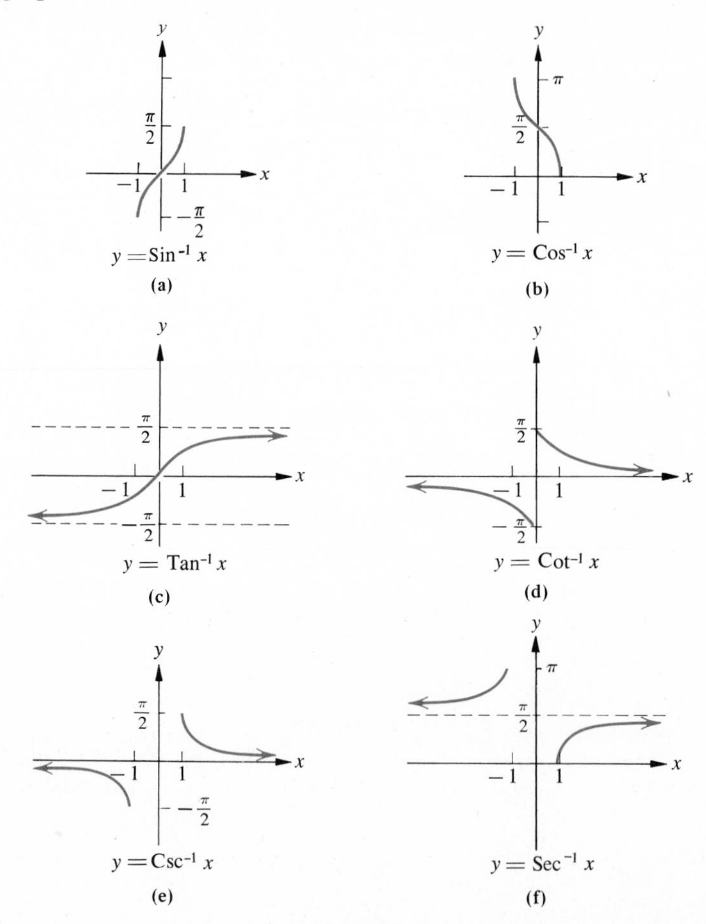

Figure 3.12

graphing ordered pairs or by reflecting the graphs of the respective functions over the appropriate domain, are shown in Figure 3.12.

EXERCISE SET 3.6

A

Rewrite each expression without using inverse notation and then specify the value of y for which the statement is true.

Example $y = \text{Arccot } 1$

Solution From Definition 3.3,

$$y = \text{Arccot } 1 \quad \leftrightarrow \quad \cot y = 1, \quad 0 < y \le \frac{\pi}{2}.$$

Hence, $y = \pi/4$.

1. $y = \text{Arccos } \dfrac{\sqrt{3}}{2}$ **2.** $y = \text{Arctan } 1$ **3.** $y = \text{Sin}^{-1} 1$

4. $y = \text{Csc}^{-1} 2$ **5.** $y = \text{Cot}^{-1} \dfrac{1}{\sqrt{3}}$ **6.** $y = \text{Sec}^{-1} \sqrt{2}$

Find the value of each of the following by using Table 1.2 or an appropriate reduction formula.

Example $\text{Cos}^{-1}\left(-\dfrac{\sqrt{3}}{2}\right)$

Solution Let $y = \text{Cos}^{-1}(-\sqrt{3}/2)$. Then from Definition 3.3,

$$\cos y = -\frac{\sqrt{3}}{2}, \quad 0 \le y \le \pi,$$

from which

$$y = \text{Cos}^{-1}\left(-\frac{\sqrt{3}}{2}\right) = \frac{5\pi}{6}.$$

7. $\text{Arcsin } \dfrac{1}{\sqrt{2}}$ **8.** $\text{Arccos } \dfrac{\sqrt{3}}{2}$ **9.** $\text{Arctan } 1$

10. $\text{Arcsin } 0$ **11.** $\text{Tan}^{-1} \sqrt{3}$ **12.** $\text{Cos}^{-1} 0$

13. $\text{Cos}^{-1} 1$ **14.** $\text{Tan}^{-1} 0$ **15.** $\text{Sec}^{-1} \dfrac{2}{\sqrt{3}}$

16. $\text{Csc}^{-1} 1$ **17.** $\text{Cot}^{-1} 1$ **18.** $\text{Sec}^{-1} 1$

19. $\text{Sin}^{-1}\left(-\frac{1}{2}\right)$ **20.** $\text{Cos}^{-1}(-1)$

Find an approximation for each of the following by using Table III or a calculator.

Example $\text{Sin}^{-1} 0.8772$

Solution Let $y = \text{Sin}^{-1} 0.8772$. Then from Definition 3.3,

$$\sin y = 0.8772, \quad \frac{\pi}{2} \le y \le \frac{\pi}{2},$$

from which

$$y = \text{Sin}^{-1} 0.8772 \approx 1.07.$$

21. Arctan 0.5726 **22.** Arcsin 0.6442 **23.** Arccos 0.8961

24. Arctan 0.1104 **25.** $\text{Sin}^{-1} 0.3523$ **26.** $\text{Cos}^{-1} 0.6600$

27. $\text{Tan}^{-1} 2.912$ **28.** $\text{Sin}^{-1} 0.8724$ **29.** $\text{Sec}^{-1} 8.299$

30. $\text{Cot}^{-1} 0.6563$ **31.** $\text{Tan}^{-1}(-0.5463)$ **32.** $\text{Sin}^{-1}(-0.6518)$

33. $\text{Cos}^{-1}(-0.9759)$ **34.** $\text{Cos}^{-1}(-0.3153)$

Rewrite each equation so that x is expressed explicitly in terms of y and/or the constants.

Example $3 \sin 2x = 1$

Solution Multiplying each member by 1/3 yields $\sin 2x = 1/3$. By Definition 3.3,

$$\sin 2x = \frac{1}{3} \leftrightarrow 2x = \text{Sin}^{-1}\frac{1}{3}, \quad -\frac{\pi}{2} \le 2x \le \frac{\pi}{2}.$$

Then, multiplying each member by 1/2, we obtain

$$x = \frac{1}{2}\text{Sin}^{-1}\frac{1}{3}, \quad -\frac{\pi}{4} \le x \le \frac{\pi}{4}.$$

35. $4 \cos x = 3$ **36.** $2 \tan x = 3$ **37.** $\sin 3x = \frac{1}{4}$

38. $\cos 2x = \frac{1}{3}$ **39.** $2 \tan 2x = 5$ **40.** $3 \sin 4x = 2$

41. $y = 2 \tan 3x$ **42.** $y = 2 \tan \frac{x}{3}$

Evaluate each of the following.

Example $\sin\left(\text{Sin}^{-1} \frac{1}{2}\right)$

Solution $\text{Sin}^{-1}\left(\frac{1}{2}\right) = \frac{\pi}{6}$; therefore,

$$\sin\left(\text{Sin}^{-1} \frac{1}{2}\right) = \sin \frac{\pi}{6} = \frac{1}{2}.$$

43. $\cos\left(\text{Cos}^{-1} \frac{\sqrt{3}}{2}\right)$

44. $\tan\,(\text{Tan}^{-1}\, 1)$

45. $\sin\,[\text{Arctan}(-1)]$

46. $\sin\left[\text{Cos}^{-1}\left(-\frac{1}{2}\right)\right]$

47. $\cot\,(\text{Sin}^{-1}\, 0.8134)$

48. $\csc\,(\text{Arccot}\, 4.933)$

49. $\csc\,[\text{Arcsec}(-2.760)]$

50. $\cot\,[\text{Arcsec}(-1.005)]$

51. $\text{Sin}^{-1}\left(\sin \frac{\pi}{4}\right)$

52. $\text{Cos}^{-1}\left(\cos \frac{\pi}{3}\right)$

53. $\text{Tan}^{-1}\left(\cos \frac{\pi}{2}\right)$

54. $\text{Csc}^{-1}\,(\cos \pi)$

55. $\text{Arcsec}\left(\tan \frac{3\pi}{4}\right)$

56. $\text{Arccot}\left(\sin \frac{3\pi}{2}\right)$

B

Write each of the following as a numerical, or an algebraic, expression without using inverse notation. Assume values for x for which each function value exists.

Example $\sin\left(\text{Arccos} \frac{3}{5}\right)$

Solution Because the value of Arccos 3/5 is not evident by inspection, we let $\beta = \text{Arccos}\ 3/5$, from which $\cos \beta = 3/5$. The sketch

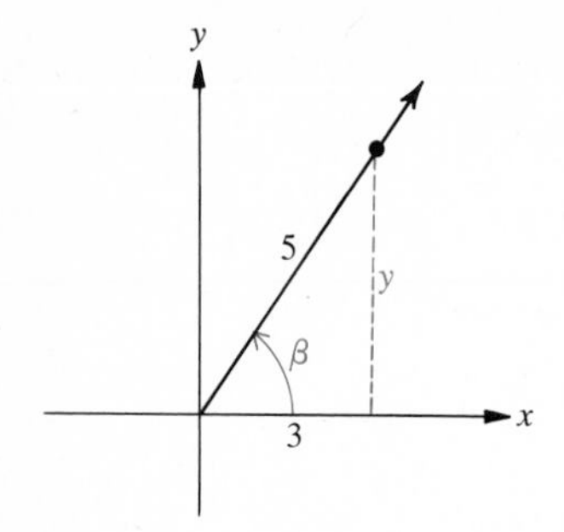

shows a right triangle in which $\cos \beta = 3/5$. From the Pythagorean theorem, we have

$$y^2 + 3^2 = 5^2$$
$$y^2 = 25 - 9 = 16$$
$$y = 4.$$

Hence,

$$\sin \beta = \sin \left(\text{Arccos } \frac{3}{5}\right) = \frac{y}{5} = \frac{4}{5}.$$

57. $\cos \left(\text{Arcsin } \frac{3}{5}\right)$

58. $\sin \left(\text{Arccos } \frac{5}{13}\right)$

59. $\sin \left(\text{Arcsin } \frac{2}{5}\right)$

60. $\cos \left(\text{Arccos } \frac{1}{3}\right)$

61. $\tan \left(\text{Arccos } \frac{4}{5}\right)$

62. $\tan \left(\text{Arcsin } \frac{5}{13}\right)$

63. $\cos \left(\text{Arcsin } \frac{x}{3}\right)$

64. $\tan \left(\text{Arccos } \frac{x}{4}\right)$

65. $\sin (\text{Arccos } x)$

66. $\sin (\text{Arctan } x)$

Chapter Summary

3.1 Graphs of the form of the graph of the sine function are called **sine waves** or **sinusoidal waves.** The part of the graph over one period is called a **cycle** of the wave. The absolute value of half the difference of the maximum and the minimum ordinates is called the **amplitude** of the wave.

The zeros of the sine function, values of x for which $\sin x = 0$, are $k\pi$, $k \in J$. These values are associated with the points where the graph of $y = \sin x$ intersects the x-axis. Zeros of the cosine function are $(\pi/2) + k\pi$, $k \in J$. These values are associated with the points where the graph of $y = \cos x$ intersects the x-axis (see Figures 3.3 and 3.5).

3.2 Functions defined by equations of the form

$$y = A \sin B(x + C) \qquad \text{or} \qquad y = A \cos B(x + C),$$

where $A, B, C, x, y \in R$ and $A, B \neq 0$, always have sine waves for their graphs. The **amplitude** of the wave or function is $|A|$, the **period** is $2\pi/|B|$, and the **phase shift** is $|C|$.

3.3 Functions defined by equations such as

$$y = 2 \sin x + \cos x$$

can be graphed by first graphing $y_1 = 2 \sin x$ and $y_2 = \cos x$, and then "adding" their ordinates graphically for arbitrary values of x.

3.4 Graphs of trigonometric functions defined by $y = \tan x$, $y = \cot x$, $y = \sec x$, and $y = \csc x$ exhibit the periodic nature of these functions. The following table shows the zeros and/or the equations for the asymptotes of these circular functions.

Function	Zeros	Equations for Asymptotes
tangent	$k\pi,\ k \in J$	$x = \frac{\pi}{2} + k\pi,\ k \in J$
cotangent	$\frac{\pi}{2} + k\pi,\ k \in J$	$x = k\pi,\ k \in J$
secant	none	$x = \frac{\pi}{2} + k\pi,\ k \in J$
cosecant	none	$x = k\pi,\ k \in J$

3.5 A pair of equations of the form

$$x = f(\alpha) \qquad \text{and} \qquad y = g(\alpha),$$

in which two variables, in this case x and y, are each expressed in terms of a third variable, are called **parametric equations.** The relation between x and y can be graphed by arbitrarily choosing real number replacements for the third variable, in this case α, and then obtaining the corresponding values for x and y.

3.6 The **inverse of a function** is the relation obtained by interchanging the components of every ordered pair in the function.

By suitably restricting the domains of the trigonometric functions, or the ranges of the respective inverse relations, an inverse function is defined for each trigonometric function (see Definition 3.3).

The function values in Arcsine, Arccosine, etc., are sometimes called the principal values of the function values in the corresponding relation. Table II or Table III in Appendix C, or a calculator, can be used to find approximations for function values of the six inverse trigonometric functions for elements in their respective domains.

Review Exercises

3.1–3.2 *Sketch the graph of the function defined by each equation over the interval* $-2\pi \le x \le 2\pi$. *Specify* **(a)** *the amplitude,* **(b)** *the period, and* **(c)** *the phase shift (if applicable).*

1. $y = 4 \sin x$

2. $y = \dfrac{1}{2} \cos 2x$

3. $y = \sin \left(x + \dfrac{\pi}{2}\right)$

4. $y = 2 \cos \left(\dfrac{1}{2}x - \dfrac{\pi}{6}\right)$

Graph two cycles of each of the following.

5. $y = \cos \dfrac{\pi}{3} x$

6. $y = -2 \sin \dfrac{\pi}{2} x$

3.3 *Using graphical addition, graph each equation over the interval* $0 \le x \le 2\pi$.

7. $y = \sin 2x - \dfrac{1}{2} \cos x$

8. $y = \dfrac{x}{2} + \sin 2x$

3.4 *Sketch the graph of the function defined by each equation over the interval* $0 \le x \le 2\pi$. *Specify* **(a)** *the range,* **(b)** *the equations of the asymptotes, and* **(c)** *the set of all zeros of the function (if applicable).*

9. $y = \tan \left(x - \dfrac{\pi}{3}\right)$

10. $y = \cot 3x$

11. $y = \csc \dfrac{1}{2} x$

12. $y = 2 \sec x$

3.5 *Find approximations to solutions of each system over the interval* $0 \le x \le 2\pi$ *by graphical methods.*

13. $y = \cos x$

$y = \dfrac{1}{3}$

14. $y = 2 \sin x$

$y = \dfrac{3}{4}$

15. $y = \sin 2x$

$y = \dfrac{1}{2} \cos x$

16. $y = \cos 2x$

$y = \dfrac{1}{2} \sin x$

17. Graph the parametric equations $x = \cos\alpha$ and $y = 4\sin\alpha$ in an x, y-coordinate system.

18. Eliminate the parameter in the equations in Exercise 17 to find an equation in x and y.

3.6 *Rewrite each expression without using inverse notation and then specify the value of y for which the statement is true.*

19. $y = \text{Arccos}\,\dfrac{1}{\sqrt{2}}$

20. $y = \text{Arccot}(-1)$

21. $y = \text{Csc}^{-1}(-2)$

22. $y = \text{Sec}^{-1}\,\dfrac{2}{\sqrt{3}}$

Find the value of each expression.

23. Arcsin 1

24. Arccot(−1)

25. $\text{Cos}^{-1}\left(-\dfrac{1}{\sqrt{2}}\right)$

26. $\text{Tan}^{-1}\,\dfrac{1}{\sqrt{3}}$

Find an approximation for each of the following by using Table III or a calculator.

27. Arccos 0.1205

28. Arctan 1.398

29. $\text{Sin}^{-1}\,0.4529$

30. $\text{Cot}^{-1}(-0.6142)$

Rewrite each equation so that x is expressed explicitly in terms of y and/or the constants.

31. $3\cos x = 5$

32. $\sin 2x = \dfrac{1}{5}$

33. $y = 3\tan 2x$

34. $y = 2\cos\dfrac{x}{2}$

Evaluate each of the following.

35. $\tan\left(\text{Cos}^{-1}\,\dfrac{1}{2}\right)$

36. cos(Arccot 1.921)

37. $\text{Sin}^{-1}\left(\tan\dfrac{\pi}{4}\right)$

38. $\text{Tan}^{-1}(\cos 0)$

4 Identities and Conditional Equations

Equations such as

$$x + y = y + x \qquad \text{and} \qquad (x + y)^2 = x^2 + 2xy + y^2,$$

which are true for all real numbers x and y, are called **identities.** Equations such as

$$x + 1 = 7 \qquad \text{and} \qquad x^2 - 3x + 2 = 0,$$

which are true for some real number replacements of the variable and false for others, are called **conditional equations** (see Appendix A.3). We shall now consider identities that involve trigonometric or circular function values. We shall simply refer to either of these identities as trigonometric identities. A knowledge of these identities will be useful in later work to simplify expressions and solve equations. In our work with such identities we shall not explicitly specify the replacement set of the variable in each case. The variables α, β, γ, and θ will represent elements of the set of all angles, and the variables x and y will be considered as representing elements either of the set of all angles or of the set of real numbers. Furthermore, it should be understood that *an equation that is specified to be an identity without noting the restrictions is in fact an identity only for all replacements of the variable for which both members of the equation are defined.*

4.1 Basic Trigonometric Identities

The trigonometric identities

$$\csc \alpha = \frac{1}{\sin \alpha}, \quad \sec \alpha = \frac{1}{\cos \alpha}, \quad \text{and} \quad \cot \alpha = \frac{1}{\tan \alpha},$$

stated in Section 1.2, and the identities

$$\tan \alpha = \frac{\sin \alpha}{\cos \alpha}, \quad \cot \alpha = \frac{\cos \alpha}{\sin \alpha}, \quad \text{and} \quad \sin^2 \alpha + \cos^2 \alpha = 1,$$

stated in Section 1.3, have been shown to be consequences of the definition of the trigonometric ratios.

Two other identities follow from the equation $\sin^2 \alpha + \cos^2 \alpha = 1$. For any α for which $\cos \alpha \neq 0$, we multiply each member by $1/\cos^2\alpha$ to obtain the equivalent equations

$$\frac{\sin^2 \alpha}{\cos^2 \alpha} + \frac{\cos^2 \alpha}{\cos^2 \alpha} = \frac{1}{\cos^2 \alpha},$$

$$\tan^2 \alpha + 1 = \sec^2 \alpha.$$

Multiplying each member of $\sin^2 \alpha + \cos^2 \alpha = 1$ by $1/\sin^2 \alpha$ for $\sin \alpha \neq 0$, we obtain

$$\frac{\sin^2 \alpha}{\sin^2 \alpha} + \frac{\cos^2 \alpha}{\sin^2 \alpha} = \frac{1}{\sin^2 \alpha},$$

$$1 + \cot^2 \alpha = \csc^2 \alpha.$$

There is no general method for proving identities. Following is a list of suggestions that may help in determining the best method to use to prove an identity.

1. If one member of an equation appears to be the more complicated, it should be transformed so that it reduces to the simpler member of the equation.
2. It is frequently helpful to change all the terms of an equation to expressions involving only sines and/or cosines, and then to simplify both members.
3. In each step of a proof, look for any possible application of algebraic procedures, including factoring, squaring an expression, and simplifying fractions.

Example Show that $\dfrac{\cos \alpha - \sin \alpha}{\cos \alpha} = 1 - \tan \alpha$ is an identity.

Solution Substituting $\sin \alpha / \cos \alpha$ for $\tan \alpha$ in the right-hand member, we obtain

$$\frac{\cos \alpha - \sin \alpha}{\cos \alpha} = 1 - \frac{\sin \alpha}{\cos \alpha}.$$

Writing the right-hand member as a single term yields

$$\frac{\cos \alpha - \sin \alpha}{\cos \alpha} = \frac{\cos \alpha - \sin \alpha}{\cos \alpha},$$

which is clearly an identity.

The basic identities that we have stated above should be recognized in equivalent forms. For example, using the symbol $\leftrightarrow$ to mean "is equivalent to," we have

$$\sec \alpha = \frac{1}{\cos \alpha} \quad \leftrightarrow \quad (\cos \alpha)(\sec \alpha) = 1 \quad \leftrightarrow \quad \cos \alpha = \frac{1}{\sec \alpha};$$

$$\csc \alpha = \frac{1}{\sin \alpha} \quad \leftrightarrow \quad (\sin \alpha)(\csc \alpha) = 1 \quad \leftrightarrow \quad \sin \alpha = \frac{1}{\csc \alpha};$$

$$\cot \alpha = \frac{1}{\tan \alpha} \quad \leftrightarrow \quad (\tan \alpha)(\cot \alpha) = 1 \quad \leftrightarrow \quad \tan \alpha = \frac{1}{\cot \alpha};$$

$$\sin^2 \alpha + \cos^2 \alpha = 1 \quad \leftrightarrow \quad \sin^2 \alpha = 1 - \cos^2 \alpha \quad \leftrightarrow \quad \cos^2 \alpha = 1 - \sin^2 \alpha;$$

$$\tan^2 \alpha + 1 = \sec^2 \alpha \quad \leftrightarrow \quad \tan^2 \alpha = \sec^2 \alpha - 1;$$

$$\cot^2 \alpha + 1 = \csc^2 \alpha \quad \leftrightarrow \quad \cot^2 \alpha = \csc^2 \alpha - 1.$$

Example Show that the equation

$$\frac{\cos x}{1 + \sin x} = \frac{1 - \sin x}{\cos x}$$

is an identity.

Solution Multiplying the left-hand member by $\dfrac{1 - \sin x}{1 - \sin x}$, we obtain

$$\frac{\cos x(1 - \sin x)}{(1 + \sin x)(1 - \sin x)} \quad \text{or} \quad \frac{\cos x(1 - \sin x)}{1 - \sin^2 x}.$$

Substituting $\cos^2 x$ for $1 - \sin^2 x$ yields

$$\frac{\cos x(1 - \sin x)}{\cos^2 x} \quad \text{or} \quad \frac{1 - \sin x}{\cos x}.$$

Thus, we have

$$\frac{1 - \sin x}{\cos x} = \frac{1 - \sin x}{\cos x},$$

which is clearly an identity.

Example Show that the equation

$$\cot \alpha + \tan \alpha = \csc \alpha \sec \alpha$$

is an identity.

Solution Substituting $\cos \alpha/\sin \alpha$ for $\cot \alpha$ and $\sin \alpha/\cos \alpha$ for $\tan \alpha$ in the left-hand member, we obtain

$$\frac{\cos \alpha}{\sin \alpha} + \frac{\sin \alpha}{\cos \alpha} \quad \text{or} \quad \frac{\cos^2 \alpha + \sin^2 \alpha}{\sin \alpha \cos \alpha} \quad \text{or} \quad \frac{1}{\sin \alpha \cos \alpha}.$$

Substituting $1/\sin \alpha$ for $\csc \alpha$ and $1/\cos \alpha$ for $\sec \alpha$ in the right-hand member, we obtain

$$\frac{1}{\sin \alpha} \cdot \frac{1}{\cos \alpha} \quad \text{or} \quad \frac{1}{\sin \alpha \cos \alpha}.$$

Hence, the two members are identical:

$$\frac{1}{\sin \alpha \cos \alpha} = \frac{1}{\sin \alpha \cos \alpha}.$$

Several other useful trigonometric relationships can be obtained from geometric considerations. In Figure 4.1a on page 142,

$$\angle AOB = \angle COB, \quad \overline{OB} = \overline{OB}, \quad \text{and} \quad \angle OBA = \angle OBC.$$

Hence (see Appendix A.9),

$$\triangle AOB \cong \triangle COB,$$

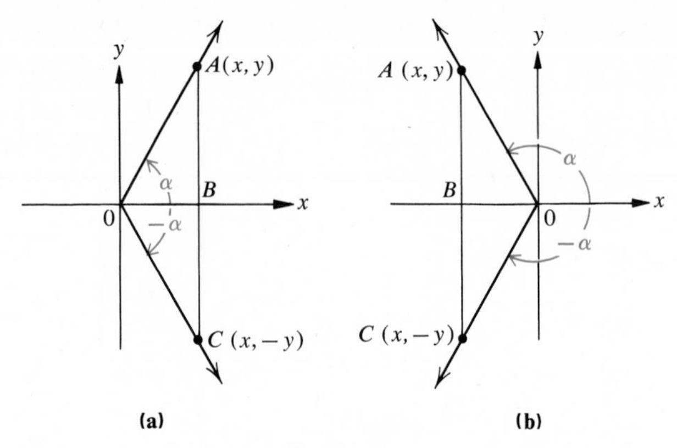

Figure 4.1

from which we have that each pair of corresponding sides are congruent and the absolute values of the respective components of the ordered pairs corresponding to points A and C are equal. Because

$$\sin(-\alpha) = \frac{-y}{r} \quad \text{and} \quad \sin \alpha = \frac{y}{r},$$

$$\mathbf{\sin(-\alpha) = -\sin \alpha.} \tag{1}$$

Also, $\cos(-\alpha) = \frac{x}{r}$ and $\cos \alpha = \frac{x}{r}$, from which

$$\mathbf{\cos(-\alpha) = \cos \alpha,} \tag{2}$$

and $\tan(-\alpha) = \frac{-y}{x}$ and $\tan \alpha = \frac{y}{x}$, from which

$$\mathbf{\tan(-\alpha) = -\tan \alpha.} \tag{3}$$

A similar argument for angles α and $-\alpha$ that are in Quadrants II and III, as shown in Figure 4.1b, leads to the results stated by Equations (1), (2), and (3).

Example Show that

$$\cos(-\alpha)\tan(-\alpha) = \sin(-\alpha). \tag{4}$$

Solution From Equations (2) and (3), we substitute $\cos \alpha$ for $\cos(-\alpha)$ and $-\tan \alpha$ for $\tan(-\alpha)$ in the left-hand member to obtain

$$\begin{aligned}\cos(-\alpha)\tan(-\alpha) &= \cos\alpha(-\tan\alpha)\\ &= \cos\alpha\left(-\frac{\sin\alpha}{\cos\alpha}\right)\\ &= -\sin\alpha\\ &= \sin(-\alpha).\end{aligned}$$

Hence, Equation (4) is an identity.

In general, it is a relatively simple matter to show that an equation is not an identity, if such is the case, by finding a *counterexample*. For example, substituting 0 for α in the equation

$$\sin\alpha = \cos\alpha$$

yields

$$\sin 0 = \cos 0,$$

$$0 = 1,$$

which is a false statement. Thus, the equation $\sin\alpha = \cos\alpha$ is not an identity.

Of course, in demonstrating a counterexample, we should select a replacement for the variable so that each member of the equation is defined.

EXERCISE SET 4.1

A

Express each of the following in terms of the function values $\sin\alpha$ *and/or* $\cos\alpha$ *and simplify the expression.*

Example $(\sin\alpha)(\cot^2\alpha)(\sec^2\alpha)$

Solution Substituting $\cos\alpha/\sin\alpha$ for $\cot\alpha$ and $1/\cos\alpha$ for $\sec\alpha$ yields

$$\begin{aligned}(\sin\alpha)(\cot^2\alpha)(\sec^2\alpha) &= (\cancel{\sin\alpha})\left(\frac{\cancel{\cos^2\alpha}}{\sin^{\cancel{2}}\alpha}\right)\left(\frac{1}{\cancel{\cos^2\alpha}}\right)\\ &= \frac{1}{\sin\alpha}.\end{aligned}$$

1. $(\sec\alpha)(\cot\alpha)$ **2.** $(\tan^2\alpha)(\csc^2\alpha)$ **3.** $1+\tan^2\alpha$

4. $1+\cot^2\alpha$ **5.** $(\sec^2\alpha-1)(\cos^2\alpha)$ **6.** $(\csc^2\alpha)(\tan\alpha)$

7. $\dfrac{\sec\alpha}{\tan\alpha}$ **8.** $\dfrac{\cot^2\alpha}{\csc^2\alpha}$ **9.** $\cos\alpha + \dfrac{\sin^2\alpha}{\cos\alpha}$

10. $\dfrac{\cos^2\alpha}{\sin\alpha} + \sin\alpha$ **11.** $\tan\alpha + \cot\alpha$ **12.** $\sec^2\alpha + \csc^2\alpha$

Show that each expression on the left is equivalent to the expression on the right for all values of the variable for which both expressions are defined.

Example $(1 - \cos^2\beta)(1 + \cot^2\beta);\ 1$

Solution Substituting $\sin^2\beta$ for $1 - \cos^2\beta$ and $\csc^2\beta$ for $1 + \cot^2\beta$ yields

$$(1 - \cos^2\beta)(1 + \cot^2\beta) = (\sin^2\beta)(\csc^2\beta).$$

Since $\csc^2\beta = 1/\sin^2\beta$,

$$(1 - \cos^2\beta)(1 + \cot^2\beta) = (\sin^2\beta)\left(\frac{1}{\sin^2\beta}\right) = 1.$$

13. $\sec\alpha\sin\alpha;\ \tan\alpha$ **14.** $\tan\alpha\cos\alpha;\ \sin\alpha$

15. $\sin\beta\cot\beta;\ \cos\beta$ **16.** $\csc\beta\tan\beta;\ \sec\beta$

17. $\dfrac{\sin\gamma}{\tan\gamma};\ \cos(-\gamma)$ **18.** $\dfrac{\sin(-\gamma)\sec\gamma}{\tan(-\gamma)};\ 1$

19. $\cot^2\gamma\sin^2\gamma;\ \cos^2\gamma$ **20.** $\tan^2\alpha\cos^2\alpha;\ \sin^2\alpha$

21. $\cos^2\alpha(1 + \tan^2\alpha);\ 1$ **22.** $(\csc^2\beta - 1)(\sin^2\beta);\ \cos^2\beta$

23. $\dfrac{(1 + \sin\beta)(1 - \sin\beta)}{\cos\beta};\ \cos\beta$ **24.** $\dfrac{(1 + \cos\gamma)(1 - \cos\gamma)}{\sin\gamma};\ \sin\gamma$

25. $\dfrac{\cos^2\gamma}{\cot^2\gamma};\ \sin^2\gamma$ **26.** $\dfrac{1 + \tan^2\gamma}{\csc^2\gamma};\ \tan^2\gamma$

27. $\dfrac{\sin^2\alpha}{\cos^2\alpha - 1 + \tan^2\alpha};\ \cot^2\alpha$ **28.** $\dfrac{\cos^2\alpha}{\sin^2\alpha - 1 + \cot^2\alpha};\ \tan^2\alpha$

29. $\dfrac{1}{\tan\alpha + \cot\alpha};\ \dfrac{\cos\alpha}{\csc\alpha}$ **30.** $\dfrac{1}{\csc\alpha - \sin\alpha};\ \dfrac{\tan\alpha}{\cos\alpha}$

Verify that each equation is an identity.

Example $\dfrac{\tan x - \sin x}{\tan x} = \dfrac{\sin^2 x}{1 + \cos x}$

Solution In this case we will simplify both members by appropriate substitutions. The left-hand member becomes

$$\frac{\tan x - \sin x}{\tan x} = \frac{\tan x}{\tan x} - \frac{\sin x}{\tan x}$$

$$= 1 - \frac{\sin x}{\frac{\sin x}{\cos x}} = 1 - \cos x.$$

The right-hand member becomes

$$\frac{\sin^2 x}{1 + \cos x} = \frac{1 - \cos^2 x}{1 + \cos x}$$

$$= \frac{(1 - \cos x)(1 + \cos x)}{1 + \cos x} = 1 - \cos x.$$

Hence, the equation is an identity.

31. $\cos x \tan x \csc x = 1$

32. $\csc x = \cot x \sec x$

33. $\sec y - \cos y = \tan y \sin y$

34. $\cos y = \sec y - \tan y \sin y$

35. $\cos^2 x - \sin^2 x = 2\cos^2 x - 1$

36. $(\sec^2 y - 1)(\csc^2 y - 1) = 1$

37. $\tan^2 y + \sec^2 y = 2\sec^2 y - 1$

38. $(\tan x + \cot x)\sin x \cos x = 1$

39. $\dfrac{\cos x + \sin x}{\sin x} = 1 + \cot x$

40. $\dfrac{1 - \tan^2 y}{\tan y} = \cot y - \tan y$

41. $\tan^2 y - \sin^2 y = \tan^2 y \sin^2 y$

42. $\dfrac{1}{1 + \sin x} + \dfrac{1}{1 - \sin x} = 2\sec^2 x$

43. $\tan y + \cot y = \sec y \csc y$

44. $\dfrac{1 + \tan^2 y}{\tan^2 y} = \csc^2 y$

45. $\cos^4 x - \sin^4 x = \cos^2 x - \sin^2 x$
Hint: Factor left-hand member.

46. $\dfrac{1 + \sec x}{\sec x} = \dfrac{\sin^2 x}{1 - \cos x}$
Hint: $\sin^2 x = (1 - \cos x)(1 + \cos x)$.

47. $\dfrac{1 - \cos x}{\sin x} = \dfrac{\sin x}{1 + \cos x}$
Hint: $\dfrac{\sin x}{1 + \cos x} = \dfrac{\sin x(1 - \cos x)}{(1 + \cos x)(1 - \cos x)}$.

48. $\dfrac{1 + \sin x}{\cos x} = \dfrac{\cos x}{1 - \sin x}$
Hint: $\dfrac{1 + \sin x}{\cos x} = \dfrac{(1 + \sin x)(1 - \sin x)}{\cos x(1 - \sin x)}$

49. $\csc y = \dfrac{\sec y + \csc y}{1 + \tan y}$

50. $\dfrac{1}{\sec y - \tan y} = \sec y + \tan y$

51. $\dfrac{\sec y - 1}{\sec y + 1} = \dfrac{1 - \cos y}{1 + \cos y}$

52. $\dfrac{\cos^2 y}{1 - \sin y} = \dfrac{\cos y}{\sec y - \tan y}$

53. $\sec^2 y + \csc^2 y = \sec^2 y \csc^2 y$

54. $\tan^4 y + \tan^2 y = \sec^4 y - \sec^2 y$

55. $\cos(-\alpha)\tan(-\alpha)\csc\alpha = -1$

56. $\csc(-\alpha) = \cot(-\alpha)\sec(-\alpha)$

57. $\dfrac{\cos(-\alpha) + \sin\alpha}{\sin(-\alpha)} = -(1 + \cot\alpha)$

58. $\dfrac{\cos(-\alpha) - \sin(-\alpha)}{\cos\alpha} = 1 + \tan\alpha$

By using counterexamples show that each equation is not an identity.

59. $\sin x + \cos x \tan x = 2$

60. $\sin^2 x + 2\cos x - \cos^2 x = 1$

61. $2\sin y\cos y + \sin y = 0$

62. $\tan^2 y - 2\tan y = 0$

B

63. Show that $\csc(-\alpha) = -\csc \alpha$.

64. Show that $\sec(-\alpha) = \sec \alpha$.

65. Show that $\cot(-\alpha) = -\cot \alpha$.

4.2 Sum and Difference Formulas for Cosine

In this section we shall establish several trigonometric identities involving the sum or difference of two angles or two real numbers. We first derive a formula for $\cos(\alpha + \beta)$. In Figure 4.2, angles α, $\alpha + \beta$,

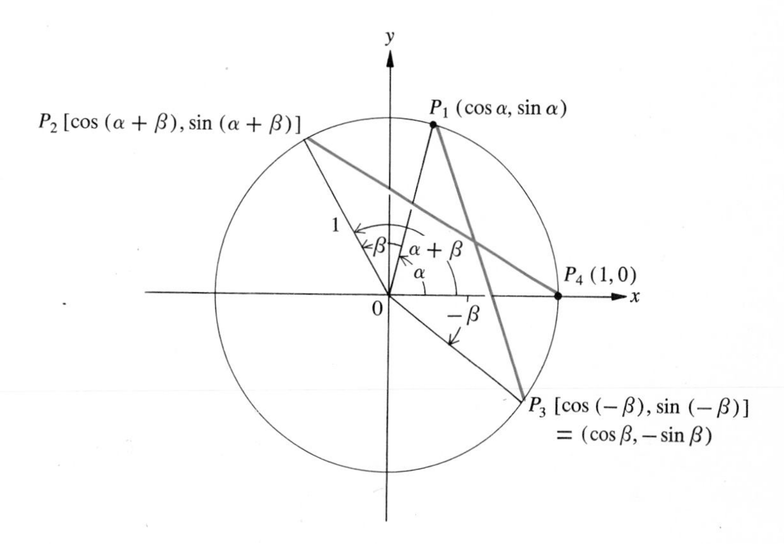

Figure 4.2

and $-\beta$ are in standard position, with their sides intersecting the circle of radius 1 at points P_1, P_2, P_3, and P_4. If P_1 has coordinates (x_1, y_1), then, since the radius r of the circle equals 1,

$$\cos \alpha = \frac{x_1}{1} \quad \text{and} \quad \sin \alpha = \frac{y_1}{1}.$$

Hence, the coordinates at P_1 are $x_1 = \cos \alpha$ and $y_1 = \sin \alpha$. We obtain the coordinates at P_2 and P_3 in a similar way. Note that from Equations (1) and (2) in Section 4.1, we have $\sin(-\beta) = -\sin \beta$ and $\cos(-\beta) = \cos \beta$. Since all radii are equal and the central angles $\angle P_4OP_2$ and $\angle P_3OP_1$ are equal (see Appendix A.9),

$$\triangle P_4OP_2 \cong \triangle P_3OP_1$$

and

$$P_4P_2 = P_3P_1.$$

From the distance formula (see Appendix A.4), we have

$$\begin{aligned}(P_3P_1)^2 &= [\cos\alpha - \cos\beta]^2 + [\sin\alpha - (-\sin\beta)]^2\\ &= (\cos\alpha - \cos\beta)^2 + (\sin\alpha + \sin\beta)^2\\ &= \cos^2\alpha - 2\cos\alpha\cos\beta + \cos^2\beta\\ &\qquad + \sin^2\alpha + 2\sin\alpha\sin\beta + \sin^2\beta.\end{aligned}$$

Since $\cos^2\alpha + \sin^2\alpha = 1$ and $\cos^2\beta + \sin^2\beta = 1$,

$$(P_3P_1)^2 = 2 - 2\cos\alpha\cos\beta + 2\sin\alpha\sin\beta.$$

Again using the distance formula, we have

$$\begin{aligned}(P_4P_2)^2 &= [\cos(\alpha+\beta) - 1]^2 + [\sin(\alpha+\beta) - 0]^2\\ &= \cos^2(\alpha+\beta) - 2\cos(\alpha+\beta) + 1 + \sin^2(\alpha+\beta).\end{aligned}$$

Since $\cos^2(\alpha+\beta) + \sin^2(\alpha+\beta) = 1$,

$$(P_4P_2)^2 = 2 - 2\cos(\alpha+\beta).$$

Since $P_4P_2 = P_3P_1$, we have

$$2 - 2\cos(\alpha+\beta) = 2 - 2\cos\alpha\cos\beta + 2\sin\alpha\sin\beta,$$

which gives

$$\boldsymbol{\cos(\alpha+\beta) = \cos\alpha\cos\beta - \sin\alpha\sin\beta.} \tag{1}$$

Substituting $-\beta$ for β in Equation (1), we have

$$\cos[\alpha + (-\beta)] = \cos\alpha\cos(-\beta) - \sin\alpha\sin(-\beta).$$

Since

$$\cos(-\beta) = \cos\beta \qquad \text{and} \qquad \sin(-\beta) = -\sin\beta,$$

we obtain

$$\boldsymbol{\cos(\alpha-\beta) = \cos\alpha\cos\beta + \sin\alpha\sin\beta.} \tag{2}$$

These very important relationships for $\cos(\alpha + \beta)$ and $\cos(\alpha - \beta)$ are called the **sum formula** and the **difference formula,** respectively, for the cosine function.

We can, as illustrated in the following examples, find some function values using the sum and difference formulas. Although using Table II, Table III, or a calculator is a more efficient way of finding these function values, such examples give us practice in applying the formulas.

Example Compute $\cos 15°$.

Solution Because we know the exact function values of 45° and 30°, we replace 15° with 45° − 30° in Equation (2) and obtain

$$\begin{aligned}\cos 15° &= \cos(45° - 30°)\\ &= \cos 45° \cos 30° + \sin 45° \sin 30°\\ &= \frac{1}{\sqrt{2}} \cdot \frac{\sqrt{3}}{2} + \frac{1}{\sqrt{2}} \cdot \frac{1}{2} = \frac{\sqrt{3}+1}{2\sqrt{2}}.\end{aligned}$$

Example Compute $\cos \frac{7\pi}{12}$.

Solution Since $\frac{1}{3} + \frac{1}{4} = \frac{7}{12}$, we have $\frac{7\pi}{12} = \frac{\pi}{3} + \frac{\pi}{4}$. By Equation (1)

$$\cos \frac{7\pi}{12} = \cos\left(\frac{\pi}{3} + \frac{\pi}{4}\right) = \cos \frac{\pi}{3} \cos \frac{\pi}{4} - \sin \frac{\pi}{3} \sin \frac{\pi}{4}.$$

The appropriate substitution for the function values yields

$$\cos \frac{7\pi}{12} = \frac{1}{2} \cdot \frac{1}{\sqrt{2}} - \frac{\sqrt{3}}{2} \cdot \frac{1}{\sqrt{2}} = \frac{1-\sqrt{3}}{2\sqrt{2}}.$$

In general we shall leave such results in radical form. However, you can, if you wish, obtain a rational number approximation. For example, since $\sqrt{2} \approx 1.414$ and $\sqrt{6} \approx 2.449$,

$$\begin{aligned}\cos \frac{7\pi}{12} &= \frac{1-\sqrt{3}}{2\sqrt{2}} = \frac{(1-\sqrt{3})\sqrt{2}}{2\sqrt{2}\sqrt{2}}\\ &= \frac{\sqrt{2}-\sqrt{6}}{4} \approx \frac{1.414 - 2.449}{4} \approx -0.2588.\end{aligned}$$

The sum and difference formulas for the cosine function can be used to establish other identities. Three identities that we will be using are

$$\cos(180° - x) = -\cos x, \tag{3}$$

$$\cos(180° + x) = -\cos x, \tag{4}$$

$$\cos(360° - x) = \cos x. \tag{5}$$

Identities such as these are known as **reduction formulas.** The proof that Equation (3) is an identity is shown here; the proofs of (4) and (5) are left as exercises.

Using the difference formula [Equation (2)] and substituting 180° for α and x for β, we obtain

$$\cos(180° - x) = (\cos 180°)(\cos x) + (\sin 180°)(\sin x).$$

Since $\cos 180° = -1$ and $\sin 180° = 0$, we have

$$\cos(180° - x) = (-1)\cos x + (0)\sin x = -\cos x.$$

Reduction formulas can be used, instead of relying on sketches of reference angles, to rewrite a function value of a larger angle as a function value of an acute angle.

Examples Write each expression as the cosine of an acute angle.

a. $\cos 150°$ b. $\cos 225°$

Solutions

a. Since $150° = 180° - 30°$, we use Equation (3) to obtain

$$\cos 150° = \cos(180° - 30°) = -\cos 30°.$$

b. Since $225° = 180° + 45°$, we use Equation (4) to obtain

$$\cos 225° = \cos(180° + 45°) = -\cos 45°.$$

EXERCISE SET 4.2

A

Find the value of each expression using Equation (1) or Equation (2) of this section.

1. $\cos 75°$ (Use: $75° = 120° - 45°$)

2. $\cos 105°$ (Use: $105° = 60° + 45°$)

3. $\cos 165°$ (Use: $165° = 120° + 45°$)

4. $\cos 195°$ (Use: $195° = 135° + 60°$)

5. $\cos \frac{\pi}{12}$ $\left(\text{Use: } \frac{\pi}{12} = \frac{\pi}{3} - \frac{\pi}{4}\right)$

6. $\cos \frac{5\pi}{12}$ $\left(\text{Use: } \frac{5\pi}{12} = \frac{\pi}{6} + \frac{\pi}{4}\right)$

Verify that each equation is correct.

Example $\cos 110° \cos 70° - \sin 110° \sin 70° = -1$

Solution From Equation (1), if we replace α with 110° and β with 70°, we obtain

$$\begin{aligned} \cos 110° \cos 70° - \sin 110° \sin 70° &= \cos(110° + 70°) \\ &= \cos 180° = -1. \end{aligned}$$

7. $\cos 40° \cos 140° - \sin 40° \sin 140° = -1$

8. $\cos 20° \cos 160° - \sin 20° \sin 160° = -1$

9. $\cos 35° \cos 55° - \sin 35° \sin 55° = 0$

10. $\cos 50° \cos 40° - \sin 50° \sin 40° = 0$

11. $\cos 100° \cos 55° + \sin 100° \sin 55° = \frac{1}{\sqrt{2}}$

12. $\cos 130° \cos 70° + \sin 130° \sin 70° = \frac{1}{2}$

Express each expression as a single function of kx, k an integer.

Example $\cos 2x \cos 3x - \sin 2x \sin 3x$

Solution From Equation (1),

$$\begin{aligned} \cos 2x \cos 3x - \sin 2x \sin 3x &= \cos(2x + 3x) \\ &= \cos 5x. \end{aligned}$$

13. $\cos 4x \cos 5x - \sin 4x \sin 5x$

14. $\cos x \cos 2x - \sin x \sin 2x$

15. $\cos 8x \cos 6x + \sin 8x \sin 6x$

16. $\cos 5x \cos x + \sin 5x \sin x$

17. $\cos \frac{2}{3} x \cos \frac{1}{3} x - \sin \frac{2}{3} x \sin \frac{1}{3} x$

18. $\cos \frac{3}{4} x \cos \frac{5}{4} x - \sin \frac{3}{4} x \sin \frac{5}{4} x$

19. $\cos(x + 70°) \cos(x - 70°) - \sin(x + 70°) \sin(x - 70°)$

20. $\cos(65° + x) \cos(65° - x) + \sin(65° + x) \sin(65° - x)$

If angles α and β have the given function values and are in the indicated quadrant, use Equation (1) to find $\cos(\alpha + \beta)$.

Example $\cos \alpha = -\frac{1}{\sqrt{5}}$ and $\sin \beta = \frac{2}{\sqrt{13}}$; α in Quadrant II and β in I

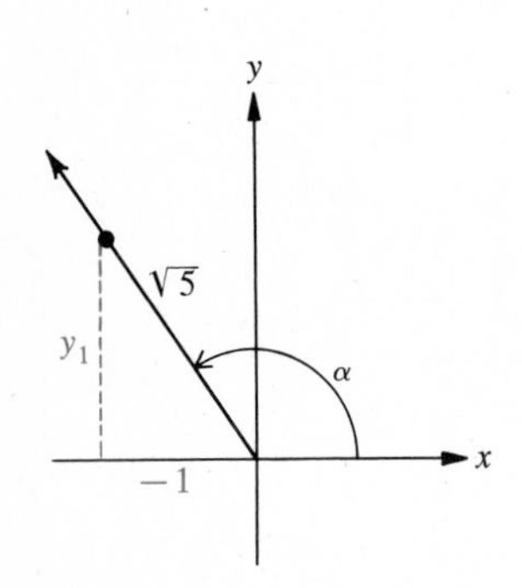

Solution Since α is in Quadrant II, as shown in the sketch, using the Pythagorean theorem, we have

$$y_1^2 + (-1)^2 = (\sqrt{5})^2,$$
$$y_1^2 + 1 = 5,$$
$$y_1^2 = 4,$$
$$y_1 = 2 \quad (-2 \text{ does not apply}).$$

Hence, $\sin \alpha = \frac{y_1}{\sqrt{5}} = \frac{2}{\sqrt{5}}$.

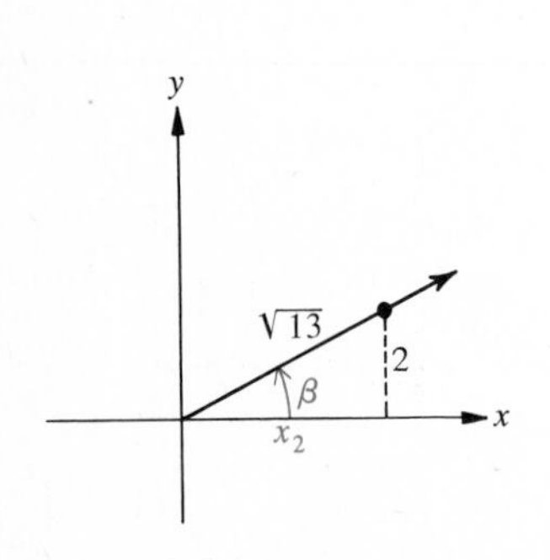

Since β is in Quadrant I, from the sketch we have

$$x_2^2 + 2^2 = (\sqrt{13})^2,$$
$$x_2^2 + 4 = 13,$$
$$x_2^2 = 9,$$
$$x_2 = 3 \quad (-3 \text{ does not apply}).$$

Hence, $\cos \beta = \frac{x_2}{\sqrt{13}} = \frac{3}{\sqrt{13}}$. Therefore,

$$\begin{aligned}\cos(\alpha + \beta) &= \cos \alpha \cos \beta - \sin \alpha \sin \beta \\ &= -\frac{1}{\sqrt{5}} \cdot \frac{3}{\sqrt{13}} - \frac{2}{\sqrt{5}} \cdot \frac{2}{\sqrt{13}} \\ &= -\frac{3}{\sqrt{65}} - \frac{4}{\sqrt{65}} = -\frac{7}{\sqrt{65}}.\end{aligned}$$

21. $\cos \alpha = \frac{3}{5}$ and $\sin \beta = \frac{5}{13}$; α and β in Quadrant I

22. $\cos \alpha = -\frac{4}{5}$ and $\sin \beta = -\frac{7}{25}$; α in Quadrant III and β in IV

23. $\sin \alpha = \frac{1}{\sqrt{10}}$ and $\cos \beta = -\frac{4}{\sqrt{17}}$; α in Quadrant II and β in III

24. $\sin \alpha = -\frac{5}{\sqrt{29}}$ and $\cos \beta = \frac{1}{\sqrt{26}}$; α in Quadrant IV and β in I

Use one of the reduction formulas [Equation (3), (4), or (5)] of this section to write each expression as a function of an acute angle.

Example $\cos 135°$

Solution Using Equation (3), we find that

$$\cos 135° = \cos(180° - 45°) = -\cos 45°.$$

25. $\cos 120°$ **26.** $\cos 240°$ **27.** $\cos 210°$

28. $\cos 300°$ **29.** $\cos 315°$ **30.** $\cos 330°$

Verify that each equation in Exercises 31–35 is an identity.

31. $\cos(180° + x) = -\cos x$

32. $\cos(360° - x) = \cos x$

33. $\cos(90° - x) = \sin x$

34. $\cos(90° + x) = -\sin x$

35. $\cos 2x = \cos^2 x - \sin^2 x$. *Hint:* $\cos 2x = \cos(x + x)$.

36. Use the results of Exercise 35 to show that

$$\cos 2x = 2\cos^2 x - 1 = 1 - 2\sin^2 x.$$

37. Show by counterexample that $\cos(\alpha + \beta) = \cos \alpha + \cos \beta$ is not an identity.

38. Show by counterexample that $\cos(\alpha - \beta) = \cos \alpha - \cos \beta$ is not an identity.

39. Show that $\cos\left(\frac{\pi}{2} - \alpha\right) = \sin \alpha$.

40. Use the results of Exercise 39 to show that $\sin\left(\frac{\pi}{2} - \alpha\right) = \cos \alpha$.

B

Evaluate each of the following.

Example $\cos\left(\text{Arccot}\,\frac{1}{\sqrt{3}} + \text{Arcsin}\,\frac{1}{\sqrt{2}}\right)$

Solution Let

$$\alpha = \text{Arccot}\,\frac{1}{\sqrt{3}} \quad \text{and} \quad \beta = \text{Arcsin}\,\frac{1}{\sqrt{2}}. \tag{1}$$

Then, using the symbols α and β and the sum formula on page 148, we have

$$\cos(\alpha + \beta) = \cos\alpha\cos\beta - \sin\alpha\sin\beta.$$

By inspection of Equations (1), $\alpha = \pi/3$. Hence, $\cos\alpha = 1/2$ and $\sin\alpha = \sqrt{3}/2$. Also by inspection of (1), $\beta = \pi/4$. Thus, we have that $\sin\beta = 1/\sqrt{2}$ and $\cos\beta = 1/\sqrt{2}$. Therefore,

$$\cos\left(\text{Arccot}\,\frac{1}{\sqrt{3}} + \text{Arcsin}\,\frac{1}{\sqrt{2}}\right) = \frac{1}{2}\cdot\frac{1}{\sqrt{2}} - \frac{\sqrt{3}}{2}\cdot\frac{1}{\sqrt{2}}$$

$$= \frac{1-\sqrt{3}}{2\sqrt{2}}.$$

41. $\cos\left(\text{Arcsin}\,\frac{1}{2} + \text{Arcsec}\,2\right)$

42. $\cos\left(\text{Arccos}\,\frac{1}{2} + \text{Arccsc}\,2\right)$

43. $\cos\left(\text{Arctan}\,1 - \text{Arccos}\,\frac{\sqrt{3}}{2}\right)$

44. $\cos\left(\text{Arcsin}\,\frac{1}{\sqrt{2}} - \text{Arcsec}\,1\right)$

45. $\cos[\text{Arccsc}(-2) - \text{Arcsec}\,\sqrt{2}]$

46. $\cos\left[\text{Arccos}\left(-\frac{1}{\sqrt{2}}\right) - \text{Arcsin}(-1)\right]$

4.3 Sum and Difference Formulas for Sine

Sum and difference formulas for the sine function can be obtained from the sum and difference formulas for the cosine function. First, however, we consider two relationships between function values for the cosine and sine functions. From the difference formula for the cosine function we have

$$\cos\left(\frac{\pi}{2} - \theta\right) = \cos\frac{\pi}{2}\cos\theta + \sin\frac{\pi}{2}\sin\theta.$$

Replacing $\cos(\pi/2)$ with 0 and $\sin(\pi/2)$ with 1, we have

$$\cos\left(\frac{\pi}{2} - \theta\right) = 0 \cdot \cos\theta + 1 \cdot \sin\theta,$$

$$\mathbf{\cos\left(\frac{\pi}{2} - \theta\right) = \sin\theta.} \tag{1}$$

Replacing θ with $\pi/2 - \theta$ in Equation (1), we obtain

$$\cos\left[\frac{\pi}{2} - \left(\frac{\pi}{2} - \theta\right)\right] = \sin\left(\frac{\pi}{2} - \theta\right),$$

$$\cos\left[\left(\frac{\pi}{2} - \frac{\pi}{2}\right) + \theta\right] = \sin\left(\frac{\pi}{2} - \theta\right),$$

$$\cos\theta = \sin\left(\frac{\pi}{2} - \theta\right),$$

$$\mathbf{\sin\left(\frac{\pi}{2} - \theta\right) = \cos\theta.} \tag{2}$$

Examples

a. $\sin\frac{\pi}{3} = \cos\left(\frac{\pi}{2} - \frac{\pi}{3}\right)$

$= \cos\frac{\pi}{6}$

b. $\cos 46° = \sin(90° - 46°)$

$= \sin 44°$

In Section 2.1 we noted that the sine and cosine are cofunctions. Equations (1) and (2) also establish this fact because in each case $(\pi/2 - \theta) + \theta = \pi/2$.

Observe that secant and cosecant are also cofunctions because

$$\sec\left(\frac{\pi}{2} - \theta\right) = \frac{1}{\cos\left(\frac{\pi}{2} - \theta\right)} = \frac{1}{\sin\theta} = \csc\theta, \tag{3}$$

and

$$\csc\left(\frac{\pi}{2} - \theta\right) = \frac{1}{\sin\left(\frac{\pi}{2} - \theta\right)} = \frac{1}{\cos\theta} = \sec\theta, \tag{4}$$

and, as above, $(\pi/2 - \theta) + \theta = \pi/2$.

Examples

a. $\csc 14° = \sec(90° - 14°)$
$= \sec 76°$

b. $\sec 23°\ 10' = \csc(90° - 23°\ 10')$
$= \csc 66°\ 50'$

Now we consider the sum and difference formulas for the sine function. If we replace θ with $\alpha + \beta$ in Equation (1), we have

$$\begin{aligned}\sin(\alpha + \beta) &= \cos\left[\frac{\pi}{2} - (\alpha + \beta)\right]\\ &= \cos\left[\left(\frac{\pi}{2} - \alpha\right) - \beta\right],\end{aligned}$$

which by the difference formula for the cosine can be written as

$$\sin(\alpha + \beta) = \cos\left(\frac{\pi}{2} - \alpha\right)\cos\beta + \sin\left(\frac{\pi}{2} - \alpha\right)\sin\beta.$$

By Equations (1) and (2),

$$\cos\left(\frac{\pi}{2} - \alpha\right) = \sin\alpha \quad \text{and} \quad \sin\left(\frac{\pi}{2} - \alpha\right) = \cos\alpha.$$

Hence, we have

$$\boldsymbol{\sin(\alpha + \beta) = \sin\alpha\cos\beta + \cos\alpha\sin\beta.} \tag{5}$$

Replacing β with $-\beta$ in Equation (5) yields

$$\sin[\alpha + (-\beta)] = \sin\alpha\cos(-\beta) + \cos\alpha\sin(-\beta).$$

Substituting $\cos\beta$ for $\cos(-\beta)$ and $(-\sin\beta)$ for $\sin(-\beta)$, we have

$$\sin(\alpha - \beta) = \sin\alpha\cos\beta + \cos\alpha(-\sin\beta),$$

from which

$$\boldsymbol{\sin(\alpha - \beta) = \sin\alpha\cos\beta - \cos\alpha\sin\beta.} \tag{6}$$

Example Compute $\sin 15^\circ$.

Solution Since $15^\circ = 45^\circ - 30^\circ$, by Equation (6),

$$\begin{aligned}\sin 15^\circ &= \sin(45^\circ - 30^\circ)\\ &= \sin 45^\circ \cos 30^\circ - \cos 45^\circ \sin 30^\circ\\ &= \frac{1}{\sqrt{2}} \cdot \frac{\sqrt{3}}{2} - \frac{1}{\sqrt{2}} \cdot \frac{1}{2} = \frac{\sqrt{3}-1}{2\sqrt{2}}.\end{aligned}$$

Example Compute $\sin \frac{\pi}{12}$.

Solution Observe that $\frac{\pi}{12} = \frac{\pi}{3} - \frac{\pi}{4}$. By using Equation (6), we have

$$\sin \frac{\pi}{12} = \sin\left(\frac{\pi}{3} - \frac{\pi}{4}\right) = \sin \frac{\pi}{3} \cos \frac{\pi}{4} - \cos \frac{\pi}{3} \sin \frac{\pi}{4}.$$

Making the appropriate substitutions for the function values yields

$$\sin \frac{\pi}{12} = \frac{\sqrt{3}}{2} \cdot \frac{1}{\sqrt{2}} - \frac{1}{2} \cdot \frac{1}{\sqrt{2}} = \frac{\sqrt{3}-1}{2\sqrt{2}}.$$

The same results are obtained if $\pi/12$ is written as $(\pi/4) - (\pi/6)$. Again we note that in practical applications, approximations for function values such as $\sin(\pi/12)$ can be found quite readily by using Table III or a calculator.

The sum and difference formulas for the sine function can be used to establish the following reduction formulas.

$$\sin(180^\circ - x) = \sin x, \tag{7}$$

$$\sin(180^\circ + x) = -\sin x, \tag{8}$$

$$\sin(360^\circ - x) = -\sin x. \tag{9}$$

The proof that Equation (7) is an identity is presented here; the proofs of (8) and (9) are left as exercises.

Using the difference formula, Equation (6), and substituting 180° for α and x for β, we obtain

$$\sin(180^\circ - x) = \sin 180^\circ \cos x - \cos 180^\circ \sin x.$$

Since $\sin 180^\circ = 0$ and $\cos 180^\circ = -1$, we have

$$\sin(180° - x) = 0 \cdot \cos x - (-1)\sin x$$
$$= \sin x.$$

EXERCISE SET 4.3

A

Find the value of each expression using Equation (5) or (6).

1. $\sin 105°$ (Use: $105° = 60° + 45°$)

2. $\sin 75°$ (Use: $75° = 120° - 45°$)

3. $\sin 165°$ (Use: $165° = 210° - 45°$)

4. $\sin 195°$ (Use: $195° = 135° + 60°$)

5. $\sin \frac{5\pi}{12}$ $\left(\text{Use: } \frac{5\pi}{12} = \frac{\pi}{6} + \frac{\pi}{4}\right)$

6. $\sin \frac{\pi}{12}$ $\left(\text{Use: } \frac{\pi}{12} = \frac{\pi}{3} - \frac{\pi}{4}\right)$

Verify that each equation is correct.

Example $\sin 100° \cos 80° + \cos 100° \sin 80° = 0$

Solution From Equation (5) of this section, if we replace α with $100°$ and β with $80°$, we obtain

$$\sin 100° \cos 80° + \cos 100° \sin 80° = \sin(100° + 80°)$$
$$= \sin 180° = 0.$$

7. $\sin 20° \cos 160° + \cos 20° \sin 160° = 0$

8. $\sin 40° \cos 140° + \cos 40° \sin 140° = 0$

9. $\sin 50° \cos 40° + \cos 50° \sin 40° = 1$

10. $\sin 35° \cos 55° + \cos 35° \sin 55° = 1$

11. $\sin 110° \cos 80° - \cos 110° \sin 80° = \frac{1}{2}$

12. $\sin 95° \cos 50° - \cos 95° \sin 50° = \frac{1}{\sqrt{2}}$

Write each expression as a single function of kx, where k is an integer.

Example $\sin 2x \cos 3x + \cos 2x \sin 3x$

Solution Using Equation (5), we obtain

$$\begin{aligned}\sin 2x \cos 3x + \cos 2x \sin 3x &= \sin(2x + 3x)\\ &= \sin 5x.\end{aligned}$$

13. $\sin x \cos 2x + \cos x \sin 2x$

14. $\sin 4x \cos 5x + \cos 4x \sin 5x$

15. $\sin 5x \cos x - \cos 5x \sin x$

16. $\sin 8x \cos 6x - \cos 8x \sin 6x$

17. $\sin \frac{3}{4}x \cos \frac{5}{4}x + \cos \frac{3}{4}x \sin \frac{5}{4}x$

18. $\sin \frac{4}{3}x \cos \frac{1}{3}x - \cos \frac{4}{3}x \sin \frac{1}{3}x$

19. $\sin(50° + x) \cos(50° - x) - \cos(50° + x) \sin(50° - x)$

20. $\sin(x + 60°) \cos(x - 60°) + \cos(x + 60°) \sin(x - 60°)$

If angles α and β have the given function values and are in the indicated quadrants, use Equation (5) to find $\sin(\alpha + \beta)$.

Example $\cos \alpha = -\frac{1}{\sqrt{5}}$, $\sin \beta = \frac{2}{\sqrt{13}}$, α in Quadrant II, β in I

Solution From the example on page 152, we obtain $\sin \alpha = 2/\sqrt{5}$ and $\cos \beta = 3/\sqrt{13}$. Therefore, from Equation (5),

$$\begin{aligned}\sin(\alpha + \beta) &= \sin \alpha \cos \beta + \cos \alpha \sin \beta\\ &= \frac{2}{\sqrt{5}} \cdot \frac{3}{\sqrt{13}} + \left(-\frac{1}{\sqrt{5}}\right) \cdot \frac{2}{\sqrt{13}}\\ &= \frac{6}{\sqrt{65}} - \frac{2}{\sqrt{65}} = \frac{4}{\sqrt{65}}.\end{aligned}$$

21. $\cos \alpha = -\frac{4}{5}$, $\sin \beta = -\frac{7}{25}$, α in Quadrant III, β in IV

22. $\cos \alpha = \frac{3}{5}$, $\sin \beta = \frac{5}{13}$, α and β in Quadrant I

23. $\sin \alpha = -\dfrac{5}{\sqrt{29}}$, $\cos \beta = \dfrac{1}{\sqrt{26}}$, α in Quadrant IV, β in I

24. $\sin \alpha = \dfrac{1}{\sqrt{10}}$, $\cos \beta = -\dfrac{4}{\sqrt{17}}$, α in Quadrant II, β in III

Use one of the reduction formulas [Equation (7), (8), or (9)] of this section to write each expression as a function of an acute angle.

Example $\sin 135°$

Solution By Equation (7),

$$\sin 135° = \sin(180° - 45°) = \sin 45°.$$

25. $\sin 240°$ **26.** $\sin 120°$ **27.** $\sin 150°$

28. $\sin 210°$ **29.** $\sin 330°$ **30.** $\sin 315°$

Verify that each equation is an identity.

31. $\sin(180° + x) = -\sin x$ **32.** $\sin(360° - x) = -\sin x$

33. $\sin(270° - x) = -\cos x$ **34.** $\sin(270° + x) = -\cos x$

35. $\sin 2x = 2 \sin x \cos x$. *Hint:* $\sin 2x = \sin(x + x)$.

36. $\sin(\alpha - 30°) = \dfrac{\sqrt{3} \sin \alpha - \cos \alpha}{2}$

37. Show by a counterexample that $\sin(\alpha - \beta) = \sin \alpha - \sin \beta$ is not an identity.

38. Show by a counterexample that $\sin(\alpha + \beta) = \sin \alpha + \sin \beta$ is not an identity.

B

Evaluate each of the following.

Example $\sin\left(\text{Arcsin}\,\dfrac{1}{\sqrt{2}} + \text{Arccos}\,\dfrac{1}{2}\right)$

Solution Let

$$\alpha = \text{Arcsin}\,\frac{1}{\sqrt{2}} \quad \text{and} \quad \beta = \text{Arccos}\,\frac{1}{2}. \tag{1}$$

Then, by using the symbols α and β, and by the sum formula on page 156, we have

$$\sin(\alpha + \beta) = \sin \alpha \cos \beta + \cos \alpha \sin \beta.$$

By inspection of Equations (1), $\alpha = \pi/4$, and we have $\sin \alpha = 1/\sqrt{2}$ and $\cos \alpha = 1/\sqrt{2}$. Also, by inspection of (1), $\beta = \pi/3$, and we have that $\cos \beta = 1/2$ and $\sin \beta = \sqrt{3}/2$. Therefore,

$$\sin\left(\text{Arcsin}\,\frac{1}{\sqrt{2}} + \text{Arccos}\,\frac{1}{2}\right) = \frac{1}{\sqrt{2}} \cdot \frac{1}{2} + \frac{1}{\sqrt{2}} \cdot \frac{\sqrt{3}}{2}$$

$$= \frac{1}{2\sqrt{2}} + \frac{\sqrt{3}}{2\sqrt{2}} = \frac{1 + \sqrt{3}}{2\sqrt{2}}.$$

39. $\sin(\text{Arcsin } 0 + \text{Arccos } 1)$

40. $\sin(\text{Arccos } 0 + \text{Arcsin } 1)$

41. $\sin\left(\text{Arcsin } 0 - \text{Arcsec}\,\frac{2}{\sqrt{3}}\right)$

42. $\sin\left(\text{Arccos } 1 - \text{Arcsin}\,\frac{1}{\sqrt{2}}\right)$

43. $\sin\left(\text{Arctan } 1 - \text{Arccos}\,\frac{\sqrt{3}}{2}\right)$

44. $\sin\left(\text{Arcsin}\,\frac{1}{2} - \text{Arcsec } 2\right)$

4.4 Sum and Difference Formulas for Tangent

Sum and difference formulas for the tangent function follow from the sum and difference formulas for the sine and cosine functions. Since

$$\tan(\alpha + \beta) = \frac{\sin(\alpha + \beta)}{\cos(\alpha + \beta)},$$

by substituting appropriately, we have

$$\tan(\alpha + \beta) = \frac{\sin \alpha \cos \beta + \cos \alpha \sin \beta}{\cos \alpha \cos \beta - \sin \alpha \sin \beta}.$$

Multiplying each term of the numerator and denominator of the right-hand member by $1/\cos \alpha \cos \beta$ yields

$$\tan(\alpha + \beta) = \frac{\dfrac{\sin \alpha \cos \beta}{\cos \alpha \cos \beta} + \dfrac{\cos \alpha \sin \beta}{\cos \alpha \cos \beta}}{\dfrac{\cos \alpha \cos \beta}{\cos \alpha \cos \beta} - \dfrac{\sin \alpha \sin \beta}{\cos \alpha \cos \beta}},$$

from which, by simplifying the right-hand member, we obtain

$$\boldsymbol{\tan(\alpha + \beta) = \frac{\tan \alpha + \tan \beta}{1 - \tan \alpha \tan \beta}}. \tag{1}$$

By substituting $-\beta$ for β in Equation (1), we have

$$\tan[\alpha + (-\beta)] = \frac{\tan \alpha + \tan(-\beta)}{1 - \tan \alpha \tan(-\beta)}.$$

Substituting $-\tan \beta$ for $\tan(-\beta)$ yields

$$\tan[\alpha + (-\beta)] = \frac{\tan \alpha + (-\tan \beta)}{1 - \tan \alpha(-\tan \beta)},$$

$$\boldsymbol{\tan(\alpha - \beta) = \frac{\tan \alpha - \tan \beta}{1 + \tan \alpha \tan \beta}}. \tag{2}$$

Example Compute $\tan 75°$

Solution Observe that $75° = 45° + 30°$. Then, using Equation (1) and replacing α with $45°$ and β with $30°$, we obtain

$$\tan 75° = \tan(45° + 30°) = \frac{\tan 45° + \tan 30°}{1 - \tan 45° \tan 30°}$$

$$= \frac{1 + \frac{1}{\sqrt{3}}}{1 - 1\left(\frac{1}{\sqrt{3}}\right)} = \frac{\left(1 + \frac{1}{\sqrt{3}}\right) \cdot \sqrt{3}}{\left(1 - \frac{1}{\sqrt{3}}\right) \cdot \sqrt{3}} = \frac{\sqrt{3} + 1}{\sqrt{3} - 1}.$$

By using Equations (1) and (2) of Section 4.3, it can be shown that tangent and cotangent are cofunctions. Thus,

$$\tan(90° - \theta) = \cot \theta, \tag{3}$$

and

$$\cot(90° - \theta) = \tan \theta, \tag{4}$$

where $(90° - \theta) + \theta = 90°$. The proofs of these identities are left as exercises (see Exercises 39 and 40).

Examples

a. $\cot 43° = \tan(90° - 47°)$

$= \tan 47°$

b. $\tan \frac{\pi}{6} = \cot\left(\frac{\pi}{2} - \frac{\pi}{6}\right)$

$= \cot \frac{\pi}{3}$

The sum and difference formulas established in this section can be used to establish the following reduction formulas.

$$\tan(180° - x) = -\tan x, \tag{5}$$

$$\tan(180° + x) = \tan x, \tag{6}$$

$$\tan(360° - x) = -\tan x. \tag{7}$$

The proof that Equation (5) is an identity is presented here; the proofs of (6) and (7) are left as exercises.

Using the difference formula [Equation (2)] and substituting 180° for α and x for β, we obtain

$$\begin{aligned} \tan(180° - x) &= \frac{\tan 180° - \tan x}{1 + \tan 180° \tan x} \\ &= \frac{0 - \tan x}{1 + (0)\tan x} \\ &= -\tan x. \end{aligned}$$

EXERCISE SET 4.4

A

Use Equation (1) or (2) to find the value of each expression.

1. $\tan 105°$ (Use: $105° = 45° + 60°$)

2. $\tan 15°$ (Use: $15° = 45° - 30°$)

3. $\tan 195°$ (Use: $195° = 240° - 45°$)

4. $\tan 165°$ (Use: $165° = 120° + 45°$)

5. $\tan \frac{\pi}{12}$ $\left(\text{Use: } \frac{\pi}{12} = \frac{\pi}{3} - \frac{\pi}{4}\right)$ **6.** $\tan \frac{3\pi}{4}$ $\left(\text{Use: } \frac{3\pi}{4} = \pi - \frac{\pi}{4}\right)$

Write each expression as a function of the complementary angle.

Examples a. $\tan 65°$ b. $\cot \frac{\pi}{6}$

Solutions

a. Using Equation (4) of this section and replacing θ with 65°,

$$\tan 65° = \cot(90° - 65°) = \cot 25°.$$

b. Using Equation (3) of this section and replacing θ with $\pi/6$,

$$\cot\frac{\pi}{6} = \tan\left(\frac{\pi}{2} - \frac{\pi}{6}\right) = \tan\frac{\pi}{3}.$$

7. $\tan 36^\circ$ **8.** $\tan 40^\circ$ **9.** $\cot 66^\circ$ **10.** $\cot 75^\circ$

11. $\tan 12^\circ\ 30'$ **12.** $\cot 30^\circ\ 20'$ **13.** $\tan\frac{\pi}{3}$ **14.** $\cot\frac{\pi}{8}$

15. $\cot\frac{\pi}{4}$ **16.** $\tan\frac{\pi}{10}$ **17.** $\cot\frac{\pi}{12}$ **18.** $\tan\frac{5\pi}{12}$

Verify that each equation is correct.

Example $\dfrac{\tan 20^\circ + \tan 25^\circ}{1 - \tan 20^\circ \tan 25^\circ} = 1$

Solution From Equation (1), if we replace α with 20° and β with 25°, we obtain

$$\frac{\tan 20^\circ + \tan 25^\circ}{1 - \tan 20^\circ \tan 25^\circ} = \tan(20^\circ + 25^\circ)$$

$$= \tan 45^\circ = 1.$$

19. $\dfrac{\tan 50^\circ + \tan 130^\circ}{1 - \tan 50^\circ \tan 130^\circ} = 0$ **20.** $\dfrac{\tan 100^\circ + \tan 50^\circ}{1 - \tan 100^\circ \tan 50^\circ} = -\dfrac{1}{\sqrt{3}}$

21. $\dfrac{\tan 140^\circ + \tan 100^\circ}{1 - \tan 140^\circ \tan 100^\circ} = \sqrt{3}$ **22.** $\dfrac{\tan 95^\circ + \tan 40^\circ}{1 - \tan 95^\circ \tan 40^\circ} = -1$

23. $\dfrac{\tan 110^\circ - \tan 50^\circ}{1 + \tan 110^\circ \tan 50^\circ} = \sqrt{3}$ **24.** $\dfrac{\tan 115^\circ - \tan 70^\circ}{1 + \tan 115^\circ \tan 70^\circ} = 1$

Write each expression as a single function of kx where k is an integer.

Example $\dfrac{\tan 2x + \tan 3x}{1 - \tan 2x \tan 3x}$

Solution Using Equation (1), we obtain

$$\frac{\tan 2x + \tan 3x}{1 - \tan 2x \tan 3x} = \tan(2x + 3x)$$

$$= \tan 5x.$$

25. $\dfrac{\tan 4x + \tan 5x}{1 - \tan 4x \tan 5x}$ **26.** $\dfrac{\tan 2x + \tan x}{1 - \tan 2x \tan x}$

27. $\dfrac{\tan 8x - \tan 6x}{1 + \tan 8x \tan 6x}$ **28.** $\dfrac{\tan 5x - \tan x}{1 + \tan 5x \tan x}$

29. $\dfrac{\tan \frac{4}{3}x - \tan \frac{1}{3}x}{1 + \tan \frac{4}{3}x \tan \frac{1}{3}x}$ **30.** $\dfrac{\tan \frac{5}{4}x + \tan \frac{3}{4}x}{1 - \tan \frac{5}{4}x \tan \frac{3}{4}x}$

Use one of the reduction formulas [Equation (5), (6), or (7)] of this section to write each expression as a function of an acute angle.

Example $\tan 135°$

Solution Using Equation (5), we find

$$\tan 135° = \tan(180° - 45°) = -\tan 45°.$$

31. $\tan 120°$ **32.** $\tan 240°$ **33.** $\tan 210°$

34. $\tan 150°$ **35.** $\tan 315°$ **36.** $\tan 330°$

Verify that each equation is an identity.

37. $\tan(180° + x) = \tan x$ **38.** $\tan(360° - x) = -\tan x$

39. $\tan(90° - x) = \cot x$ **40.** $\cot(90° - x) = \tan x$

41. $\tan 2x = \dfrac{2 \tan x}{1 - \tan^2 x}$. *Hint:* $\tan 2x = \tan(x + x)$.

42. $\tan\left(x + \dfrac{\pi}{4}\right) = \dfrac{1 + \tan x}{1 - \tan x}$

B

43. $\dfrac{\tan(45° - \alpha)}{\tan(45° + \alpha)} = \dfrac{(1 - \tan \alpha)^2}{(1 + \tan \alpha)^2}$ **44.** $\tan \beta = \dfrac{\tan(45° + \beta) - 1}{1 + \tan(45° + \beta)}$

45. $\cot(\alpha + \beta) = \dfrac{\cot \alpha \cot \beta - 1}{\cot \beta + \cot \alpha}$ **46.** $\cot(\alpha - \beta) = \dfrac{\cot \alpha \cot \beta + 1}{\cot \beta - \cot \alpha}$

4.5 Double-Angle Formulas

Additional identities follow from the sum and difference formulas developed in Sections 4.2 and 4.3. For example, if α is substituted for β in

$$\cos(\alpha + \beta) = \cos \alpha \cos \beta - \sin \alpha \sin \beta,$$

we obtain

$$\cos(\alpha + \alpha) = \cos \alpha \cos \alpha - \sin \alpha \sin \alpha,$$

$$\mathbf{\cos 2\alpha = \cos^2 \alpha - \sin^2 \alpha.} \tag{1}$$

The right-hand member can be written in a variety of convenient forms. Thus, if $1 - \sin^2 \alpha$ is substituted for $\cos^2 \alpha$, we have

$$\cos 2\alpha = (1 - \sin^2 \alpha) - \sin^2 \alpha,$$

$$\mathbf{\cos 2\alpha = 1 - 2 \sin^2 \alpha.} \tag{2}$$

If $1 - \cos^2 \alpha$ is substituted for $\sin^2 \alpha$ in Equation (1), we have

$$\cos 2\alpha = \cos^2 \alpha - (1 - \cos^2 \alpha),$$

$$\mathbf{\cos 2\alpha = 2 \cos^2 \alpha - 1.} \tag{3}$$

Identities (1), (2), and (3), and similar identities that follow for the sine and tangent functions, called **double-angle formulas,** are often helpful for rewriting expressions in more useful forms. You will prove additional identities to become familiar with these important relationships.

Example Show that $\dfrac{1 - \cos 2x}{2 \sin x \cos x} = \tan x$ is an identity.

Solution From Equation (2), we can substitute $1 - 2 \sin^2 x$ for $\cos 2x$ in the left-hand member to obtain

$$\frac{1 - (1 - 2 \sin^2 x)}{2 \sin x \cos x} = \frac{2 \sin^2 x}{2 \sin x \cos x}$$

$$= \frac{\sin x}{\cos x} = \tan x.$$

Hence, the original equation is an identity.

Example Show that $\dfrac{5 + \cos 2\beta}{1 + \cos 2\beta} - 1 = 2 \sec^2 \beta$ is an identity.

Solution From Equation (3), we substitute $2 \cos^2 \beta - 1$ for $\cos 2\beta$ in the left-hand member to obtain

$$\frac{5 + (2\cos^2\beta - 1)}{1 + (2\cos^2\beta - 1)} - 1 = \frac{4 + 2\cos^2\beta}{2\cos^2\beta} - 1$$

$$= \frac{4 + 2\cos^2\beta - 2\cos^2\beta}{2\cos^2\beta}$$

$$= \frac{4}{2\cos^2\beta} = 2\sec^2\beta.$$

Hence, the original equation is an identity.

If α is substituted for β in

$$\sin(\alpha + \beta) = \sin\alpha\cos\beta + \cos\alpha\sin\beta,$$

we obtain

$$\sin(\alpha + \alpha) = \sin\alpha\cos\alpha + \cos\alpha\sin\alpha,$$

$$\mathbf{\sin 2\alpha = 2\sin\alpha\cos\alpha.} \tag{4}$$

Example Show that $\dfrac{2}{\sin 2\theta} = \tan\theta + \cot\theta$ is an identity.

Solution Rewriting the right-hand member of the equation in terms of $\sin\theta$ and $\cos\theta$, we obtain

$$\tan\theta + \cot\theta = \frac{\sin\theta}{\cos\theta} + \frac{\cos\theta}{\sin\theta} = \frac{\sin^2\theta + \cos^2\theta}{\sin\theta\cos\theta}$$

$$= \frac{1}{\sin\theta\cos\theta} = \frac{2}{2\sin\theta\cos\theta}$$

$$= \frac{2}{\sin 2\theta}.$$

Hence, the original equation is an identity.

If α is substituted for β in

$$\tan(\alpha + \beta) = \frac{\tan\alpha + \tan\beta}{1 - \tan\alpha\tan\beta},$$

we obtain

$$\tan(\alpha + \alpha) = \frac{\tan \alpha + \tan \alpha}{1 - \tan \alpha \tan \alpha},$$

$$\mathbf{\tan 2\alpha = \frac{2 \tan \alpha}{1 - \tan^2 \alpha}.} \tag{5}$$

Example Show that $\tan 2x = \dfrac{2 \tan x}{2 - \sec^2 x}$ is an identity.

Solution We can substitute $\tan^2 x + 1$ for $\sec^2 x$ to obtain

$$\begin{aligned} \tan 2x &= \frac{2 \tan x}{2 - (\tan^2 x + 1)} \\ &= \frac{2 \tan x}{1 - \tan^2 x}, \end{aligned}$$

which is an identity by Equation (5).

Relationships similar to those developed in this section for sine, cosine, and tangent can be derived for cotangent, secant, and cosecant. However, they are of less importance and will not be considered.

EXERCISE SET 4.5

A

Express each of the following as a single function of kx, or $k\alpha$, where k is a positive integer. Assume that x or α takes on no value for which the expression is undefined.

Examples a. $1 - 2 \sin^2 x$ b. $\dfrac{\tan 3x}{1 - \tan^2 3x}$

Solutions

a. From Equation (2),

$$1 - 2 \sin^2 x = \cos 2x.$$

b. From Equation (5),

$$\frac{\tan 3x}{1 - \tan^2 3x} = \frac{1}{2} \cdot \frac{2 \tan 3x}{1 - \tan^2 x} = \frac{1}{2} \tan 2(3x) = \frac{1}{2} \tan 6x.$$

1. $2 \sin x \cos x$

2. $2 \sin 2x \cos 2x$

3. $\cos^2 x - \sin^2 x$

4. $2 \cos^2 x - 1$

5. $\dfrac{2 \tan x}{1 - \tan^2 x}$

6. $\dfrac{4 \tan 2x}{1 - \tan^2 2x}$

7. $\sin \dfrac{x}{2} \cos \dfrac{x}{2}$

8. $4 \sin 3x \cos 3x$

9. $\cos^2 5\alpha - \sin^2 5\alpha$

10. $1 - 2 \sin^2 4\alpha$

11. $4 \cos^2 6\alpha - 2$

12. $2 \sin 5\alpha \cos 5\alpha$

13. $\sin 2\alpha \cos \alpha - \cos 2\alpha \sin \alpha$

14. $\sin 3\alpha \cos 2\alpha + \cos 3\alpha \sin 2\alpha$

15. $\cos 3x \cos 5x + \sin 3x \sin 5x$

16. $\cos 4x \cos 2x - \sin 4x \sin 2x$

Verify that each equation is an identity.

17. $\cos^2 \alpha = \dfrac{1 + \cos 2\alpha}{2}$

18. $\sin^2 \alpha = \dfrac{1 - \cos 2\alpha}{2}$

19. $\cos 2x + 2 \sin^2 x = 1$

20. $\sin 2x \csc x = 2 \cos x$

21. $\dfrac{\sin 2x}{1 + \cos 2x} = \tan x$

22. $\sin 2x = \dfrac{2 \tan x}{1 + \tan^2 x}$

23. $\sec^2 x = \dfrac{2}{1 + \cos 2x}$

24. $\cot x = \dfrac{1 + \cos 2x}{\sin 2x}$

25. $\cos 2\beta = \dfrac{1 - \tan^2 \beta}{1 + \tan^2 \beta}$

26. $\cot^2 \beta = \dfrac{1 + \cos 2\beta}{1 - \cos 2\beta}$

27. $\cot 4x = \dfrac{1 - \tan^2 2x}{2 \tan 2x}$

28. $\cos^4 x - \sin^4 x = \cos 2x$

29. $\sin 3x = 3 \sin x - 4 \sin^3 x$
Hint: First rewrite $\sin 3x$ as $\sin(2x + x)$.

30. $\cos 3x = 4 \cos^3 x - 3 \cos x$

31. $\sin 4\alpha = (\cos \alpha)(4 \sin \alpha - 8 \sin^3 \alpha)$

32. $\cos 4\alpha = 1 - 8 \sin^2 \alpha \cos^2 \alpha$

Example If $\sin x = \dfrac{3}{5}$ and $\cos x < 0$, find $\sin 2x$.

Solution To evaluate

$$\sin 2x = 2 \sin x \cos x, \tag{1}$$

the values for $\cos x$ for those values of x for which $\sin x = 3/5$ and $\cos x < 0$ are needed. Substituting $3/5$ for $\sin x$ in

$$\sin^2 x + \cos^2 x = 1$$

(continued)

gives

$$\left(\frac{3}{5}\right)^2 + \cos^2 x = 1,$$

$$\cos^2 x = 1 - \frac{9}{25} = \frac{16}{25}.$$

Because $\cos x < 0$, $\cos x = -4/5$. Substituting $3/5$ for $\sin x$ and $-4/5$ for $\cos x$ in $\sin 2x = 2 \sin x \cos x$ yields

$$\sin 2x = 2\left(\frac{3}{5}\right)\left(-\frac{4}{5}\right) = -\frac{24}{25}.$$

33. If $\sin x = \frac{4}{5}$ and $\cot x > 0$, find $\sin 2x$.

34. If $\sin x = \frac{12}{13}$ and $\sec x < 0$, find $\cos 2x$.

35. If $\cos 2\alpha = -\frac{8}{17}$ and $180° \leq 2\alpha \leq 270°$, find $\sin \alpha$.

36. If $\cos 2a = \frac{5}{13}$ and $270° \leq 2\alpha \leq 360°$, find $\cos \alpha$.

37. If $\tan \beta = -\frac{3}{4}$, find $\tan 2\beta$.

38. If $\tan \beta = \frac{5}{12}$, find $\cot 2\beta$.

4.6 Additional Identities

The identities in this section are less important than some of the identities introduced in the preceding sections. However, they are sometimes useful in rewriting expressions involving trigonometric function values.

In Section 4.5 we observed that

$$\cos 2\alpha = 2 \cos^2 \alpha - 1.$$

Solving this for $\cos^2 \alpha$, we have

$$1 + \cos 2\alpha = 2 \cos^2 \alpha,$$

$$\frac{1 + \cos 2\alpha}{2} = \cos^2 \alpha,$$

$$\boldsymbol{\cos^2 \alpha = \frac{1 + \cos 2\alpha}{2}}. \qquad (1)$$

We also noted in Section 4.5 that

$$\cos 2\alpha = 1 - 2 \sin^2 \alpha.$$

Solving this for $\sin^2 \alpha$, we have

$$2 \sin^2 \alpha = 1 - \cos 2\alpha,$$

$$\boldsymbol{\sin^2 \alpha = \frac{1 - \cos 2\alpha}{2}}. \tag{2}$$

These identities enable us to write the second-degree terms, $\cos^2 \alpha$ and $\sin^2 \alpha$, in terms of first-degree terms. They also enable us to write additional useful identities.

Half-Angle Formulas

Substituting $\alpha/2$ for α in Equation (1) yields

$$\cos^2 \frac{\alpha}{2} = \frac{1 + \cos 2(\alpha/2)}{2},$$

from which

$$\boldsymbol{\cos^2 \frac{\alpha}{2} = \frac{1 + \cos \alpha}{2}}. \tag{3}$$

Substituting $\alpha/2$ for α in Equation (2) yields

$$\sin^2 \frac{\alpha}{2} = \frac{1 - \cos 2(\alpha/2)}{2},$$

from which

$$\boldsymbol{\sin^2 \frac{\alpha}{2} = \frac{1 - \cos \alpha}{2}}. \tag{4}$$

Values for $\cos(\alpha/2)$ and $\sin(\alpha/2)$ are positive or negative, depending on the value of $\alpha/2$. Table 1.1 (page 11) can be used to help you select the appropriate sign.

Example Use Equation (4) to find $\sin \frac{\pi}{8}$.

Solution With $\alpha/2 = \pi/8$ and $\alpha = 2(\pi/8) = \pi/4$, we obtain

(continued)

$$\sin^2 \frac{\pi}{8} = \frac{1 - \cos \dfrac{\pi}{4}}{2}.$$

Because $0 < \pi/8 < \pi/2$, we use the positive square root for $\sin(\pi/8)$. Then, since $\cos(\pi/4) = 1/\sqrt{2}$,

$$\sin \frac{\pi}{8} = \sqrt{\frac{1 - \dfrac{1}{\sqrt{2}}}{2}} = \sqrt{\frac{\sqrt{2} - 1}{2\sqrt{2}}}.$$

Formulas for $\tan(\alpha/2)$ also follow directly from identities we have developed. Substituting $\alpha/2$ for α in

$$\tan \alpha = \frac{\sin \alpha}{\cos \alpha},$$

we have

$$\tan \frac{\alpha}{2} = \frac{\sin \dfrac{\alpha}{2}}{\cos \dfrac{\alpha}{2}}.$$

Multiplying both the numerator and the denominator of the right-hand member by $2 \sin(\alpha/2)$ yields

$$\tan \frac{\alpha}{2} = \frac{2 \sin^2 \dfrac{\alpha}{2}}{2 \sin \dfrac{\alpha}{2} \cos \dfrac{\alpha}{2}}.$$

Substituting $(1 - \cos \alpha)/2$ for $\sin^2(\alpha/2)$ and substituting $\sin 2(\alpha/2)$ for $2 \sin(\alpha/2) \cos(\alpha/2)$, we obtain

$$\tan \frac{\alpha}{2} = \frac{2\left(\dfrac{1 - \cos \alpha}{2}\right)}{\sin 2\left(\dfrac{\alpha}{2}\right)},$$

$$\boldsymbol{\tan \frac{\alpha}{2} = \frac{1 - \cos \alpha}{\sin \alpha}}. \qquad (5)$$

By multiplying the numerator and the denominator of the right-hand member of Equation (5) by $1 + \cos \alpha$, we obtain

$$\tan\frac{\alpha}{2} = \frac{(1-\cos\alpha)(1+\cos\alpha)}{\sin\alpha(1+\cos\alpha)} = \frac{\sin^2\alpha}{\sin\alpha(1+\cos\alpha)},$$

$$\mathbf{\tan\frac{\alpha}{2} = \frac{\sin\alpha}{1+\cos\alpha}}. \tag{6}$$

Example Use Equation (5) to find $\tan\frac{\pi}{8}$.

Solution If $\pi/8 = \alpha/2$, then $\alpha = \pi/4$ and

$$\tan\frac{\pi}{8} = \frac{1-\cos\frac{\pi}{4}}{\sin\frac{\pi}{4}} = \frac{1-\frac{1}{\sqrt{2}}}{\frac{1}{\sqrt{2}}} = \sqrt{2}-1.$$

Just as identities such as $\sin 2\alpha = 2\sin\alpha\cos\alpha$ are sometimes called double-angle formulas, the identities developed in this section are sometimes called **half-angle formulas.**

Relationships similar to those developed above for sine, cosine, and tangent can be derived for cotangent, secant, and cosecant. They are of less importance and will not be considered.

Product, Sum, and Difference Formulas

The following identities are derived from the sum and difference formulas for the cosine and sine that were developed in Sections 4.2 and 4.3. The proofs are left as exercises.

$$\cos\alpha\cos\beta = \tfrac{1}{2}[\cos(\alpha+\beta) + \cos(\alpha-\beta)]; \tag{7}$$

$$\sin\alpha\sin\beta = -\tfrac{1}{2}[\cos(\alpha+\beta) - \cos(\alpha-\beta)]; \tag{8}$$

$$\sin\alpha\cos\beta = \tfrac{1}{2}[\sin(\alpha+\beta) + \sin(\alpha-\beta)]; \tag{9}$$

$$\cos\alpha\sin\beta = \tfrac{1}{2}[\sin(\alpha+\beta) - \sin(\alpha-\beta)]. \tag{10}$$

Example Express $\cos 5x \sin 2x$ as a sum or difference.

Solution From Equation (10),

$$\begin{aligned}\cos 5x\sin 2x &= \tfrac{1}{2}[\sin(5x+2x) - \sin(5x-2x)]\\ &= \tfrac{1}{2}(\sin 7x - \sin 3x).\end{aligned}$$

The following four identities enable us to express sums and differences of function values in terms of products.

$$\cos \alpha + \cos \beta = 2 \cos \left(\frac{\alpha + \beta}{2}\right) \cos \left(\frac{\alpha - \beta}{2}\right); \qquad (11)$$

$$\cos \alpha - \cos \beta = -2 \sin \left(\frac{\alpha + \beta}{2}\right) \sin \left(\frac{\alpha - \beta}{2}\right); \qquad (12)$$

$$\sin \alpha + \sin \beta = 2 \sin \left(\frac{\alpha + \beta}{2}\right) \cos \left(\frac{\alpha - \beta}{2}\right); \qquad (13)$$

$$\sin \alpha - \sin \beta = 2 \cos \left(\frac{\alpha + \beta}{2}\right) \sin \left(\frac{\alpha - \beta}{2}\right). \qquad (14)$$

Example Express $\sin 4\alpha + \sin 2\alpha$ as a product.

Solution From Equation (13),

$$\sin 4\alpha + \sin 2\alpha = 2 \sin \left(\frac{4\alpha + 2\alpha}{2}\right) \cos \left(\frac{4\alpha - 2\alpha}{2}\right)$$

$$= 2 \sin 3\alpha \cdot \cos \alpha.$$

EXERCISE SET 4.6

A

Use the half-angle formulas to find the exact value for each of the following.

Example $\sin 105°$

Solution From Equation (4), with $\alpha/2 = 105°$ and $\alpha = 210°$

$$\sin 105° = \sqrt{\frac{1 - \cos 210°}{2}},$$

where the positive square root is chosen because 105° is in Quadrant II and $\sin 105°$ is therefore positive. Since $\cos 210° = -\sqrt{3}/2$.

$$\sin 105° = \sqrt{\frac{1 - (-\sqrt{3}/2)}{2}} = \sqrt{\frac{2 + \sqrt{3}}{4}} = \frac{\sqrt{2 + \sqrt{3}}}{2}.$$

1. $\sin 15°$ **2.** $\cos 15°$ **3.** $\tan 15°$ **4.** $\sin 75°$

5. $\cos 75°$ **6.** $\tan 75°$ **7.** $\cos \frac{\pi}{8}$ **8.** $\sin \frac{\pi}{12}$

9. $\tan \frac{5\pi}{12}$ **10.** $\cos \frac{\pi}{12}$ **11.** $\cos \frac{7\pi}{12}$ **12.** $\sin \frac{7\pi}{12}$

Verify each identity.

13. $\tan\frac{x}{2} = \csc x - \cot x$

14. $\sin\frac{\theta}{2}\cos\frac{\theta}{2} = \frac{\sin\theta}{2}$

15. $\left(\cos\frac{\alpha}{2} - \sin\frac{\alpha}{2}\right)^2 = 1 - \sin\alpha$

16. $\sin\beta - \cos\beta\tan\frac{\beta}{2} = \tan\frac{\beta}{2}$

17. $\sin^2\frac{\gamma}{2} = \frac{\sec\gamma - 1}{2\sec\gamma}$

18. $\tan\frac{x}{2}\sin x = \frac{\tan x - \sin x}{\sin x \sec x}$

In Exercises 19–24, assume $0° \leq x < 360°$.

Example If $\tan x = -\frac{3}{4}$ and $\cos x$ is negative, find $\sin\frac{x}{2}$.

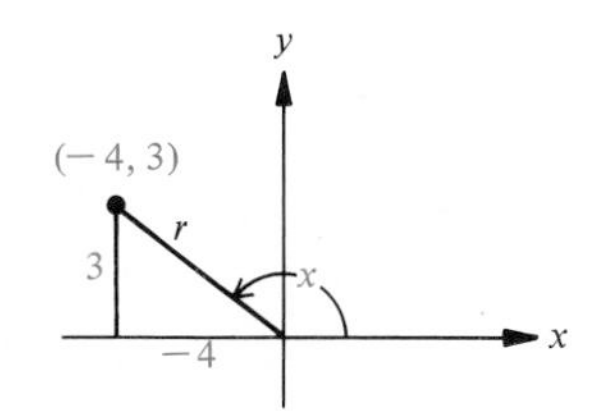

Solution Since $\cos x$ and $\tan x$ are both negative, x is an angle in Quadrant II, and because $\tan x = -3/4$, a point on the terminal ray has the x-coordinate -4 and the y-coordinate 3. From the Pythagorean formula (see Appendix A.8),

$$r = \sqrt{x^2 + y^2} = \sqrt{(-4)^2 + 3^2} = 5.$$

From Equation (4) with $\cos x = -4/5$, we obtain

$$\sin\frac{x}{2} = \sqrt{\frac{1 - (-4/5)}{2}} = \sqrt{\frac{9}{10}} = \frac{3}{\sqrt{10}},$$

where the positive square root is taken because x is in Quadrant II and therefore $x/2$ must be in Quadrant I.

19. If $\sin x = \frac{\sqrt{3}}{2}$ and $\cos x > 0$, find $\tan\frac{x}{2}$.

20. If $\sin\theta = \frac{1}{2}$ and $\sec\theta > 0$, find $\sin\frac{\theta}{2}$.

21. If $\cos\alpha = -\frac{1}{3}$ and $\tan\alpha > 0$, find $\sin\frac{\alpha}{2}$.

22. If $\tan\beta = 3$ and $\sin\beta > 0$, find $\cos\frac{\beta}{2}$.

23. If $\tan x = -1$ and $\sin x < 0$, find $\cos\frac{x}{2}$.

24. If $\sin\theta = \frac{1}{\sqrt{2}}$ and $\cos\theta < 0$, find $\tan\frac{\theta}{2}$.

Write each expression as a sum or difference of function values.

Example $\sin 3x \sin x$

Solution From Equation (8),

$$\begin{aligned}\sin 3x \sin x &= -\tfrac{1}{2}[\cos(3x + x) - \cos(3x - x)] \\ &= -\tfrac{1}{2}(\cos 4x - \cos 2x).\end{aligned}$$

25. $\sin 5° \cos 12°$ **26.** $\cos 7° \cos 20°$ **27.** $\cos 2x \sin x$

28. $\sin 2x \sin 5x$ **29.** $\cos 2x \cos 5x$ **30.** $\sin 6x \cos x$

Write each expression as a product.

Example $\cos 3\alpha - \cos \alpha$

Solution From Equation (12),

$$\begin{aligned}\cos 3\alpha - \cos \alpha &= -2 \sin \left(\frac{3\alpha + \alpha}{2}\right) \sin \left(\frac{3\alpha - \alpha}{2}\right) \\ &= -2 \sin 2\alpha \sin \alpha.\end{aligned}$$

31. $\cos 20° + \cos 4°$ **32.** $\sin 18° - \sin 6°$ **33.** $\sin 3\alpha + \sin 5\alpha$

34. $\cos \alpha + \cos 5\alpha$ **35.** $\sin 5\alpha - \sin \alpha$ **36.** $\cos 6\alpha - \cos 2\alpha$

B

Verify each of the following identities that are listed in this section.

37. Equation (7) **38.** Equation (8)

39. Equation (9) **40.** Equation (10)

41. Equation (11) **42.** Equation (12)

43. Equation (13) **44.** Equation (14)

4.7 Conditional Equations

In Sections 4.1–4.6 we considered identities—equations that are satisfied by all angle or all real number replacements for which both of the members are defined. We now consider equations that are conditional—that is, they are satisfied only for particular replacements for the variable.

We have already solved many simple conditional equations, such as

$$\sin x = \frac{1}{2},$$

by inspection and by using Table 1.2, Table II or Table III in Appendix C, or a calculator. In this case,

for $0 \leq x \leq \pi/2$, the solution set is $\{\pi/6\}$;

for $0 \leq x < 2\pi$, the solution set is $\{\pi/6, 5\pi/6\}$;

for $x \in A$ (the set of all angles), in degree units, the solution set is

$$\{x \mid x = (30 + k \cdot 360)^\circ \text{ or } x = (150 + k \cdot 360)^\circ\}, \quad k \in J.$$

Perhaps you can determine the solution set of

$$\sin x = \cos x, \quad 0 < x < \frac{\pi}{2}, \tag{1}$$

by inspection—perhaps not. However, you can generate equivalent equations and hopefully you will generate one from which the solution(s) will be evident by inspection. Generally, solutions can be determined most easily from equations that involve *values of one function only*. The identities developed in the preceding sections are useful in obtaining such equivalent equations from equations involving values of more than one function. For example, in Equation (1), for $0 < x < \pi/2$, $\cos x \neq 0$. Thus, we can multiply each member by $1/\cos x$ to obtain

$$\frac{\sin x}{\cos x} = 1.$$

Substituting $\tan x$ for $\sin x/\cos x$, we obtain

$$\tan x = 1.$$

Since, in this case, values of x are restricted to $0 < x < \pi/2$, we find the solution set by inspection to be $\{\pi/4\}$.

Because function values ($\cos x$, $\sin x$, etc.) are real numbers, equations that are quadratic in a given trigonometric function value, such as

$$\cos^2 x + \cos x - 2 = 0, \quad x \in A, \tag{2}$$

can be solved by first solving the quadratic equation for the function value involved. To emphasize that we are using a standard algebraic method for solving a quadratic equation, we also show each step of the process, where we let $\cos x = u$ and write Equation (2) as

$$u^2 + u - 2 = 0. \tag{3}$$

Since the left-hand members of Equations (2) and (3) are factorable, the equations can be written equivalently as

$$(\cos x + 2)(\cos x - 1) = 0 \quad \text{or} \quad (u + 2)(u - 1) = 0.$$

Since $\cos x + 2$ cannot equal zero (there is no real number x such that $\cos x = -2$), we seek values of x for which

$$\cos x - 1 = 0 \quad \text{or} \quad u - 1 = 0.$$

From either equation, we have

$$\cos x = 1.$$

By inspection the solution set of $\cos x = 1$ and hence of Equation (2) is

$$\{x \mid x = k \cdot 360°, \quad k \in J\} \quad \text{or} \quad \{x \mid x = k \cdot 2\pi, \quad k \in J\}.$$

EXERCISE SET 4.7

Solve each equation. Find all solutions in the interval $0 \leq x \leq \pi/2$.

Examples

a. $(2 \cos x - \sqrt{3})(\sin x - 1) = 0$ b. $\sqrt{2} \sin x \cos x - \cos x = 0$

Solutions

a. The left-hand member equals 0 if

$$2 \cos x - \sqrt{3} = 0 \quad \text{or} \quad \sin x - 1 = 0.$$

From these equations we have

$$\cos x = \frac{\sqrt{3}}{2} \quad \text{or} \quad \sin x = 1,$$

and the solution set is $\{\pi/6, \pi/2\}$.

b. Factoring the left-hand member we have

$$\cos x(\sqrt{2} \sin x - 1) = 0.$$

The left-hand member equals 0 if

$$\cos x = 0 \qquad \text{or} \qquad \sin x = \frac{1}{\sqrt{2}},$$

and the solution set is $\{\pi/2, \pi/4\}$.

1. $2 \sin x - 1 = 0$
2. $2 \cos x - 1 = 0$
3. $2 \cos^2 x = 1$
4. $4 \sin^2 x = 3$
5. $3 \csc^2 x - 3 = 0$
6. $\sec^2 x = 4$
7. $(2 \cos^2 x - 1)(\sin x - 1) = 0$
8. $(\sin^2 x - 1)(\sec x - 2) = 0$
9. $\cos x \sin x = 0$
10. $3 \tan x \sec x = 0$
11. $2 \cos^3 x - \cos x = 0$
12. $2 \sin^3 x - \sin x = 0$
13. $2 \sin x \cos x + \sin x = 0$
14. $2 \sin x \cos x - \cos x = 0$

Solve each equation. Find all solutions in the interval $0 \leq x < 2\pi$.

Example $2 \cos^2 x + \cos x - 1 = 0$

Solution Factoring the left-hand member, we obtain

$$(2 \cos x - 1)(\cos x + 1) = 0,$$

$$\cos x = \frac{1}{2} \qquad \text{or} \qquad \cos x = -1,$$

and the solution set is $\{\pi/3, 5\pi/3, \pi\}$.

15. $2 \cos^2 x - \cos x - 1 = 0$
16. $2 \cos^2 x + 3 \cos x - 2 = 0$
17. $9 \cos^2 x - 21 \cos x - 8 = 0$
18. $25 \sin^2 x - 20 \sin x + 4 = 0$

Solve each equation **(a)** *in the interval* $0° \leq x < 360°$ *and* **(b)** *for all* $x \in A$.

Example $\sec x - 2 \cos x = 1$

Solution

a. We must first write each term using the same function value. Hence, we substitute $1/\cos x$ for $\sec x$ to obtain

$$\frac{1}{\cos x} - 2 \cos x = 1.$$

(continued)

Multiplying each member by $\cos x$ ($\cos x \neq 0$), we have

$$1 - 2\cos^2 x = \cos x,$$

$$2\cos^2 x + \cos x - 1 = 0.$$

This equation is the same as in the preceding example. By the same procedure, then, the solution set over the interval $0° \leq x < 360°$ is

$$\{60°, 300°, 180°\}.$$

b. For all $x \in A$, the solution set is

$$\{x \mid x = 60° + k \cdot 360°\} \cup \{x \mid x = 180° + k \cdot 360°\} \cup \{x \mid x = 300° + k \cdot 360°\}, \quad k \in J,$$

19. $\csc x + \sin x + 2 = 0$

20. $2\cos x + \sec x - 3 = 0$

21. $2\sin^2 x - \cos x - 1 = 0$

22. $2\cos^2 x - 3\sin x - 3 = 0$

23. $2\sin x + \csc x - 3 = 0$

24. $\tan x + \cot x + 2 = 0$

4.8 Conditional Equations for Multiples

Conditional equations such as

$$\tan 3x = 1, \tag{1}$$

$$2\sin x \cos x = \frac{\sqrt{3}}{2}, \tag{2}$$

and

$$\cos 2x = \cos^2 x - 1 \tag{3}$$

need further discussion.

First consider the solution set of

$$\tan 3x = 1 \tag{1}$$

over the interval $0 \leq x < 2\pi$. By inspection we have that

$$3x = \frac{\pi}{4} + k \cdot \pi,$$

and, dividing each term by 3,

$$x = \frac{\pi}{12} + k \cdot \frac{\pi}{3}, \quad k \in J.$$

To find the solution set over the interval $0 \le x < 2\pi$, we take 0, 1, 2, 3, 4, and 5 as replacements for k and obtain

$$\left\{\frac{\pi}{12}, \frac{5\pi}{12}, \frac{3\pi}{4}, \frac{13\pi}{12}, \frac{17\pi}{12}, \frac{7\pi}{4}\right\}.$$

The identities involving multiples of the variable are often useful to help solve certain kinds of equations. For example, the solution set of

$$2 \sin x \cos x = \frac{\sqrt{3}}{2}, \tag{2}$$

for $x \in R$ over the interval $0 \le x \le \pi/2$, can readily be found if $\sin 2x$ is substituted for the left-hand member, $2 \sin x \cos x$, to obtain

$$\sin 2x = \frac{\sqrt{3}}{2}.$$

By inspection we have that

$$2x = \frac{\pi}{3} \quad \text{or} \quad 2x = \frac{2\pi}{3},$$

$$x = \frac{\pi}{6} \quad \text{or} \quad x = \frac{\pi}{3},$$

and the solution set of Equation (2) is $\{\pi/6, \pi/3\}$.

One method of solving

$$\cos 2x = \cos^2 x - 1, \tag{3}$$

for $x \in R$, consists of substituting $2 \cos^2 x - 1$ for $\cos 2x$ to obtain the equivalent equations

$$2 \cos^2 x - 1 = \cos^2 x - 1,$$

$$\cos^2 x = 0.$$

Then, we have as the solution set,

$$\left\{x \mid x = \frac{\pi}{2} + k\pi, \ k \in J\right\}.$$

Alternatively, we could have substituted $(1 + \cos 2x)/2$ (page 170) for $\cos^2 x$ in the right-hand member of Equation (3) to obtain the equivalent equations

$$\cos 2x = \frac{1 + \cos 2x}{2} - 1,$$

$$2 \cos 2x = 1 + \cos 2x - 2,$$

$$\cos 2x = -1.$$

By inspection, we have that

$$2x = \pi + k \cdot 2\pi, \quad k \in J,$$

from which the solution set is given by

$$\left\{x \mid x = \frac{\pi}{2} + k\pi,\ k \in J\right\},$$

as we found above using the other method.

EXERCISE SET 4.8

A

Solve each equation for all x, such that $0 \le x \le \pi/2$.

Example $\cos 3x = \dfrac{1}{2}$

Solution Using Table 1.2 and the fact that the cosine of an angle in Quadrant I or IV is positive, we first consider all possible solutions and obtain

$$3x = \frac{\pi}{3} + k \cdot 2\pi \quad \text{or} \quad 3x = \frac{5\pi}{3} + k \cdot 2\pi.$$

From which, dividing each term by 3, we have

$$x = \frac{\pi}{9} + k \cdot \frac{2\pi}{3} \quad \text{or} \quad x = \frac{5\pi}{9} + k \cdot \frac{2\pi}{3}.$$

Since the problem restricts x to the interval $0 \le x \le \pi/2$, we need only the value 0 for k. if $k = 0$, $x = \pi/9$ or $x = 5\pi/9$. Furthermore, since $5\pi/9$ is not in the required interval, the solution set is $\{\pi/9\}$.

1. $\sin 3x = 0$ **2.** $\cos 3x = 1$

3. $\tan 4x = -\sqrt{3}$ **4.** $\cot 4x = -1$

5. $\sec 5x = 2$ **6.** $\csc 5x = \sqrt{2}$

7. $\cos 2x - \sin x = 0$ **8.** $\cos 2x + \cos x = 0$

9. $\sin 2x + \sin x = 0$ **10.** $\sin 2x - \cos x = 0$

11. $\sin 2x = \cos 2x$ **12.** $\sin \frac{x}{2} = \tan \frac{x}{2}$

13. $\tan \frac{x}{2} - \cos x = -1$ **14.** $\cos x = \sin \frac{x}{2}$

Solve each equation in the interval $0° \leq x < 360°$.

Example $\sin 4x = \frac{\sqrt{3}}{2}$

Solution By inspection we have $4x = 60°$. However, since $\sin \theta$ is also positive for angles in Quadrant II, we know that $4x = 120°$ will also satisfy the equation. Thus for $k \in J$, we have

$$4x = 60° + k \cdot 360° \quad \text{or} \quad 4x = 120° + k \cdot 360°.$$

Dividing each member of both equations by 4, we obtain

$$x = 15° + k \cdot 90° \quad \text{or} \quad x = 30° + k \cdot 90°.$$

Since the problem restricts x to the interval $0° \leq x < 360°$, we need only the values of 0, 1, 2, and 3 for k.

If $k = 0$, $x = 15°$ or $x = 30°$,
if $k = 1$, $x = 105°$ or $x = 120°$,
if $k = 2$, $x = 195°$ or $x = 210°$,
if $k = 3$, $x = 285°$ or $x = 300°$,

and the solution set is $\{15°, 30°, 105°, 120°, 195°, 210°, 285°, 300°\}$.

15. $\sin 3x = \frac{1}{2}$ **16.** $\cos 4x = -1$

17. $\sin 3x = \cos 3x$ **18.** $\tan 2x = \cot 2x$

19. $\sin^2 2x = 1$ **20.** $\cos^2 2x = 1$

21. $\tan^2 3x = 1$ **22.** $\cot^2 3x = 3$

23. $\left(1 + \cos \frac{x}{2}\right) \cos \frac{x}{2} = 0$

24. $\sin^2 \frac{x}{4} - \frac{1}{2} \sin \frac{x}{4} = 0$

25. $5 \sec 3x = 2 \cos 3x - 3$

26. $\tan x \tan 2x = 1$

B

Solve each equation for all x, such that $x \in R$.

27. $\sin 2x = \cos x$

28. $2 \cos 2x - 1 = -3 \sin^2 x$

29. $\sin x \cos 3x = 0$

30. $(\sin 2x - 1)(\cos 3x + 1) = 0$

31. $\tan 2x - 3 \sec^2 2x = -5$

32. $\cos^2 2x + 3 \sin 2x - 3 = 0$

33. $\sin 2x \cos x + \cos 2x \sin x = \frac{1}{\sqrt{2}}$

34. $\sin 3x \cos x - \cos 3x \sin x = \frac{\sqrt{3}}{2}$

35. Use graphical methods to approximate a positive solution for

$$\cos 2x = x, \quad x \in R.$$

Hint: Graph $y = \cos 2x$ and $y = x$ on the same axes.

36. Use graphical methods to approximate a positive solution for

$$x^2 - 3 \sin \frac{x}{2} = 0, \quad x \in R.$$

Chapter Summary

4.1–4.6 Equations are **identities** if they are satisfied for all real number replacements of the variable for which both members of the equation are defined.

For convenience, the basic identities that have been established are listed inside the back cover.

An equation can be shown *not* to be an identity by replacing the variable with a real number for which the resulting statement is false.

4.7–4.8 **Conditional equations** involving circular function values can be solved by inspection or by generating equivalent equations from which the solutions are evident by inspection. The identities introduced in this chapter are useful in obtaining such equivalent equations.

Review Exercises

A

4.1–4.5 *Express each of the following in terms of the function values* $\sin\alpha$ *and/or* $\cos\alpha$ *and simplify the expression.*

1. $\csc\alpha\tan\alpha$

2. $\cot^2\alpha\sec^2\alpha$

3. $\sin^2\alpha(\csc^2\alpha - 1)$

4. $\cot\alpha\cos\alpha\sec^2\alpha$

Verify that each equation is an identity.

5. $\cos\alpha\csc\alpha = \cot\alpha$

6. $\dfrac{\cos\alpha}{\cot\alpha} = \sin\alpha$

7. $\sin^2\alpha(1 + \cot^2\alpha) = 1$

8. $\dfrac{\cos\alpha\csc\alpha}{\cot\alpha} = 1$

9. $\tan\alpha = \dfrac{1}{\cos\alpha\csc\alpha}$

10. $1 + \tan^2\alpha = \csc^2\alpha\tan^2\alpha$

11. $\sin(-\alpha) = -\dfrac{\tan\alpha}{\sec\alpha}$

12. $\tan(-\alpha) = \dfrac{\cos(-\alpha)\sin(-\alpha)}{1 - \sin^2\alpha}$

13. $\cos(270° + x) = \sin x$

14. $\cos(270° - x) = -\sin x$

15. $\sin(90° + x) = \cos x$

16. $\sin(\alpha - 45°) = \dfrac{\sin\alpha - \cos\alpha}{\sqrt{2}}$

17. $\sin(\alpha - 60°) = \dfrac{\sin\alpha - \sqrt{3}\cos\alpha}{2}$

18. $\tan(\alpha - 45°) = \dfrac{\tan\alpha - 1}{1 + \tan\alpha}$

19. $\cos 2x + 1 = 2\cos^2 x$

20. $\sin 2x = 2\sin^2 x\cos x\csc x$

21. $\csc x = 2\cos x\csc 2x$

22. $\cos 2x = \cos^4 x - \sin^4 x$

By counterexamples, show that each equation is not an identity.

23. $\sin\alpha\sec\alpha = 2\tan\alpha$

24. $\dfrac{1 - \cos\alpha}{\sin\alpha} = \dfrac{1 + \sin\alpha}{\cos\alpha}$

4.6 *Use half-angle formulas to find the exact value for each of the following.*

25. $\cos 105°$

26. $\tan 165°$

Verify the identities in Exercises 27 and 28.

27. $\cos^2 \frac{x}{2} = \frac{\sin^2 x}{2(1 - \cos x)}$

28. $\sin^2 \frac{x}{2} = \frac{\csc x - \cot x}{2 \csc x}$

29. Express $\cos x \cos 5x$ in terms of sums of function values.

30. Express $\sin 6x + \sin 4x$ in terms of products of function values.

4.7 *Solve each equation in the interval* $0 \leq x \leq \frac{\pi}{2}$.

31. $2 \sin x - \sqrt{3} = 0$

32. $\csc^2 x = 4$

33. $\sec x - 2 \cos x = 0$

34. $\cot x \csc x = 0$

Solve each equation in the interval $0° \leq x < 360°$.

35. $\cot^2 x - 2 \csc^2 x + 5 = 0$

36. $\sec^2 x - 4 \cos^2 x = 0$

4.8 *Solve each equation where* $0° \leq x < 360°$.

37. $\cos 3x = \frac{1}{\sqrt{2}}$

38. $\sin 4x = \frac{\sqrt{3}}{2}$

39. $\sin 2x - \tan x = 0$

40. $\tan 2x = \tan x$

5 Complex Numbers

The set of real numbers does not contain an element whose square is a negative real number. Thus, any symbol of the form $\sqrt{-b}$, $b \in R$, $b > 0$, does not represent a real number, and equations such as

$$x^2 + 4 = 0 \qquad \text{equivalent to} \qquad x = \pm\sqrt{-4} \tag{1}$$

and

$$x^2 - 2x + 5 = 0 \qquad \text{equivalent to} \qquad x = \frac{2 \pm \sqrt{-16}}{2} \tag{2}$$

have no solutions in this set. In this chapter we consider a set of numbers, called the **set of complex numbers** and designated by C, that does provide solutions to such equations.

5.1 Sums and Products

The elements in the set C that we now consider are represented by various symbols. In this section we consider one form that is frequently used. In the following sections we consider other symbolic forms for these numbers.

Definition 5.1 *For all a, $b \in R$, a* **complex number** *is*

$$z = a + bi, \qquad \textit{where} \quad i^2 = -1.$$

In this chapter we shall restrict the use of the variable z to represent elements in the set C. It is sometimes convenient to refer to a as the **real part** of the complex number and b as the **imaginary part,** with i the **imaginary unit.**

Properties of Equality

Equality in the set of complex numbers is established as follows.

Definition 5.2 *For every pair of complex numbers* $z_1 = a_1 + b_1 i$ *and* $z_2 = a_2 + b_2 i$,

$$z_1 = z_2 \quad \textit{if and only if} \quad a_1 = a_2 \quad \textit{and} \quad b_1 = b_2.$$

Example $8 + 9i = 2x + y^2 i$ if and only if $2x = 8$ and $y^2 = 9$ or, equivalently, $x = 4$ and $y = \pm 3$.

Properties similar to the equality axioms in the set of real numbers R (see Appendix A.2) and the substitution law are valid for the set of complex numbers. These laws follow from the corresponding properties for real numbers. Thus, if $z_1, z_2, z_3 \in C$, then

a. $z_1 = z_1$ — Reflexive law
b. If $z_1 = z_2$, then $z_2 = z_1$ — Symmetric law
c. If $z_1 = z_2$ and $z_2 = z_3$, then $z_1 = z_3$ — Transitive law

Computing Sums and Products

If complex numbers are treated as if i were a variable and $a + bi$ were a binomial in the variable i, then the sum or product of two complex numbers can be found by applying the ordinary rules of the algebra of real numbers. Thus,

$$(2 + 3i) + (5 - 2i) = (2 + 5) + (3 - 2)i$$
$$= 7 + i;$$
$$(2 + 3i)(5 - 2i) = 10 + 11i - 6i^2.$$

However, replacing i^2 with -1, we have

$$(2 + 3i)(5 - 2i) = 10 + 11i - 6(-1)$$
$$= 16 + 11i.$$

Using such an approach for sums and products, we find that

1. C will contain elements $a + 0i$, or a, that can be identified with the real numbers for these operations, and
2. C will contain elements $0 + bi$, or bi, whose squares are negative real numbers.

For example,

$$\begin{aligned}(2) + (3) &= (2 + 0i) + (3 + 0i)\\ &= 5 + 0i = 5,\\ (2)(3) &= (2 + 0i)(3 + 0i)\\ &= 6 + 0i + 0i^2 = 6,\\ (3i)^2 &= 9i^2 = -9.\end{aligned}$$

Because the complex number $a + bi$, where $b = 0$, behaves like the real number a, we shall simply write a for $a + 0i$ as convenient and refer to a or $a + 0i$ as a real number.

If $b \neq 0$, the complex number $a + bi$ is called an **imaginary number** and for the special case where $a = 0$ and $b \neq 0$, the complex number is called a **pure imaginary number.** The relationship of the set C and other sets of numbers is shown in Figure 5.1. For example, 3 and -5 are elements of C that are real; $2 + 5i$ and $5i$ are elements of C that are imaginary numbers. Furthermore, $5i$ is a pure imaginary number.

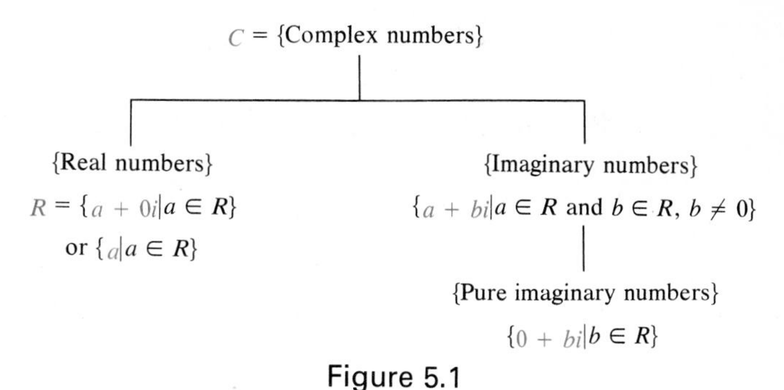

Figure 5.1

Periodic Property of i

Powers of the imaginary unit i have an interesting periodic property. Observe that

$$\begin{aligned}i^1 &= i,\\ i^2 &= -1,\end{aligned}$$

$$i^3 = i^2 \cdot i = -1 \cdot i = -i,$$
$$i^4 = i^2 \cdot i^2 = (-1)(-1) = 1,$$
$$i^5 = i^4 \cdot i = 1 \cdot i = i,$$
$$i^6 = i^4 \cdot i^2 = 1 \cdot (-1) = -1,$$
$$i^7 = i^4 \cdot i^3 = 1 \cdot (-i) = -i,$$

etc.

Thus, any integral power of i can be expressed as one of the numbers i, -1, $-i$, or 1.

Radical Notation

Another notation used to represent a complex number involves the square root symbol. In the set C, we make the following definition.

Definition 5.3 *For all $b \in R$, $b > 0$,*

$$\sqrt{-b} = \sqrt{-1}\sqrt{b} = i\sqrt{b}.$$

Observe that for the special case where $b = 1$,

$$\sqrt{-1} = \sqrt{-1}\sqrt{1} = i \cdot 1 = i.$$

Note that for a and b positive real numbers,

$$\sqrt{-a}\sqrt{-b} = (i\sqrt{a})(i\sqrt{b}) = i^2\sqrt{a}\sqrt{b} = -\sqrt{ab},$$

a negative real number. Hence, the property of real numbers

$$\sqrt{a}\sqrt{b} = \sqrt{ab}, \quad a,\ b \geq 0$$

is not applicable in the set of complex numbers. For example,

$$\sqrt{-4}\sqrt{-9} \neq \sqrt{(-4)(-9)}.$$

In this case,

$$\sqrt{-4}\sqrt{-9} = 2i \cdot 3i = 6i^2 = -6.$$

To avoid difficulty in rewriting products, expressions of the form $\sqrt{-b}$ $(b > 0)$ should first be expressed in the form $\sqrt{b}\,i$ or $i\sqrt{b}$.

Examples

a. $\sqrt{-3}(5 + \sqrt{-3})$
$= i\sqrt{3}(5 + i\sqrt{3})$
$= 5i\sqrt{3} + i^2\sqrt{9})$
$= 5i\sqrt{3} + (-1)(3)$
$= -3 + 5i\sqrt{3}$

b. $(3 - \sqrt{-2})(3 + \sqrt{-2})$
$= (3 - i\sqrt{2})(3 + i\sqrt{2})$
$= 9 - i^2(2)$
$= 9 - (-1)(2)$
$= 11$

EXERCISE SET 5.1

A

Find real numbers x and y for which each statement is true.

Example By Definition 5.2, $x + 2yi = -5 + 8i$ if and only if

$$x = -5 \quad \text{and} \quad 2y = 8,$$

or, equivalently, if

$$x = -5 \quad \text{and} \quad y = 4.$$

1. $-5x + yi = -5 - 3i$

2. $x + 4i = 2yi$

3. $y + 4xi = 8 - 2i$

4. $x - 4i = 2y - x^2i$

Write each expression in the form a + bi.

Example $(3 + 2i) + (1 - 3i) = (3 + 1) + [2 + (-3)]i$
$= 4 - i.$

5. $(4 + 8i) + (-3 + 2i)$

6. $(-6 + 6i) + (4 + 2i)$

7. $(-9 - 5i) + (-1 - 6i)$

8. $(-4 + 3i) + (5 - i)$

9. $(3 + 3i) + (-4 + 3i)$

10. $(-1 + 6i) + (2 - 2i)$

11. $(a + bi) + (-a - bi)$

12. $(a + bi) + (a - bi)$

Example $2(1 - 5i) = 2(1) + 2(-5)i$
$= 2 - 10i.$

13. $4(3 - 3i)$

14. $5(2 + 4i)$

15. $-6(6 + 2i)$

16. $-4(8 - i)$

17. $\frac{1}{2}(4 + 6i)$

18. $\frac{2}{3}(9 - 6i)$

Example $(-3 + 4i)(1 + 6i) = -3 + 4i - 18i + 24i^2$
$= -3 - 14i + 24(-1) = -27 - 14i.$

19. $(-3 + 8i)(1 + 2i)$

20. $(4 + 2i)(-8 - 2i)$

21. $(-6 - i)(-9 + 8i)$

22. $(-3 - 9i)(3 + 4i)$

23. $(5 - 6i)(4 + i)$

24. $(5 + i)(-1 + i)$

25. $(a + bi)(a - bi)$

26. $(a + bi)(a + bi)$

Write each expression as one of the complex numbers, i, −1, −i, or 1.

27. i^5 **28.** i^8 **29.** i^{10} **30.** i^{11}

31. i^{14} **32.** i^{16} **33.** i^{19} **34.** i^{23}

Write each expression in the form bi or a + bi.

Examples

a. $\sqrt{-16} = \sqrt{16}\sqrt{-1}$
$= 4i$

b. $4 + \sqrt{-8} = 4 + \sqrt{4}\sqrt{2}\sqrt{-1}$
$= 4 + 2\sqrt{2}i$

35. $\sqrt{-25}$ **36.** $\sqrt{-49}$ **37.** $\sqrt{-12}$ **38.** $\sqrt{-27}$

39. $1 + \sqrt{-9}$ **40.** $2 - \sqrt{-64}$ **41.** $3 - \sqrt{-20}$ **42.** $5 + \sqrt{-48}$

Examples

a. $\sqrt{-9}(2 - 3\sqrt{-9}) = (\sqrt{9}\sqrt{-1})(2 - 3\sqrt{9}\sqrt{-1})$
$= 3i(2 - 3 \cdot 3i)$
$= 3i(2 - 9i)$
$= 6i - 27i^2 = 27 + 6i$

b. $(1 + \sqrt{-5})(1 - 2\sqrt{-5}) = (1 + \sqrt{5}i)(1 - 2\sqrt{5}i)$
$= 1 + \sqrt{5}i - 2\sqrt{5}i - 2(\sqrt{5})^2 i^2$
$= 1 - \sqrt{5}i - 2 \cdot 5(-1)$
$= 11 - \sqrt{5}i$

43. $\sqrt{-4}(1 - 2\sqrt{-4})$

44. $\sqrt{-8}(3 + 4\sqrt{-2})$

45. $\sqrt{-16}(2 + \sqrt{-4})$

46. $\sqrt{-9}(3 - \sqrt{-25})$

47. $(\sqrt{-50} + 2)\sqrt{-10}$

48. $(\sqrt{-48} - 1)\sqrt{-21}$

49. $(5 + 2\sqrt{-4})(3 - \sqrt{-4})$

50. $(1 - 4\sqrt{-9})(7 + 2\sqrt{-9})$

51. $(2 + \sqrt{-3})(3 - \sqrt{-3})$

52. $(3 - \sqrt{-7})(3 + \sqrt{-7})$

53. Show that $2i$ and $-2i$ satisfy $x^2 + 4 = 0$.

54. Show that $1 + 2i$ and $1 - 2i$ satisfy $x^2 - 2x + 5 = 0$.

B

55. Show that $\left(-\frac{1}{2} + i\frac{\sqrt{3}}{2}\right)^3 = 1$ and $\left(-\frac{1}{2} - i\frac{\sqrt{3}}{2}\right)^3 = 1$.

56. Show that $\left(\frac{1}{2} + i\frac{\sqrt{3}}{2}\right)^3 = -1$ and $\left(\frac{1}{2} - i\frac{\sqrt{3}}{2}\right)^3 = -1$.

Find a and b for which each statement in Exercises 57 and 58 is true.

57. $(a + bi)^2 = -1$. *Hint:* First express the given equation as $a^2 - b^2 + 2abi = -1$ and then use Definition 5.2.

58. $(a + bi)^2 = 1$.

The formula $e^{ix} = \cos x + i \sin x$, *where* $e = 2.718\ldots$, *is called* **Euler's formula** (*it is usually derived in a course in calculus*). *Use this formula to show that the following equations are identities.*

59. $e^{\pi i} = -1$

60. $e^{-ix} = \cos x - i \sin x$

61. $\sin x = \dfrac{e^{ix} - e^{-ix}}{2i}$

62. $\cos x = \dfrac{e^{ix} + e^{-ix}}{2}$

63. $\sin^2 x + \cos^2 x = 1$. *Hint:* Use Exercises 61 and 62.

5.2 Differences and Quotients

The operations of addition and multiplication of complex numbers have been defined so that many of the properties of real numbers are valid in the set of complex numbers.

Two other useful operations, subtraction and division, are defined in terms of addition and multiplication. These definitions are similar to the definitions for a difference and a quotient in the set of real numbers.

Definition 5.4 *For all* $z_1, z_2 \in C$, *the* **difference**

$$z_1 - z_2 = z_1 + (-z_2).$$

Example $(3 + 4i) - (5 + 2i) = (3 + 4i) + (-5 - 2i)$

$$= (3 - 5) + (4 - 2)i = -2 + 2i.$$

Definition 5.5 *For all* $z_1, z_2 \in C$ *where* $z_2 \neq 0$, *the* **quotient**

$$\frac{z_1}{z_2} = z_1\left(\frac{1}{z_2}\right).$$

Since the axioms used for the real numbers together with equality axioms are also valid in C, for each property that follows from these axioms in the set R (see Appendix A.2), there is a corresponding property in the set C. One such property,

If $z_1, z_2, z_3 \in C$, *such that* $z_2, z_3 \neq 0$, *then*

$$\frac{z_1}{z_2} = \frac{z_1 z_3}{z_2 z_3}, \tag{1}$$

is particularly useful in rewriting quotients in the set C. We first make the following definition.

Definition 5.6 *For all* $z = a + bi$, *the* **conjugate** *of* z *is*

$$\bar{z} = a - bi.$$

Examples

a. The conjugate of $5 + 4i$ is $5 - 4i$.

b. The conjugate of $2 - 3i$ is $2 + 3i$.

A quotient of complex numbers, z_1/z_2, can be expressed in the form $a + bi$ by first multiplying both the numerator and the denominator by the conjugate of the denominator. If the denominator is a pure imaginary number, then it is necessary only to multiply the numerator and denominator by i. If the denominator is a real number, then it is necessary only to write the quotient as two terms.

Examples

a. $\dfrac{3}{1 + 2i} = \dfrac{3(1 - 2i)}{(1 + 2i)(1 - 2i)} = \dfrac{3 - 6i}{1 - 4i^2} = \dfrac{3 - 6i}{5} = \dfrac{3}{5} - \dfrac{6}{5}i$

b. $\dfrac{4 - i}{2 - 3i} = \dfrac{(4 - i)(2 + 3i)}{(2 - 3i)(2 + 3i)} = \dfrac{8 + 10i - 3i^2}{4 - 9i^2} = \dfrac{11 + 10i}{13} = \dfrac{11}{13} + \dfrac{10}{13}i$

c. $\dfrac{3 - i}{3i} = \dfrac{(3 - i)i}{3i \cdot i} = \dfrac{3i - i^2}{3i^2} = \dfrac{3i + 1}{-3} = -\dfrac{1}{3} - i$

d. $\dfrac{4 + 5i}{7} = \dfrac{4}{7} + \dfrac{5}{7}i$

In Example a, note that the product of the complex number $1 + 2i$ and its conjugate $1 - 2i$ is the real number 5. In Example b, note that the product of the complex number $2 - 3i$ and its conjugate $2 + 3i$ is the real number 13. In fact, observe that for all $a + bi \in C$,

$$(a + bi)(a - bi) = a^2 - b^2i^2 = a^2 + b^2,$$

a real number.

As we noted in Section 5.1, to avoid difficulty in rewriting products, expressions of the form $\sqrt{-b}$ $(b > 0)$ should first be expressed in the form $\sqrt{b}\,i$ or $i\sqrt{b}$. Hence, quotients of this form should also be rewritten before using Equation (1).

Examples

a. $\dfrac{5}{\sqrt{-4}} = \dfrac{5}{2i} = \dfrac{5 \cdot i}{2i \cdot i} = \dfrac{5i}{2(-1)} = -\dfrac{5i}{2}$

b. $\dfrac{4}{3 - \sqrt{-2}} = \dfrac{4(3 + i\sqrt{2})}{(3 - i\sqrt{2})(3 + i\sqrt{2})} = \dfrac{12 + 4i\sqrt{2}}{9 - i^2(2)} = \dfrac{12 + 4i\sqrt{2}}{9 - (-1)(2)} = \dfrac{12}{11} + \dfrac{4\sqrt{2}}{11}i$

EXERCISE SET 5.2

A

Express each difference in the form $a + bi$.

Example $(1 - 4i) - (-1 + 10i) = (1 - 4i) + (1 - 10i)$
$= (1 + 1) + (-4 - 10)i = 2 - 14i$

1. $(6 + i) - (-7 + 5i)$

$6 + i + (7 - 5i)$
$(6+7)+(-5+1)i$
$13 + -4i$

2. $(-7 - 3i) - (-5 + 5i)$

$(-7 - 3i) + (5 - 5i)$
$(-7+5)+(-3-5)i$
$-2 + -8i$

3. $(3 - i) - (1 - 3i)$ **4.** $(3 + 2i) - (-5 + 8i)$

5. $(5 + 2i) - 3i$ **6.** $4 - (2 - 3i)$

Write the conjugate of each complex number in the form $a + bi$.

Examples a. $z = -3 + 4i$ b. $z = 5 - 2i$

Solutions a. $\bar{z} = -3 - 4i$ b. $\bar{z} = 5 + 2i$

7. $6 + 3i$ **8.** $5 - 5i$ **9.** $-3 - 3i$

10. $-4 + i$ **11.** $3i$ **12.** 7

Express each quotient in the form $a + bi$.

Examples

a. $$\frac{8 - 2i}{8i} = \frac{(8 - 2i)i}{8i \cdot i} = \frac{8i - 2i^2}{8i^2} = \frac{8i + 2}{-8} = -\frac{1}{4} - i$$

b. $$\frac{-4}{-3 + i} = \frac{(-4)(-3 - i)}{(-3 + i)(-3 - i)} = \frac{12 + 4i}{9 - i^2} = \frac{6}{5} + \frac{2}{5}i$$

13. $\dfrac{-9 - 6i}{-4i}$ **14.** $\dfrac{-2 + 3i}{3i}$ **15.** $\dfrac{4}{2 - 9i}$ **16.** $\dfrac{2}{-8 + 5i}$

17. $\dfrac{2 + i}{4 + i}$ **18.** $\dfrac{2 - 5i}{-1 + 6i}$ **19.** $\dfrac{-5 + 9i}{6 - 2i}$ **20.** $\dfrac{3 - 2i}{3 + 2i}$

Examples

a. $$\frac{3}{\sqrt{-2}} = \frac{3(i)}{i\sqrt{2}(i)} = \frac{3i}{i^2\sqrt{2}} = \frac{3i}{(-1)\sqrt{2}} = \frac{-3i}{\sqrt{2}}$$

b. $$\frac{5}{2 + \sqrt{-3}} = \frac{5(2 - i\sqrt{3})}{(2 + i\sqrt{3})(2 - i\sqrt{3})} = \frac{10 - 5i\sqrt{3}}{4 - i^2(3)} = \frac{10 - 5i\sqrt{3}}{4 - (-1)(3)} = \frac{10}{7} - \frac{5\sqrt{3}}{7}i$$

21. $\dfrac{1}{\sqrt{-5}}$ **22.** $\dfrac{2}{\sqrt{-7}}$ **23.** $\dfrac{4}{\sqrt{-8}}$

24. $\dfrac{7}{\sqrt{-12}}$ **25.** $\dfrac{3}{4 + \sqrt{-5}}$ **26.** $\dfrac{7}{5 - \sqrt{-3}}$

27. $\dfrac{6}{\sqrt{-2} + 3}$ **28.** $\dfrac{4}{\sqrt{-7} - 6}$ **29.** $\dfrac{6}{\sqrt{2} + \sqrt{-5}}$

30. $\dfrac{10}{\sqrt{3} - \sqrt{-6}}$ **31.** $\dfrac{2}{\sqrt{5} + \sqrt{-8}}$ **32.** $\dfrac{3}{\sqrt{6} - \sqrt{-12}}$

5.3 Graphical Representation

The elements in $\{a + bi \mid a, b \in R\}$ can be placed in a one-to-one correspondence with the set of points in a plane by associating a with the abscissa of a point and b with the ordinate. In this way each point in a plane can be viewed as the graph of a complex number. A plane on which complex numbers are thus represented is called a **complex plane.** The elements in $\{a + 0i \mid a \in R\}$, or simply $\{a \mid a \in R\}$, are associated with the points on the x-axis, called the **real axis** and the elements in $\{0 + bi \mid b \in R\}$, or simply $\{bi \mid b \in R\}$, are associated with points on the y-axis, called the **imaginary axis.**

Examples Graph each number z, its negative, and its conjugate.

a. $z = 3 - 2i$ b. $z = -2 - 4i$

Solutions

a. The negative of $3 - 2i$ is $-3 + 2i$; the conjugate is $3 + 2i$.

b. The negative of $-2 - 4i$ is $2 + 4i$; the conjugate is $-2 + 4i$.

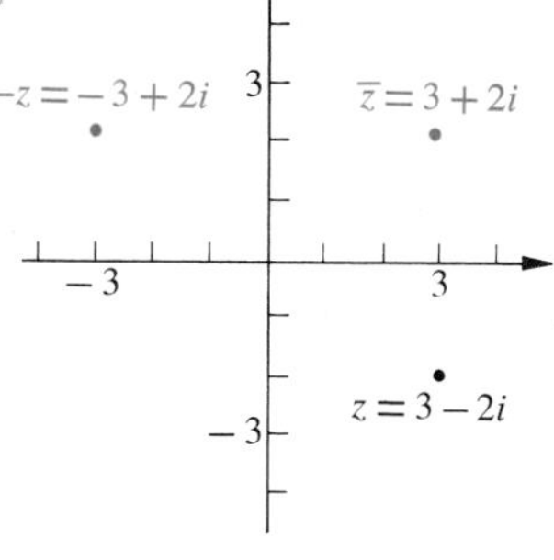

a.

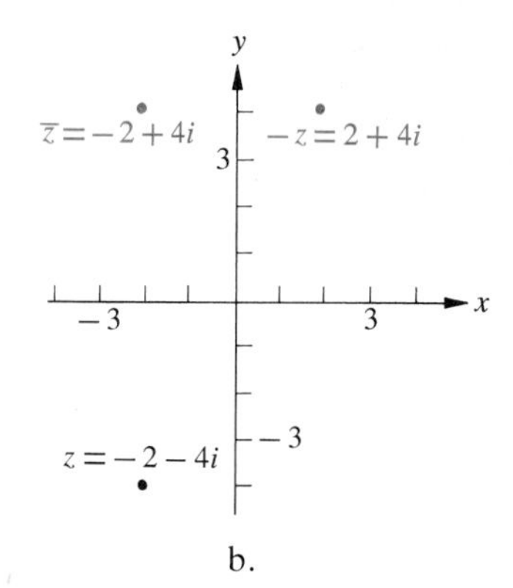

b.

Sometimes the graph of a complex number is shown as a geometric vector with its initial point located at the origin. With this representation, complex numbers can be added graphically by the parallelogram law applicable to geometric vectors.

Examples Represent each sum graphically using geometric vectors and check the results by analytic methods.

a. $(2 + 3i) + (4 - 5i)$ b. $(2 + 2i) + (-4 + 5i)$

Solutions

a.

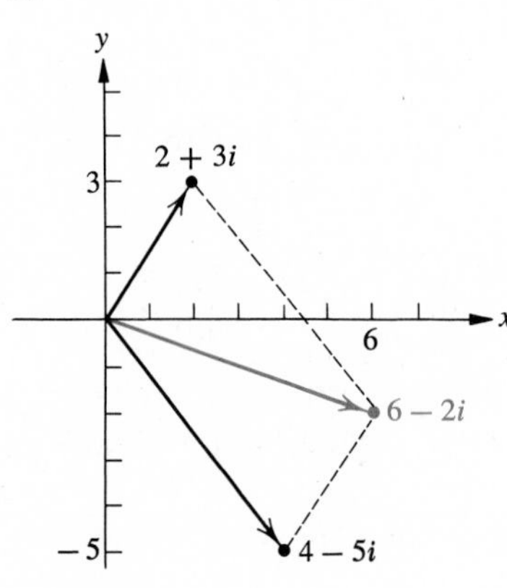

b.

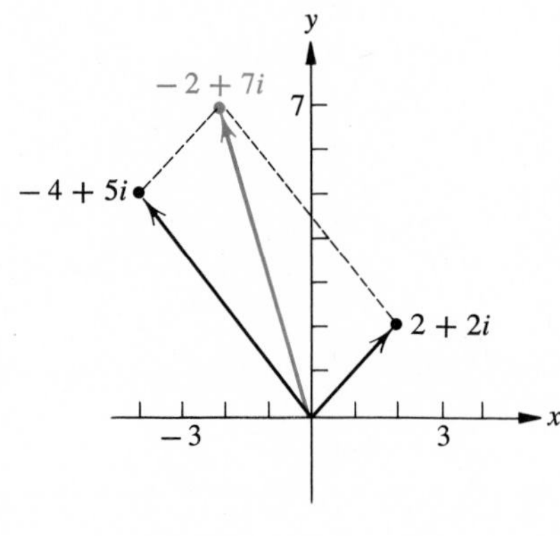

a. Check: $(2 + 3i) + (4 - 5i) = 6 - 2i$

b. Check: $(2 + 2i) + (-4 + 5i) = -2 + 7i$

Complex Numbers as Ordered Pairs

The elements in both the set $C = \{a + bi \mid a, b \in R\}$ and the elements in the set $\{(a, b) \mid a, b \in R\}$ have now been associated with the points in a plane. Hence, elements in C can also be associated with, and represented by, ordered pairs of real numbers (a, b).

Examples

a. $2 + 3i = (2, 3)$

b. $-5 = -5 + 0i = (-5, 0)$

c. $6i = 0 + 6i = (0, 6)$

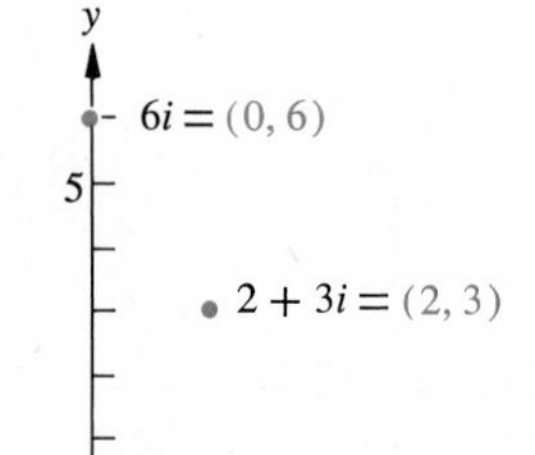

Absolute Value and Modulus

Corresponding to the norm $\|\vec{\mathbf{v}}\|$ and the direction angle α for each geometric vector $\vec{\mathbf{v}}$, there are a real number and a *set* of angles associated with each complex number.

Definition 5.7 *For all* $z \in C$, *the* **absolute value** *or* **modulus** *of* z, *where* $z = a + bi = (a, b)$, *is*

$$\rho = |z| = |a + bi|$$
$$= |(a, b)| = \sqrt{a^2 + b^2}.$$

The modulus ρ (the Greek letter "rho") is the distance from the origin to the graph of (a, b), as shown in Figure 5.2.

Definition 5.8 *For all* $z \in C$, *an* **argument** *of* $z = a + bi = (a, b)$, *denoted by* $\arg(a + bi)$ *or* $\arg(a, b)$, *is an angle* θ *(theta) such that*

$$\cos \theta = \frac{a}{\sqrt{a^2 + b^2}} = \frac{a}{\rho}$$

and

$$\sin \theta = \frac{b}{\sqrt{a^2 + b^2}} = \frac{b}{\rho} \quad (a^2 + b^2 \neq 0).$$

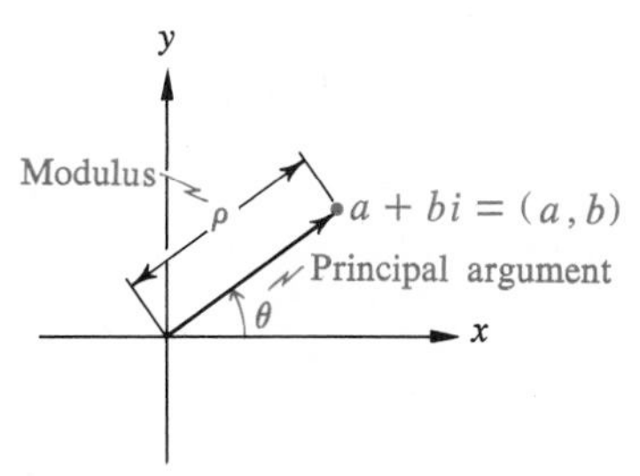

Figure 5.2

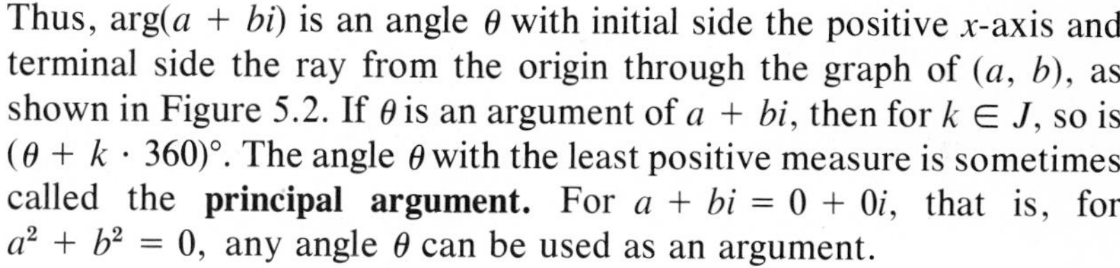

Thus, $\arg(a + bi)$ is an angle θ with initial side the positive x-axis and terminal side the ray from the origin through the graph of (a, b), as shown in Figure 5.2. If θ is an argument of $a + bi$, then for $k \in J$, so is $(\theta + k \cdot 360)°$. The angle θ with the least positive measure is sometimes called the **principal argument.** For $a + bi = 0 + 0i$, that is, for $a^2 + b^2 = 0$, any angle θ can be used as an argument.

Example Find the absolute value and principal argument of $\sqrt{3} + i$.

Solution $\rho = |\sqrt{3} + i| = \sqrt{(\sqrt{3})^2 + 1^2} = \sqrt{3 + 1} = 2$;

$$\cos \theta = \frac{a}{\rho} = \frac{\sqrt{3}}{2} \quad \text{and} \quad \sin \theta = \frac{b}{\rho} = \frac{1}{2}.$$

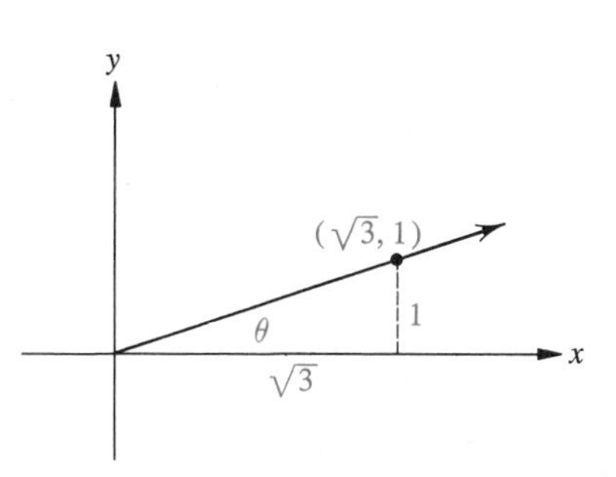

Because both $\cos \theta$ and $\sin \theta$ are positive, θ is in Quadrant I, and hence $\theta = 30°$.

There is another method that is often used to determine θ. Consider the number $a + bi = \sqrt{3} + i$ from the example above. First plot the point $(a, b) = (\sqrt{3}, 1)$ and form angle θ. From the sketch, θ is in Quadrant I, and $\tan \theta = 1/\sqrt{3}$. Hence $\theta = 30°$.

EXERCISE SET 5.3

A

Express each complex number as an ordered pair.

1. $-1 + 6i$ **2.** $5 - 9i$ **3.** $8 + i$

4. $-2 + 7i$ **5.** $6i$ **6.** -5

Express each complex number in the form $a + bi$.

7. $(-8, 7)$ **8.** $(0, 5)$ **9.** $(-4, 0)$

10. $(6, -3)$ **11.** $(7, -\sqrt{2})$ **12.** $(-\sqrt{3}, 1)$

Graph each complex number z, its negative $-z$, and its conjugate $\bar{z}$.

Examples a. $-5 - 4i$ b. $(3, 2)$

Solutions

a. The negative is $5 + 4i$; the conjugate is $-5 + 4i$.

b. The negative is $(-3, -2)$; the conjugate is $(3, -2)$.

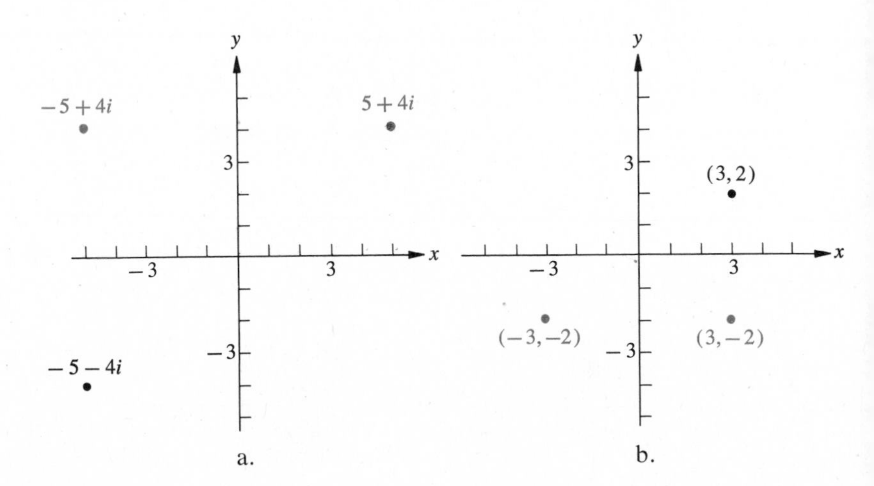

a. b.

13. $2 - 3i$ **14.** $-6 + 6i$ **15.** 5 **16.** $2i$

17. $(4, 6)$ **18.** $(-5, -3)$ **19.** $(-1, 0)$ **20.** $(0, 3)$

Represent each sum graphically. Check the results by algebraic methods.

Example $(8 + 3i) + (1 - 5i)$

Solution

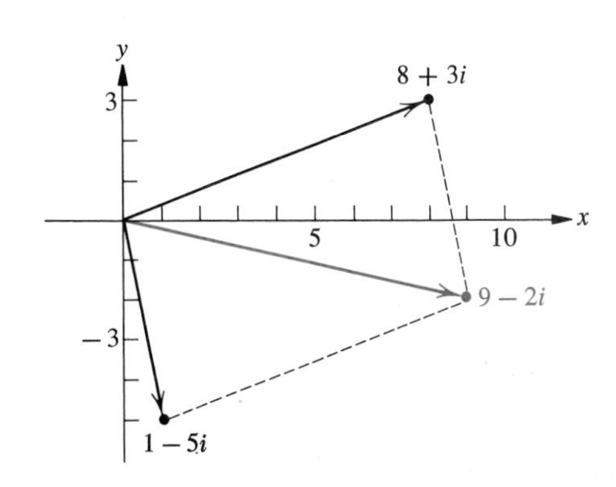

Check: $(8 + 3i) + (1 - 5i) = (8 + 1) + (3 - 5)i = 9 - 2i.$

21. $2 + (-3 + 6i)$

22. $(7 - i) + 2i$

23. $(4 - 7i) + (-8 + 2i)$

24. $(2 + 4i) + (1 - 5i)$

25. $(-6 + 7i) + (5 - 3i)$

26. $(-3 - 5i) + (3 - 2i)$

Find the absolute value and principal argument of each complex number.

Examples a. $-7 + 5i$ b. $(3, 9)$

Solutions From Definitions 5.7 and 5.8,

a. $\rho = |-7 + 5i| = \sqrt{(-7)^2 + 5^2} = \sqrt{74}$;

$$\cos \theta = -\frac{7}{\sqrt{74}} \approx -0.8137 \quad \text{and} \quad \sin \theta = \frac{5}{\sqrt{74}} \approx 0.5812.$$

Because $\cos \theta$ is negative and $\sin \theta$ is positive, θ is in Quadrant II; hence, $\theta \approx 144.5°$.

b. $\rho = |(3, 9)| = \sqrt{3^2 + 9^2} = \sqrt{90} = 3\sqrt{10}$;

$$\cos \theta = \frac{3}{3\sqrt{10}} \approx 0.3162 \quad \text{and} \quad \sin \theta = \frac{9}{3\sqrt{10}} \approx 0.9487.$$

(continued)

Because cos θ and sin θ are both positive, θ is in Quadrant I; hence, $\theta \approx 71.6°$.

27. $5 + 2i$	**28.** $6 + 8i$	**29.** $-2 - 2i$	**30.** $8 - 3i$
31. 7	**32.** -4	**33.** $-5i$	**34.** $6i$
35. $(3, -9)$	**36.** $(-6, -2)$	**37.** $(4, 0)$	**38.** $(0, -7)$

39. Determine the conditions that a and b must satisfy if the graph of the complex number $a + bi$ is
a. on the real axis. **b.** on the imaginary axis.
c. above the real axis. **d.** below the real axis.

40. What condition does $a^2 + b^2$ satisfy if the graph of $a + bi$ is
a. on a circle with radius of length 5.
b. inside a circle with radius of length 5.
c. outside a circle with radius of length 5.

B

Graphically, the multiplication or division of any complex number $a + bi$ by the complex number i produces a 90° rotation of the geometric vector corresponding to $a + bi$, counterclockwise for multiplication and clockwise for division. Graph **(a)** *the given complex number $a + bi$,* **(b)** *$i \cdot (a + bi)$, and* **(c)** *$\frac{a + bi}{i}$.*

Example $2 + 3i$

Solution

a.

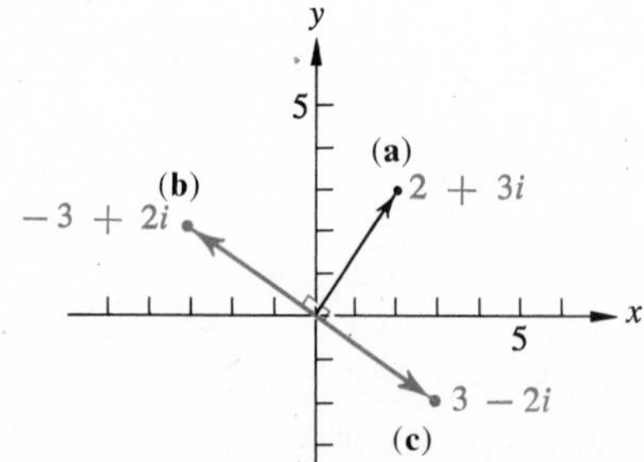

b. $i \cdot (2 + 3i) = 2i + 3i^2 = -3 + 2i$

c. $$\frac{2 + 3i}{i} = \frac{(2 + 3i) \cdot i}{i \cdot i} = \frac{2i + 3i^2}{i^2}$$

$$= \frac{2i - 3}{-1} = 3 - 2i$$

41. $5 + 4i$ **42.** $1 - 2i$ **43.** $-6 + i$

44. $-3 - 4i$ **45.** 2 **46.** $-5i$

5.4 Trigonometric Form

In previous sections we have represented complex numbers in the form $a + bi$ and (a, b). It is sometimes convenient to represent a complex number in terms of its absolute value $\rho = \sqrt{a^2 + b^2}$ and its argument θ. From Definition 5.8,

$$\cos \theta = \frac{a}{\rho} \quad \text{and} \quad \sin \theta = \frac{b}{\rho}.$$

Therefore, as shown in Figure 5.3,

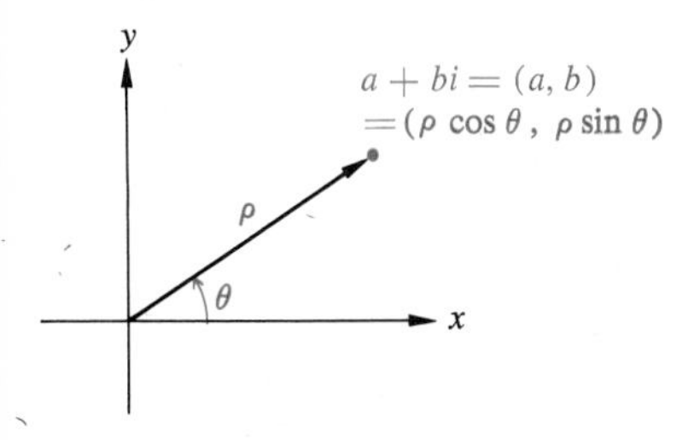

Figure 5.3

$$a = \rho \cos \theta \quad \text{and} \quad b = \rho \sin \theta,$$

from which we have

$$a + bi = \rho \cos \theta + (\rho \sin \theta)i,$$

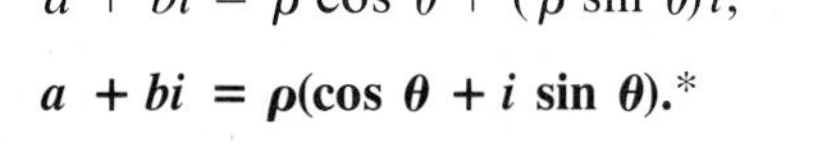

$$\mathbf{a + bi = \rho(\cos \theta + i \sin \theta).*}$$

The sine and cosine functions are periodic with period of 360°. Hence, it is also true that

$$\boldsymbol{\alpha + bi = \rho[\cos(\theta + k \cdot 360°) + i \sin(\theta + k \cdot 360°)], \quad k \in J,}$$

where θ is the angle of smallest nonnegative measure. Either equation is called the **trigonometric form** of a complex number.

Example Represent $-2 + 2i$ in trigonometric form.

Solution From Definition 5.7,

$$\rho = |-2 + 2i| = \sqrt{(-2)^2 + 2^2} = \sqrt{8} = 2\sqrt{2}.$$

From Definition 5.8,

$$\cos \theta = \frac{-2}{2\sqrt{2}} = -\frac{1}{\sqrt{2}} \quad \text{and} \quad \sin \theta = \frac{2}{2\sqrt{2}} = \frac{1}{\sqrt{2}}.$$

(continued)

* The expression $\cos \theta + i \sin \theta$ is sometimes abbreviated as cis θ.

Thus, θ is in Quadrant II and $\theta = 135°$. For $k \in J$,

$$-2 + 2i = 2\sqrt{2}[\cos(135° + k \cdot 360°) + i \sin(135° + k \cdot 360°)].$$

For $k = 0$,

$$-2 + 2i = 2\sqrt{2}(\cos 135° + i \sin 135°).$$

Example Represent $3(\cos 60° + i \sin 60°)$ graphically and express the number in the form $a + bi$.

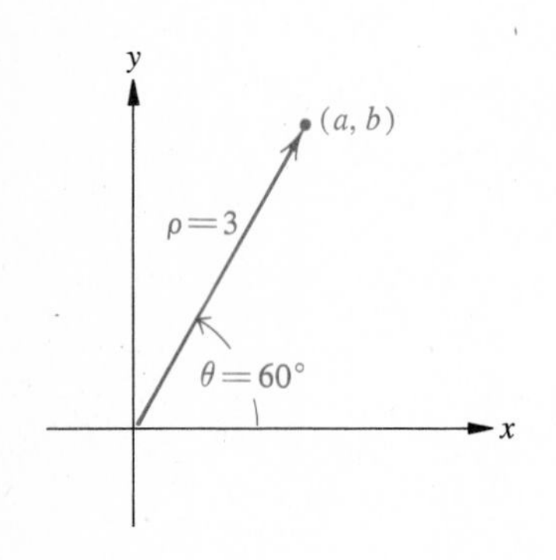

Solution Since

$$a = \rho \cos \theta = 3 \cos 60° = 3\left(\frac{1}{2}\right) = \frac{3}{2},$$

and

$$b = \rho \sin \theta = 3 \sin 60° = 3\left(\frac{\sqrt{3}}{2}\right) = \frac{3\sqrt{3}}{2},$$

we have

$$3(\cos 60° + i \sin 60°) = \frac{3}{2} + \frac{3\sqrt{3}}{2} i.$$

Products and Quotients

Products and quotients of complex numbers expressed in trigonometric form can be found quite readily. If

$$z_1 = \rho_1(\cos \theta_1 + i \sin \theta_1) \quad \text{and} \quad z_2 = \rho_2(\cos \theta_2 + i \sin \theta_2),$$

then

$$\begin{aligned} z_1 \cdot z_2 &= \rho_1(\cos \theta_1 + i \sin \theta_1) \cdot \rho_2(\cos \theta_2 + i \sin \theta_2) \\ &= \rho_1\rho_2[(\cos \theta_1 \cos \theta_2 - \sin \theta_1 \sin \theta_2) \\ &\qquad + i(\cos \theta_1 \sin \theta_2 + \sin \theta_1 \cos \theta_2)]. \end{aligned}$$

Substituting $\cos(\theta_1 + \theta_2)$ for $\cos \theta_1 \cos \theta_2 - \sin \theta_1 \sin \theta_2$ and then $\sin(\theta_1 + \theta_2)$ for $\cos \theta_1 \sin \theta_2 + \sin \theta_1 \cos \theta_2$, we have

$$\boldsymbol{z_1 \cdot z_2 = \rho_1\rho_2[\cos(\theta_1 + \theta_2) + i \sin(\theta_1 + \theta_2)].} \tag{1}$$

It can also be shown in a similar way that

$$\boldsymbol{\frac{z_1}{z_2} = \frac{\rho_1}{\rho_2}[\cos(\theta_1 - \theta_2) + i \sin(\theta_1 - \theta_2)], \quad z_2 \neq (0, 0).} \tag{2}$$

Example Express $6(\cos 75° + i \sin 75°) \cdot 3(\cos 15° + i \sin 15°)$ in the form $a + bi$.

Solution From Equation (1),

$$\begin{aligned} 6(\cos 75° + i \sin 75°) \cdot 3(\cos 15° + i \sin 15°) &= 6 \cdot 3[\cos(75° + 15°) + i \sin(75° + 15°)] \\ &= 18(\cos 90° + i \sin 90°) \\ &= 18(0 + i) = 18i. \end{aligned}$$

Example Express $\dfrac{6(\cos 75° + i \sin 75°)}{3(\cos 15° + i \sin 15°)}$ in the form $a + bi$.

Solution From Equation (2),

$$\begin{aligned} \frac{6(\cos 75° + i \sin 75°)}{3(\cos 15° + i \sin 15°)} &= \frac{6}{3}[\cos(75° - 15°) + i \sin(75° - 15°)] \\ &= 2(\cos 60° + i \sin 60°) \\ &= 2\left(\frac{1}{2} + i\frac{\sqrt{3}}{2}\right) = 1 + \sqrt{3}i. \end{aligned}$$

EXERCISE SET 5.4

A

Represent each complex number in trigonometric form, using the angle with least positive measure.

Example $\sqrt{3} - i$

Solution From Definition 5.7, $\rho = |\sqrt{3} - i| = \sqrt{3 + 1} = 2$. From Definition 5.8, $\cos \theta = \sqrt{3}/2$ and $\sin \theta = -1/2$. Hence θ is in Quadrant IV and $\theta = 330°$. Therefore,

$$\sqrt{3} - i = 2[\cos(330° + k \cdot 360°) + i \sin(330° + k \cdot 360°)], \quad k \in J.$$

For $k = 0$, we obtain

$$\sqrt{3} - i = 2(\cos 330° + i \sin 330°).$$

1. $-2 + 2\sqrt{3}i$ **2.** $6 + 6i$ **3.** $-5\sqrt{3} - 5i$ **4.** $3\sqrt{3} + 3i$

5. $2 - 2i$ **6.** $-4 - 4i$ **7.** -8 **8.** $-9i$

Represent each complex number graphically and write in the form $a + bi$.

Example $6(\cos 315° + i \sin 315°)$

Solution Since

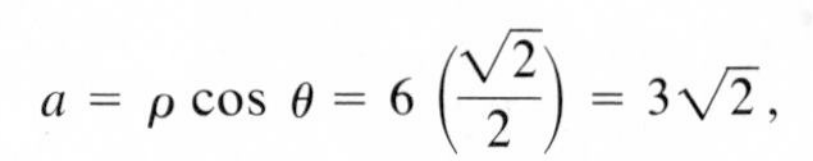

$$a = \rho \cos \theta = 6\left(\frac{\sqrt{2}}{2}\right) = 3\sqrt{2},$$

and

$$b = \rho \sin \theta = 6\left(\frac{-\sqrt{2}}{2}\right) = -3\sqrt{2},$$

we have

$$a + bi = 3\sqrt{2} - 3\sqrt{2}\, i.$$

9. $2(\cos 240° + i \sin 240°)$

10. $2(\cos 90° + i \sin 90°)$

11. $2(\cos 135° + i \sin 135°)$

12. $3(\cos 30° + i \sin 30°)$

13. $\cos(-45°) + i \sin(-45°)$

14. $\cos(-210°) + i \sin(-210°)$

15. $\frac{1}{2}(\cos 150° + i \sin 150°)$

16. $\frac{3}{4}(\cos 330° + i \sin 330°)$

Express each product in the form $a + bi$.

Example $2(\cos 140° + i \sin 140°) \cdot 5(\cos 70° + i \sin 70°)$

Solution From Equation (1) in this section,

$$\begin{aligned} 2(\cos 140° + i \sin 140°) &\cdot 5(\cos 70° + i \sin 70°) \\ &= 2 \cdot 5[\cos(140° + 70°) + i \sin(140° + 70°)] \\ &= 10(\cos 210° + i \sin 210°) \\ &= 10\left[\frac{-\sqrt{3}}{2} + i\left(-\frac{1}{2}\right)\right] = -5\sqrt{3} - 5i. \end{aligned}$$

17. $5(\cos 100° + i \sin 100°) \cdot 2(\cos 35° + i \sin 35°)$

18. $3(\cos 130° + i \sin 130°) \cdot \frac{2}{3}(\cos 140° + i \sin 140°)$

19. $4(\cos 85° + i \sin 85°) \cdot \frac{3}{4}(\cos 245° + i \sin 245°)$

20. $4(\cos 20° + i \sin 20°) \cdot \frac{1}{2}(\cos 40° + i \sin 40°)$

21. $2(\cos 40° + i \sin 40°) \cdot (\cos 95° + i \sin 95°)$

22. $2(\cos 5° + i \sin 5°) \cdot 3(\cos 40° + i \sin 40°)$

Express each quotient in the form $a + bi$.

Example $\dfrac{6(\cos 85^\circ + i \sin 85^\circ)}{2(\cos 40^\circ + i \sin 40^\circ)}$

Solution From Equation (2) in this section,

$$\begin{aligned}\frac{6(\cos 85^\circ + i \sin 85^\circ)}{2(\cos 40^\circ + i \sin 40^\circ)} &= \frac{6}{2}[\cos(85^\circ - 40^\circ) + i \sin(85^\circ - 40^\circ)] \\ &= 3(\cos 45^\circ + i \sin 45^\circ) \\ &= \frac{3}{\sqrt{2}} + \frac{3}{\sqrt{2}} i.\end{aligned}$$

23. $\dfrac{8(\cos 260^\circ + i \sin 260^\circ)}{2(\cos 80^\circ + i \sin 80^\circ)}$

24. $\dfrac{6(\cos 215^\circ + i \sin 215^\circ)}{3(\cos 80^\circ + i \sin 80^\circ)}$

25. $\dfrac{2(\cos 255^\circ + i \sin 255^\circ)}{5(\cos 15^\circ + i \sin 15^\circ)}$

26. $\dfrac{2(\cos 280^\circ + i \sin 280^\circ)}{3(\cos 70^\circ + i \sin 70^\circ)}$

27. $\dfrac{3(\cos 70^\circ + i \sin 70^\circ)}{\cos 130^\circ + i \sin 130^\circ}$

28. $\dfrac{\cos 19^\circ + i \sin 19^\circ}{4(\cos 259^\circ + i \sin 259^\circ)}$

29. Write $[2(\cos 20^\circ + i \sin 20^\circ)]^3$ in the form $a + bi$.

30. Write $\left(\dfrac{1}{\sqrt{2}} + \dfrac{1}{\sqrt{2}} i\right)^3$ in the form $a + bi$.

31. Show that if

$$a + bi = \rho(\cos \theta + i \sin \theta),$$

then

$$(a + bi)^2 = \rho^2(\cos 2\theta + i \sin 2\theta).$$

32. Use the result of Exercise 31 to show that if

$$a + bi = \rho(\cos \theta + i \sin \theta),$$

then

$$(a + bi)^3 = \rho^3(\cos 3\theta + i \sin 3\theta).$$

33. Use the result of Exercise 32 to show that if

$$a + bi = \rho(\cos \theta + i \sin \theta),$$

then

$$(a + bi)^4 = \rho^4(\cos 4\theta + i \sin 4\theta).$$

34. Using the results of Exercises 31–33, what can you conjecture about $(a + bi)^n$, $n \in N$, if $a + bi = \rho(\cos\theta + i\sin\theta)$?

35. Show that $\cos\theta + i\sin\theta$ and $\cos(-\theta) + i\sin(-\theta)$ are conjugates of each other.

36. Show that $\cos\theta + i\sin\theta$ and $\cos(-\theta) + i\sin(-\theta)$ are reciprocals of each other.

5.5 De Moivre's Theorem; Powers and Roots

Consider the complex number $z = a + bi = \rho(\cos\theta + i\sin\theta)$. From Equation (1) in Section 5.4,

$$\begin{aligned}[\rho(\cos\theta + i\sin\theta)]^2 &= \rho(\cos\theta + i\sin\theta)\cdot\rho(\cos\theta + i\sin\theta)\\ &= \rho^2[\cos(\theta+\theta) + i\sin(\theta+\theta)]\\ &= \rho^2(\cos 2\theta + i\sin 2\theta);\end{aligned}$$

$$\begin{aligned}[\rho(\cos\theta + i\sin\theta)]^3 &= \rho^2(\cos 2\theta + i\sin 2\theta)\cdot\rho(\cos\theta + i\sin\theta)\\ &= \rho^3[\cos(2\theta+\theta) + i\sin(2\theta+\theta)]\\ &= \rho^3(\cos 3\theta + i\sin 3\theta).\end{aligned}$$

It can be shown that similar results are valid for each such power for $n \in N$. That is, if $z = \rho(\cos\theta + i\sin\theta)$ and $n \in N$, then

$$\boldsymbol{z^n = \rho^n(\cos n\theta + i\sin n\theta).} \qquad (1)$$

This statement is known as **De Moivre's theorem.** The proof of the statement and extensions of it which are presented in this section involve the process of mathematical induction and will not be shown. However, the informal argument above should make the validity of Equation (1) plausible.

Example Express $(\sqrt{3} + i)^5$ in the form $a + bi$.

Solution From Definition 5.7,

$$\rho = |\sqrt{3} + i| = \sqrt{3 + 1} = 2.$$

By Definition 5.8, $\cos\theta = \sqrt{3}/2$ and $\sin\theta = 1/2$. Therefore θ is in Quadrant I and $\theta = 30°$. Thus,

$$
\begin{aligned}
(\sqrt{3} + i)^5 &= [2(\cos 30° + i \sin 30°)]^5 \\
&= 2^5(\cos 5 \cdot 30° + i \sin 5 \cdot 30°) \\
&= 32(\cos 150° + i \sin 150°) \\
&= 32\left(-\frac{\sqrt{3}}{2} + \frac{1}{2} i\right) = -16\sqrt{3} + 16i.
\end{aligned}
$$

Equation (1) was stated for powers with natural number exponents. If this statement is to be valid for integral exponents, consistent meanings must be assigned to z^0 and z^{-n} for $n \in N$.

Substituting 0 for n in Equation (1), we have

$$
\begin{aligned}
z^0 = \rho^0(\cos 0 \cdot \theta + i \sin 0 \cdot \theta) &= 1 \cdot (\cos 0 + i \sin 0) \\
&= 1 \cdot (1 + 0) = 1.
\end{aligned}
$$

Substituting $-n$ for n in Equation (1), we have

$$z^{-n} = \rho^{-n}[\cos(-n\theta) + i \sin(-n\theta)].$$

Since $\cos(-n\theta) = \cos n\theta$ and $\sin(-n\theta) = -\sin n\theta$, we have

$$z^{-n} = \frac{1}{\rho^n} \cdot (\cos n\theta - i \sin n\theta).$$

Multiplying the numerator and the denominator of the right-hand member by the conjugate of the numerator yields

$$
\begin{aligned}
z^{-n} &= \frac{1}{\rho^n} \cdot \frac{(\cos n\theta - i \sin n\theta) \cdot (\cos n\theta + i \sin n\theta)}{\cos n\theta + i \sin n\theta} \\
&= \frac{1}{\rho^n} \cdot \frac{\cos^2 n\theta + \sin^2 n\theta}{\cos n\theta + i \sin n\theta} \\
&= \frac{1}{\rho^n} \cdot \frac{1}{\cos n\theta + i \sin n\theta} = \frac{1}{\rho^n(\cos n\theta + i \sin n\theta)}.
\end{aligned}
$$

Substituting z^n for $\rho^n(\cos n\theta + i \sin n\theta)$ yields

$$z^{-n} = \frac{1}{z^n}.$$

Thus, if we want Equation (1) to be valid for every exponent $n \in J$, z^0 and z^{-n} must be given the following interpretations.

Definition 5.9 *For all $z \neq 0 + 0i$,*

$$\text{I.} \qquad z^0 = 1 + 0i = 1.$$

$$\text{II.} \qquad z^{-n} = \frac{1}{z^n}, \quad n \in N.$$

De Moivre's theorem [Equation (1)] can now be applied to powers with $n \in J$.

Example Express $(-\sqrt{3} + i)^{-3}$ in the form $a + bi$.

Solution From Definitions 5.7 and 5.8,

$$\rho = |-\sqrt{3} + i| = \sqrt{3 + 1} = 2;$$

since

$$\cos \theta = -\frac{\sqrt{3}}{2} \quad \text{and} \quad \sin \theta = \frac{1}{2},$$

θ is in Quadrant II and $\theta = 150°$. Thus,

$$(-\sqrt{3} + i)^{-3} = [2(\cos 150° + i \sin 150°)]^{-3},$$

and from De Moivre's theorem we have

$$\begin{aligned}(-\sqrt{3} + i)^{-3} &= \frac{1}{2^3}[\cos(-3 \cdot 150°) + i \sin(-3 \cdot 150°)] \\ &= \frac{1}{8}[\cos(-450°) + i \sin(-450°)] \\ &= \frac{1}{8}[0 - 1 \cdot i] = -\frac{1}{8}i.\end{aligned}$$

Another extension of De Moivre's theorem is possible for rational number exponents $1/n$, where $n \in N$. First we define an nth root of z in the same way that an nth root of a real number a is defined, when such a number exists.

Definition 5.10 *For $z \in C$, $n \in N$, the number ω is an nth root of z if $\omega^n = z$.*

We can now state another extension of De Moivre's theorem applicable to roots of complex numbers:

If $n \in N$ and $z = \rho(\cos \theta + i \sin \theta)$, $z \neq 0$, then

$$\rho^{1/n}\left[\cos\left(\frac{\theta + k\cdot 360^\circ}{n}\right) + i\sin\left(\frac{\theta + k\cdot 360^\circ}{n}\right)\right], \quad k \in J$$

specifies all nth roots of z.

Using this extension of De Moivre's theorem, we can now find n distinct complex nth roots for each $z \in C$, where $z \neq 0$. For $k = 0$, one nth root is

$$\rho^{1/n}\left(\cos\frac{\theta}{n} + i\sin\frac{\theta}{n}\right),$$

where θ is the angle with the least positive measure. This root is called the **principal nth root** of z.

Example Express each of the five fifth roots of $z = \sqrt{2} + \sqrt{2}i$ in trigonometric form.

Solution From Definition 5.7,

$$\rho = |\sqrt{2} + \sqrt{2}i| = \sqrt{2 + 2} = 2;$$

since, from Definition 5.8,

$$\cos\theta = \frac{\sqrt{2}}{2} \quad \text{and} \quad \sin\theta = \frac{\sqrt{2}}{2},$$

the argument θ is in Quadrant I and $\theta = 45^\circ$. Thus, in trigonometric form, for $k \in J$,

$$z = 2[\cos(45^\circ + k\cdot 360^\circ) + i\sin(45^\circ + k\cdot 360^\circ)].$$

Because the sine and cosine functions are periodic, each number

$$2^{1/5}\left[\cos\left(\frac{45^\circ + k\cdot 360^\circ}{5}\right) + i\sin\left(\frac{45^\circ + k\cdot 360^\circ}{5}\right)\right], \quad k \in J,$$

is a fifth root of z. Taking $k = 0, 1, 2, 3,$ and 4 in turn yields the roots

$$2^{1/5}(\cos 9^\circ + i\sin 9^\circ),$$

$$2^{1/5}(\cos 81^\circ + i\sin 81^\circ),$$

$$2^{1/5}(\cos 153^\circ + i\sin 153^\circ),$$

$$2^{1/5}(\cos 225^\circ + i\sin 225^\circ),$$

$$2^{1/5}(\cos 297^\circ + i\sin 297^\circ).$$

(continued)

Notice that the substitution of any other integer for k will simply produce one of these five complex numbers. For example, if $k = 5$,

$$2^{1/5}[\cos(9 + 360)^\circ + i \sin(9 + 360)^\circ] = 2^{1/5}(\cos 9^\circ + i \sin 9^\circ).$$

EXERCISE SET 5.5

A

Express each power in the form $a + bi$.

Example $(\sqrt{3} + i)^{-2}$

Solution From Definition 5.7.

$$\rho = |\sqrt{3} + i| = \sqrt{(\sqrt{3})^2 + (1)^2} = \sqrt{4} = 2;$$

since, from Definition 5.8,

$$\cos \theta = \frac{\sqrt{3}}{2} \qquad \text{and} \qquad \sin \theta = \frac{1}{2},$$

the argument θ is in Quadrant I and $\theta = 30^\circ$. Thus,

$$(\sqrt{3} + i)^{-2} = [2(\cos 30^\circ + i \sin 30^\circ)]^{-2}.$$

From De Moivre's theorem,

$$(\sqrt{3} + i)^{-2} = \frac{1}{2^2}[\cos(-2 \cdot 30^\circ) + i \sin(-2 \cdot 30^\circ)]$$

$$= \frac{1}{4}[\cos(-60^\circ) + i \sin(-60^\circ)]$$

$$= \frac{1}{4}\left(\frac{1}{2} - \frac{\sqrt{3}}{2} i\right) = \frac{1}{8} - \frac{\sqrt{3}}{8} i.$$

1. $[3(\cos 20^\circ + i \sin 20^\circ)]^3$

2. $[2(\cos 42^\circ + i \sin 42^\circ)]^5$

3. $(-1 + i)^6$

4. $(2\sqrt{3} - 2i)^5$

5. $\left(\frac{1}{4} + \frac{\sqrt{3}}{4} i\right)^3$

6. $\left(-\frac{\sqrt{3}}{2} + \frac{1}{2} i\right)^6$

7. $[9(\cos 315^\circ + i \sin 315^\circ)]^{-2}$

8. $\{8[\cos(120^\circ) + i \sin(120^\circ)]\}^{-3}$

9. $(2 - 2i)^{-3}$

10. $(\sqrt{3} - i)^{-3}$

11. $(-1 - \sqrt{3}i)^{-5}$

12. $(-2 - 2i)^{-4}$

In Exercises 13–18, find the roots indicated and express each of them in trigonometric form.

Example · The two square roots of $\dfrac{\sqrt{3}}{2} - \dfrac{1}{2}i$

Solution

$$\rho = \left|\frac{\sqrt{3}}{2} - \frac{1}{2}i\right| = \sqrt{\frac{3}{4} + \frac{1}{4}} = 1.$$

Thus, $\cos\theta = \sqrt{3}/2$ and $\sin\theta = -1/2$, and θ is in Quadrant IV. Hence $\theta = 330°$ and

$$\frac{\sqrt{3}}{2} - \frac{1}{2}i = \cos(330° + k \cdot 360°) + i\sin(330° + k \cdot 360°).$$

The square roots of $\dfrac{\sqrt{3}}{2} - \dfrac{1}{2}i$, using De Moivre's theorem, are

$$1^{1/2}\left[\cos\left(\frac{330° + k \cdot 360°}{2}\right) + i\sin\left(\frac{330° + k \cdot 360°}{2}\right)\right],$$

where $k = 0$ and $k = 1$. For $k = 0$, we have

$$\cos 165° + i\sin 165°,$$

and, for $k = 1$, we have

$$\cos 345° + i\sin 345°.$$

13. The two square roots of

a. $\cos 80 + i\sin 80°$ **b.** $\dfrac{1}{2} + \dfrac{\sqrt{3}}{2}i$

14. The two square roots of

a. $9(\cos 140° + i\sin 140°)$ **b.** $-4i$

15. The three cube roots of

a. $8(\cos 120° + i\sin 120°)$ **b.** $\dfrac{1}{2} + \dfrac{\sqrt{3}}{2}i$

16. The three cube roots of

a. $2(\cos 75° + i\sin 75°)$ **b.** $-27i$

17. The four fourth roots of

a. $\cos 240° + i \sin 240°$ **b.** $\frac{1}{2} - \frac{\sqrt{3}}{2} i$

18. The four fourth roots of

a. $81(\cos 60° + i \sin 60°)$ **b.** $-16i$

Solve each equation for all $x \in C$.

19. $x^3 + 1 = 0$. *Hint:* $x^3 = -1 = -1 + 0i$.
Thus, x is a cube root of $-1 + 0i$.

20. $x^5 + 1 = 0$ **21.** $x^3 - 1 = 0$

22. $x^5 - 1 = 0$ **23.** $x^3 = \frac{\sqrt{3}}{4} - \frac{1}{4} i$

24. $x^5 = 12 - 5i$

B

25. Graph each of the following on a separate Cartesian coordinate system; make a conjecture about the graphs of all nth roots of 1.

a. The two square roots of 1 **b.** The three cube roots of 1

c. The four fourth roots of 1 **d.** The five fifth roots of 1

e. The six sixth roots of 1 **f.** The eight eighth roots of 1

5.6 Polar Coordinates

The basis that we used to graph functions in Chapter 3 was the fact that we associated a point in the plane with an ordered pair of real numbers (x, y) called the Cartesian coordinates of the point, the first component of the ordered pair being the directed distance of the point from the y-axis and the second component being the directed distance from the x-axis.

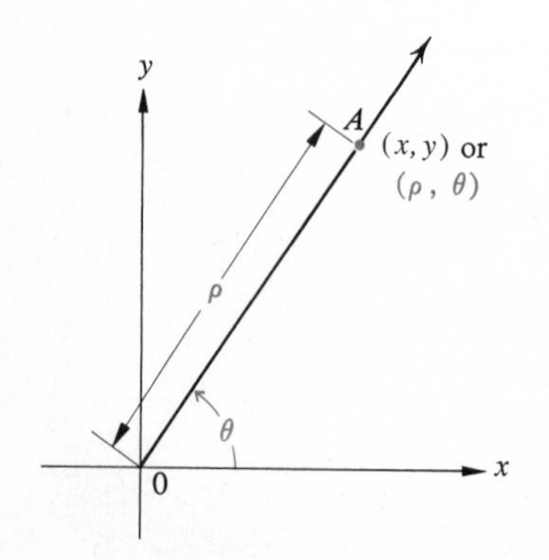

Figure 5.4

Graphing Ordered Pairs (ρ, θ)

The location of a point A in the plane can also be specified by ρ (rho), the distance (a positive number) from the origin O to A and the measure of an angle θ that the ray through point A makes with the positive x-axis, as shown in Figure 5.4. If we extend the replacement set of ρ to include the negative real numbers and the replacement set of θ to in-

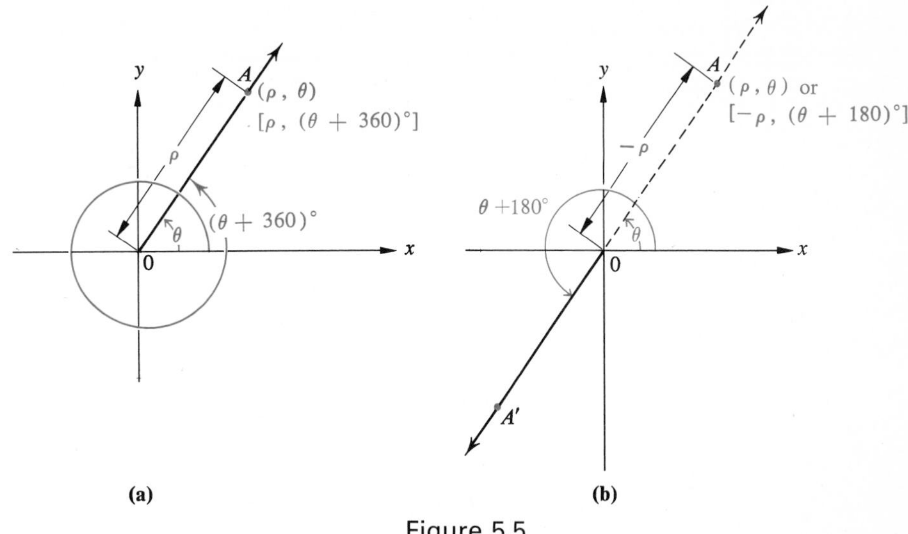

Figure 5.5

clude the set of all angles, we can specify point A by many ordered pairs of the form (ρ, θ). The two components of each of these ordered pairs are called **polar coordinates** of A. For example, in Figure 5.5a, if (ρ, θ) are polar coordinates of A, then so are $[\rho, (\theta + k \cdot 360)°]$, $k \in J$. Also, if $-\rho$ $(\rho > 0)$ denotes the directed distance from O to A along the negative extension of the ray OA', as shown in Figure 5.5b, then $[-\rho, (\theta + 180 + k \cdot 360)°]$ are also acceptable coordinates of A. For example, in Figure 5.6, the point having polar coordinates (4, 30°) also has polar coordinates (4, 390°) and (4, −330°) for positive values of ρ, and (−4, 210°) for a negative value of ρ. Of course, there are infinitely many other possible polar coordinates for this point.

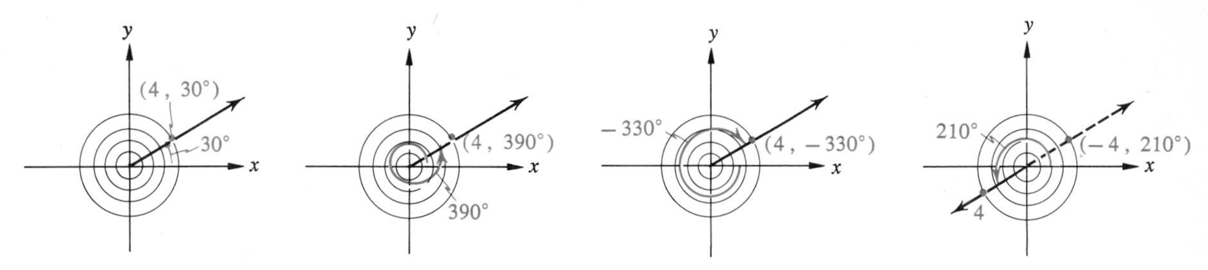

Figure 5.6

If point A is at the origin, then $\rho = 0$, and the coordinates of A are $(0, \theta)$, where θ is an arbitrary angle, not necessarily zero. For example, (0, 90°) and (0, 60°) are both coordinates of the origin.

Cartesian and Polar Coordinates

For any point on the terminal ray of angle θ with Cartesian coordinates x and y, $\cos\theta = x/\rho$ and $\sin\theta = y/\rho$. Hence, the Cartesian and polar coordinates can be related as follows:

$$\left.\begin{aligned} x &= \rho\cos\theta \\ y &= \rho\sin\theta \end{aligned}\right\} \quad (1) \qquad \left.\begin{aligned} \rho &= \pm\sqrt{x^2+y^2} \\ \cos\theta &= \frac{x}{\rho} \\ \sin\theta &= \frac{y}{\rho}, \quad \rho \neq 0 \end{aligned}\right\} \quad (2)$$

Equations (1) can be used to find the rectangular coordinates for a point if the polar coordinates are known, and Equations (2) can be used to find the polar coordinates for a point if the Cartesian coordinates are known. The positive or negative radical expression is chosen depending on the particular problem.

Example Find the Cartesian coordinates of the point with polar coordinates (3, 60°).

Solution From Equations (1), it follows that

$$x = 3\cos 60° = 3\cdot\frac{1}{2} = \frac{3}{2},$$

$$y = 3\sin 60° = 3\cdot\frac{\sqrt{3}}{2} = \frac{3\sqrt{3}}{2}.$$

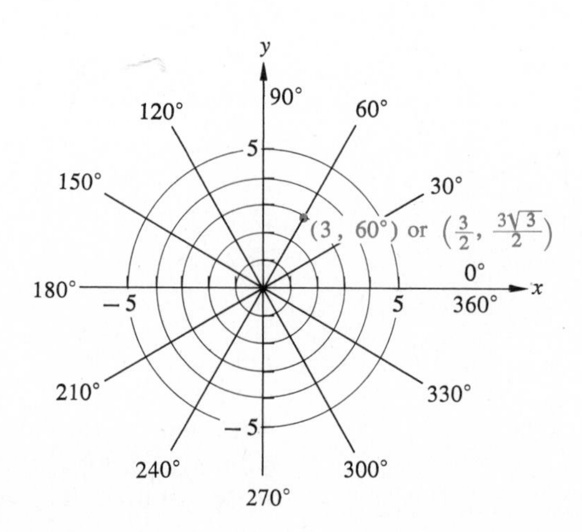

Example Find an approximation for the polar coordinates for the point with Cartesian coordinates (5, −3), where $0° < \theta < 360°$ and $\rho > 0$. Graph the point.

Solution From Equations (2) and since $\rho > 0$,

$$\rho = \sqrt{x^2 + y^2} = \sqrt{25 + 9} = \sqrt{34} \approx 5.831;$$

$$\cos \theta = \frac{5}{5.831} \approx 0.8575.$$

Since the point is in Quadrant IV, then $\theta \approx 329.0°$. Therefore, the pair of polar coordinates is $(\sqrt{34}, 329°)$. The point and the corresponding ordered pairs are shown in the figure.

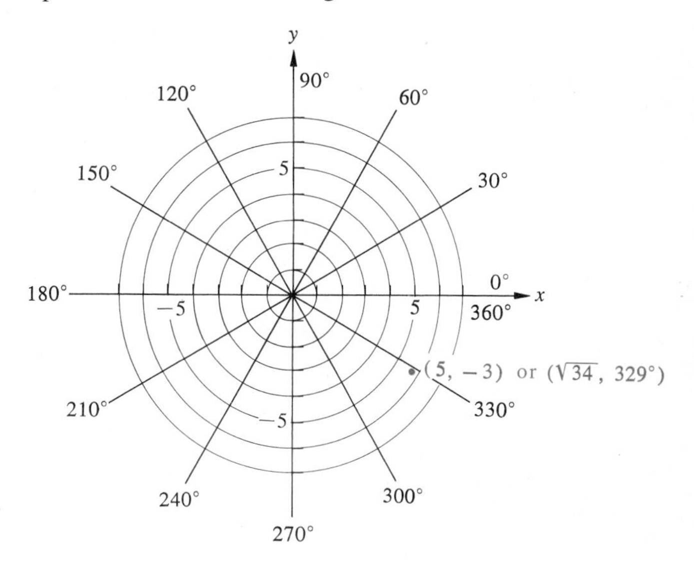

Graphing Equations in Polar Form

In our earlier work we have seen that an equation in two variables, say x and y, serves to pair elements from the replacement sets of the variables and has a graph in the geometric plane in a Cartesian coordinate system. An equation in ρ and θ also has a graph in a polar coordinate system.

Example Graph $\rho = 4 \sin \theta$ in the interval $0° \le \theta < 360°$.

Solution Some ordered pairs with commonly used angles 30°, 45°, etc., are first obtained. The tabulation of the data in the arrangement shown is sometimes helpful. These ordered pairs (ρ, θ) are shown in

θ	$\sin \theta$	ρ or $4 \sin \theta$
0°	0	0
30°	0.50	2.0
45°	0.71	2.8
60°	0.87	3.5
90°	1.00	4.0
120°	0.87	3.5
135°	0.71	2.8
150°	0.50	2.0
180°	0	0

(*continued*)

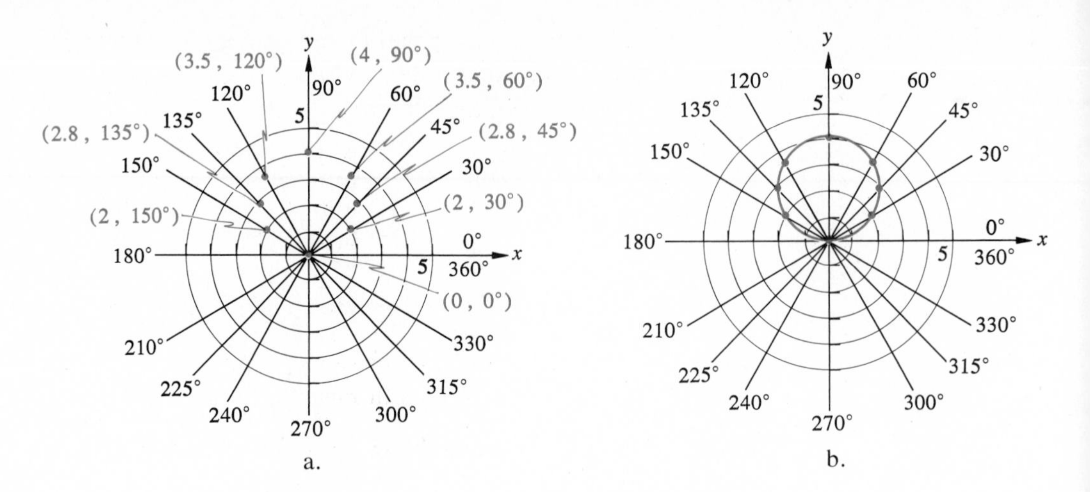

figure a. The complete curve is obtained for $0° \leq \theta < 180°$. For values of θ where $180° \leq \theta < 360°$, the components of ordered pairs (ρ, θ) are also coordinates of points on the curve shown in b.

Example Graph the equation $\rho = \dfrac{3}{\sin \theta + \cos \theta}$.

Solution Ordered pairs are first obtained as shown in the table and in figure a below. The graph, a straight line, is shown in b.

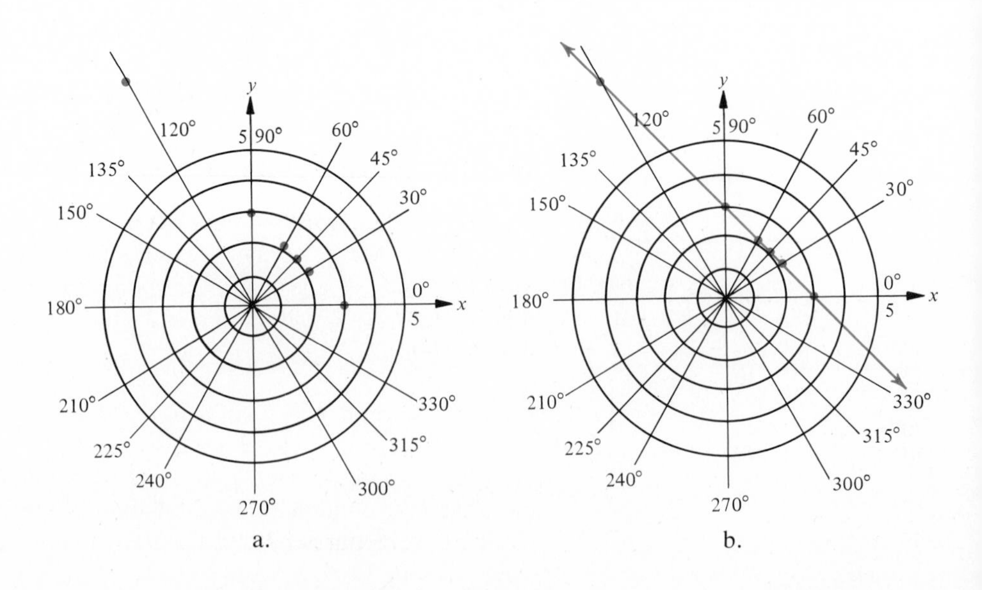

θ	$\sin\theta$	$\cos\theta$	$\sin\theta + \cos\theta$	$\rho = \dfrac{3}{\sin\theta + \cos\theta}$
0°	0	1.00	1.00	3.0
30°	0.50	0.87	1.37	2.2
45°	0.71	0.71	1.42	2.1
60°	0.87	0.50	1.37	2.2
90°	1.00	0	1.00	3.0
120°	0.87	−0.50	0.37	8.1
135°	0.71	−0.71	0	undef.

Several additional interesting types of curves that are graphs of polar equations are shown in Figures 5.7 and 5.8.

LIMAÇONS

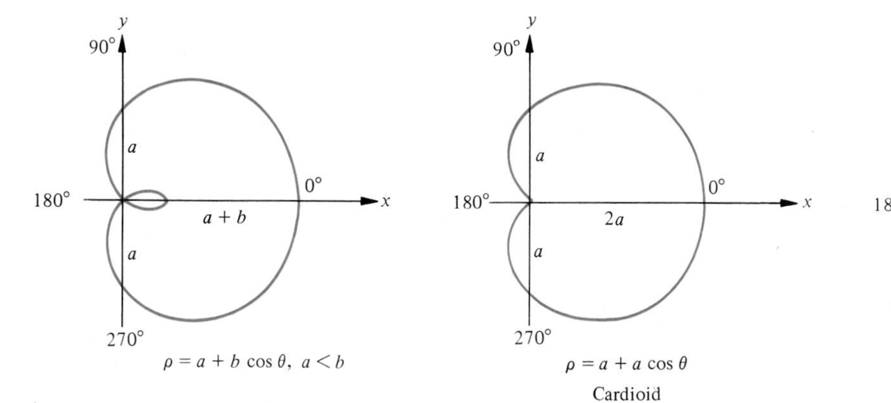

$\rho = a + b\cos\theta,\ a < b$

$\rho = a + a\cos\theta$

Cardioid

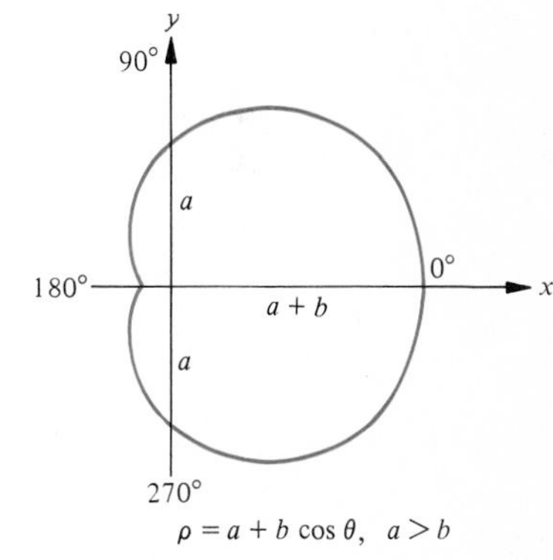

$\rho = a + b\cos\theta,\ a > b$

LEMNISCATES

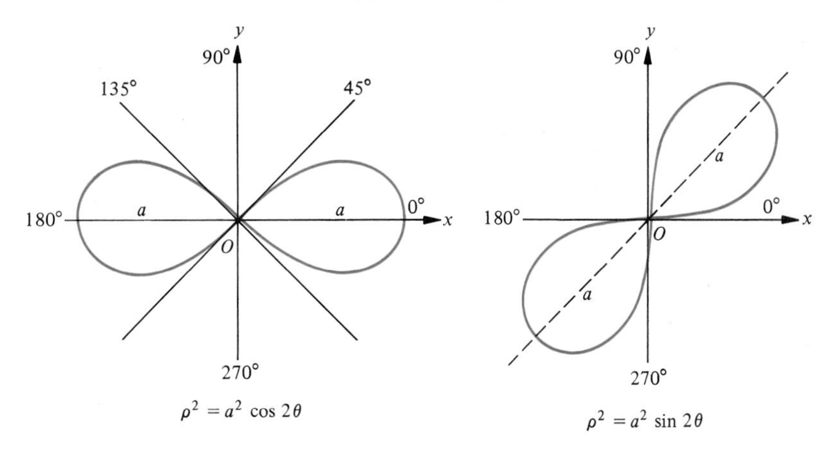

$\rho^2 = a^2\cos 2\theta$

$\rho^2 = a^2\sin 2\theta$

Figure 5.7

THREE–LEAVED ROSE

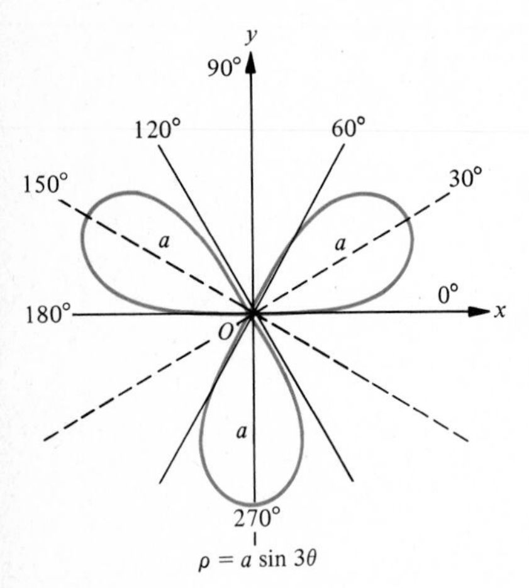

$\rho = a \sin 3\theta$

FOUR–LEAVED ROSE

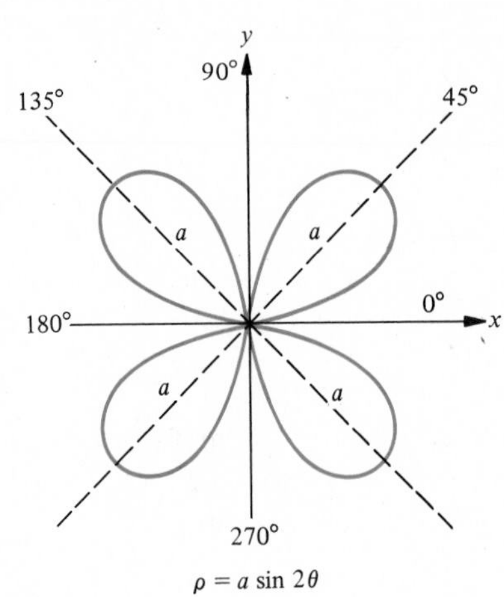

$\rho = a \sin 2\theta$

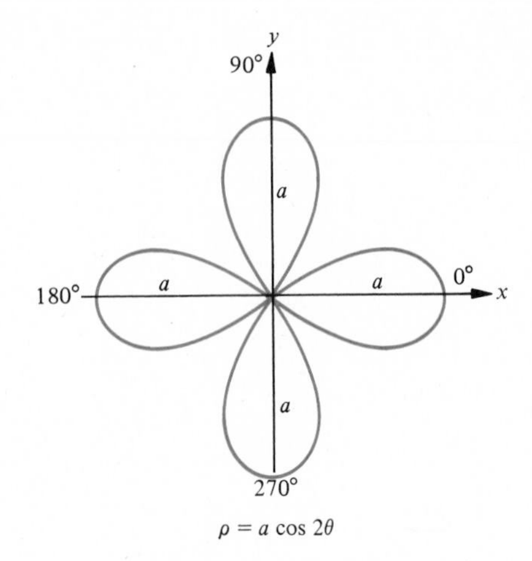

$\rho = a \cos 2\theta$

SPIRAL OF ARCHIMEDES

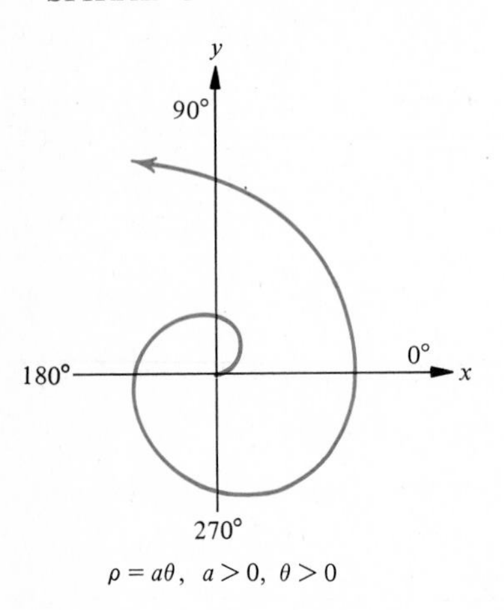

$\rho = a\theta, \ a > 0, \ \theta > 0$

HYPERBOLIC SPIRAL

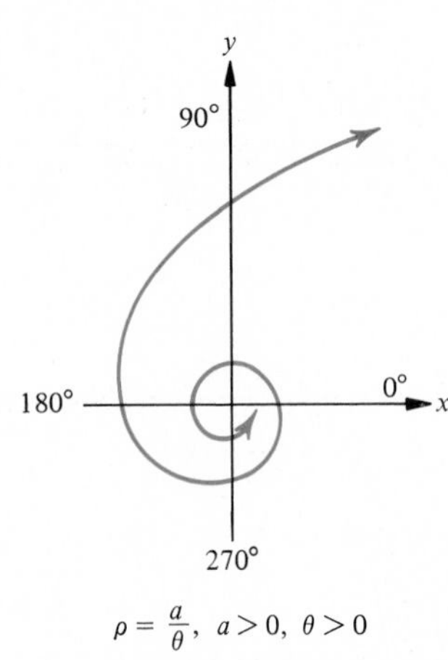

$\rho = \frac{a}{\theta}, \ a > 0, \ \theta > 0$

LOGARITHMIC SPIRAL

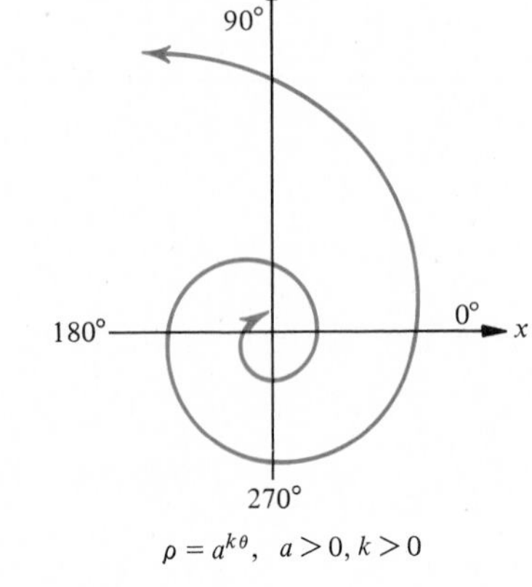

$\rho = a^{k\theta}, \ a > 0, k > 0$

Figure 5.8

Rewriting Equations

Equations (1) and (2) on page 216 can be used to transform an equation in Cartesian form to polar form, and vice versa.

Example Rewrite $x^2 + y^2 - 4x = 0$ in polar form.

Solution Substituting ρ^2 for $x^2 + y^2$ and $\rho \cos \theta$ for x in

$$x^2 + y^2 - 4x = 0$$

yields

$$\rho^2 - 4\rho \cos \theta = 0,$$

$$\rho(\rho - 4 \cos \theta) = 0,$$

$$\rho = 0 \quad \text{or} \quad \rho - 4 \cos \theta = 0.$$

Note that the graph of $\rho = 0$ is the origin and that the graph of $\rho = 4 \cos \theta$ includes the origin as one of its points (for example, for $\theta = 90°$, we have $\rho = 0$). Hence the equation $\rho = 4 \cos \theta$ corresponds to $x^2 + y^2 - 4x = 0$, and it is not necessary to write $\rho = 0$ as part of the answer.

Frequently, the polar form of an equation is more useful than the rectangular form; furthermore, its graph may be simpler to sketch from the polar representation.

EXERCISE SET 5.6

A

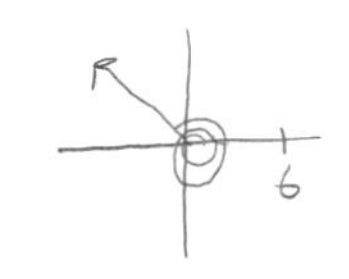

Find four sets of polar coordinates $(-360° < \theta \leq 360°)$ *for each point whose coordinates are given.*

1. $(3,\ 380°)$ **2.** $(6,\ 840°)$ **3.** $(5,\ -450°)$

4. $(4,\ -540°)$ **5.** $(-4,\ 420°)$ **6.** $(-5,\ -600°)$

Find the Cartesian coordinates of each point with polar coordinates as given.

7. $\left(4, \frac{\pi^R}{6}\right)$ **8.** $\left(6, \frac{2\pi^R}{3}\right)$ **9.** $\left(3, -\frac{3\pi^R}{4}\right)$

10. $\left(0, \frac{13\pi^R}{3}\right)$ **11.** $\left(-3, \frac{5\pi^R}{6}\right)$ **12.** $\left(-4, \frac{\pi^R}{4}\right)$

Find the pair of polar coordinates (θ in radians) using an angle α such that $0^R < \alpha < 2\pi^R$ and $\rho > 0$ for each point with Cartesian coordinates as given.

13. $(4, 0)$ **14.** $(0, -3)$ **15.** $(2, 2)$

16. $(-1, -\sqrt{3})$ **17.** $\left(-\frac{\sqrt{3}}{2}, \frac{1}{2}\right)$ **18.** $\left(\frac{1}{2}, -\frac{\sqrt{3}}{2}\right)$

Transform each Cartesian equation to an equation in polar form.

Example $y = 3x$

Solution Substituting $\rho \sin \theta$ for y and $\rho \cos \theta$ for x yields

$$\rho \sin \theta = 3 \rho \cos \theta.$$

Multiplying each member by $\dfrac{1}{\rho \cos \theta}$ $(\rho \neq 0, \theta \neq 90^\circ + k \cdot 180^\circ)$ yields

$$\frac{\sin \theta}{\cos \theta} = 3,$$

$$\tan \theta = 3.$$

Notice that for values of θ for which the equation is satisfied, $\tan \theta$ is a constant, namely 3. Thus the graph of the equation is a straight line with slope 3.

19. $x = 4$ **20.** $y = -3$ **21.** $x^2 + y^2 = 16$

22. $4x^2 + y^2 = 4$ **23.** $x^2 + y^2 - 3y = 0$ **24.** $x^2 + y^2 + 4x = 0$

Transform each polar equation to an equation in Cartesian form.

Example $\rho = 2 \sec \theta$

Solution Substituting $\dfrac{1}{\cos \theta}$ for $\sec \theta$ yields

$$\rho = \frac{2}{\cos \theta},$$

$$\rho \cos \theta = 2.$$

Because $\rho \cos \theta = x$, the Cartesian form of the equation is $x = 2$.

25. $\rho = -\csc \theta$ **26.** $\rho = 9$ **27.** $\rho = 4 \sin \theta$

28. $\rho = 3 \cos \theta$ **29.** $\rho = \dfrac{4}{1 - \sin \theta}$ **30.** $\rho = \dfrac{4}{1 + 2 \cos \theta}$

Graph each equation for $0^\circ \leq \theta < 360^\circ$.

Example $\rho = 2 + 2 \cos \theta$

θ	$\cos\theta$	$2\cos\theta$	ρ or $2 + 2\cos\theta$
0°	1.00	2.00	4.0
30°	0.87	1.74	3.7
45°	0.71	1.42	3.4
60°	0.50	1.00	3.0
90°	0	0	2.0
120°	−0.50	−1.00	1.0
135°	−0.71	−1.42	0.6
150°	−0.87	−1.74	0.3
180°	−1.00	−2.00	0
210°	−0.87	−1.74	0.3
225°	−0.71	−1.42	0.6
240°	−0.50	−1.00	1.0
270°	0	0	2.0
300°	0.50	1.00	3.0
315°	0.71	1.42	3.4
330°	0.87	1.74	3.7

Solution Ordered pairs are first obtained as shown in the table and in figure a. The graph is shown in b.

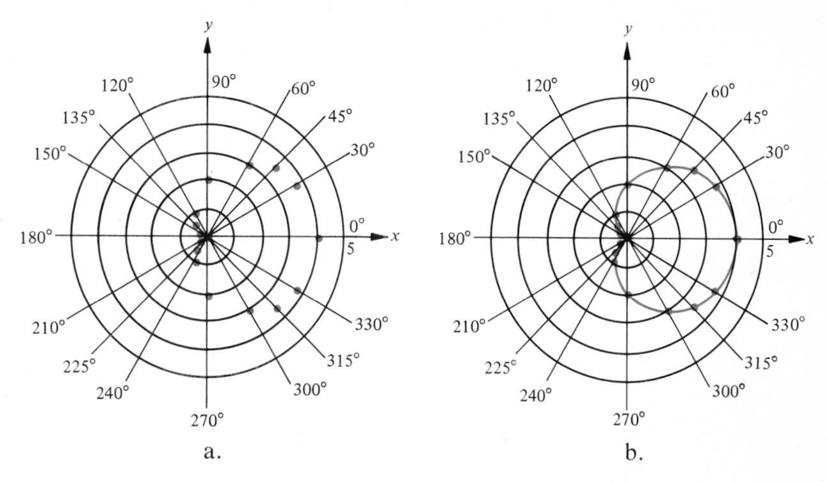

a. b.

31. $\rho = 4$. *Hint:* Consider the equation $\rho = 0 \cdot \theta + 4$.

32. $\theta = 30°$. *Hint:* Consider the equation $\theta = 0 \cdot \rho + 30°$.

33. $\rho = 4 \cos \theta$ (circle)

34. $\rho = 2 \sin 3\theta$ (three-leaved rose)

35. $\rho = \sin 2\theta$ (four-leaved rose)

36. $\rho = 1 + 2 \cos \theta$ (limaçon)

37. $\rho = 3 + 3 \cos \theta$ (cardioid)

38. $\rho = 2 + \cos \theta$ (limaçon)

39. $\rho^2 = 4 \cos 2\theta$ (lemniscate)

40. $\rho = 2\theta, \theta > 0$ (spiral of Archimedes)

41. $\rho = \dfrac{1}{\theta}, \theta > 0$ (hyperbolic spiral)

42. $\rho = 2^\theta, \theta > 0$ (logarithmic spiral)

B

Graph each of the given pairs of equations and find coordinates for the points of intersection.

43. $\rho = \sin \theta$

$\rho = \cos \theta$

44. $\rho = 1 + \sin \theta$

$\rho = 1$

45. $\rho = 1 + \sin\theta$
$\rho = 1 + \cos\theta$

46. $\rho = 2\sin\theta$
$\rho = 2\sin 2\theta$

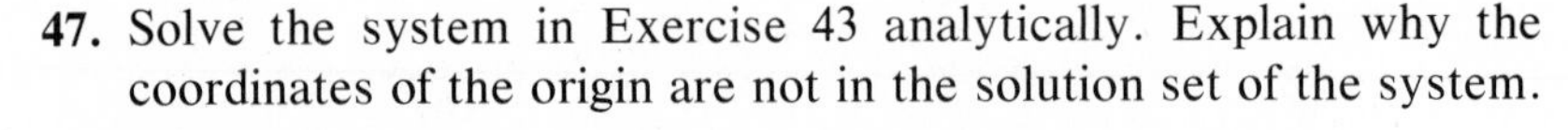

47. Solve the system in Exercise 43 analytically. Explain why the coordinates of the origin are not in the solution set of the system.

48. Solve the system in Exercise 44 analytically.

49. Solve the system in Exercise 45 analytically.

50. Solve the system in Exercise 46 analytically.

51. Show that the distance between the points with coordinates (ρ_1, θ_1) and (ρ_2, θ_2) is given by $d = \sqrt{\rho_1^2 + \rho_2^2 - 2\rho_1\rho_2\cos(\theta_1 - \theta_2)}$.

52. Find an equation in polar form for the circle with center at $(a, 0^R)$ and radius of length a. *Hint:* Recall that an angle inscribed in a semicircle is a right angle.

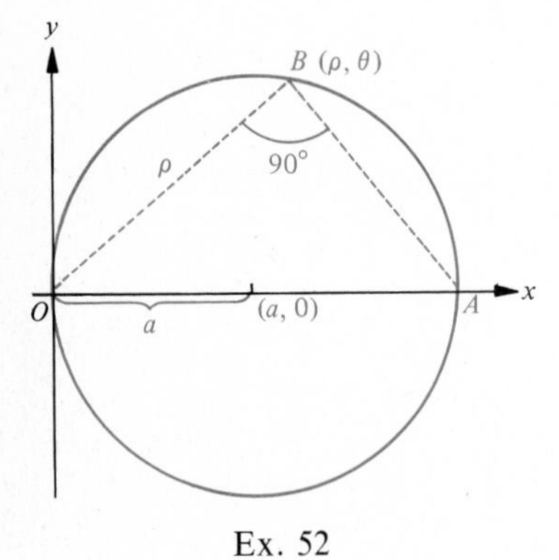

Ex. 52

Chapter Summary

5.1 A **complex number** is defined by

$$z = a + bi, \qquad a, b \in R \quad \text{and} \quad i^2 = -1.$$

If $z_1 = a_1 + b_1 i$ and $z_2 = a_2 + b_2 i$, then

$$z_1 = z_2 \quad \text{if and only if} \quad a_1 = a_2 \quad \text{and} \quad b_1 = b_2.$$

Sums and products of complex numbers are found by treating $a + bi$ as a binomial in the variable i and replacing i^2 with -1.

A complex number $a + bi$ can be identified with a **real number** if $b = 0$; if $b \neq 0$, the number is an **imaginary number,** and for the special case where $a = 0$, the number is a **pure imaginary number.**

For all $b \in R$, $b > 0$,

$$\sqrt{-b} = \sqrt{b}i = i\sqrt{b}.$$

Hence, for a and b positive real numbers,

$$\sqrt{-a}\sqrt{-b} = i^2\sqrt{ab} = -\sqrt{ab}.$$

5.2 The properties used for the set of complex numbers are similar to the corresponding properties of real numbers. Furthermore,

$$z_1 - z_2 = z_1 + (-z_2);$$

$$\frac{z_1}{z_2} = z_1 \cdot \frac{1}{z_2} \quad (z_2 \neq 0).$$

If $z_2, z_3 \neq 0$, then

$$\frac{z_1}{z_2} = \frac{z_1 z_3}{z_2 z_3}.$$

If $z = a + bi$, then the **conjugate** of z is $\bar{z} = a - bi$. The product of $a + bi$ and its conjugate equals $a^2 + b^2$, a real number.

5.3 A plane on which complex numbers are graphed is called a **complex plane.** The x-axis is called the **real axis** and the y-axis is called the **imaginary axis.**

Complex numbers can be added graphically by the parallelogram law applicable to geometric vectors.

The **absolute value** or **modulus** of the complex number z is

$$\rho = |z| = |a + bi| = |(a, b)| = \sqrt{a^2 + b^2}.$$

The modulus is the distance from the origin to the graph of (a, b). An **argument** or **amplitude** of the complex number $z = a + bi = (a, b)$, denoted by **arg($a + bi$)** or **arg(a, b),** is an angle θ such that

$$\cos \theta = \frac{a}{\rho} \quad \text{and} \quad \sin \theta = \frac{b}{\rho} \quad (a^2 + b^2 \neq 0).$$

If θ is an argument of $a + bi$, then $(\theta + k \cdot 360)°$, $k \in J$, is also an argument of $a + bi$. The angle θ with the least positive measure is called the **principal argument.**

5.4 A nonzero complex number $a + bi$ can be expressed in **trigonometric form** as

$$a + bi = \rho[\cos(\theta + k \cdot 360°) + i \sin(\theta + k \cdot 360°)], \quad k \in J.$$

If $z_1 = \rho_1(\cos \theta_1 + i \sin \theta_1)$ and $z_2 = \rho_2(\cos \theta_2 + i \sin \theta_2)$, then

$$z_1 \cdot z_2 = \rho_1 \cdot \rho_2[\cos(\theta_1 + \theta_2) + i \sin(\theta_1 + \theta_2)],$$

$$\frac{z_1}{z_2} = \frac{\rho_1}{\rho_2}[\cos(\theta_1 - \theta_2) + i \sin(\theta_1 - \theta_2)].$$

5.5 If $z = \rho(\cos\theta + i\sin\theta)$ and $n \in n$, then

$$z^n = \rho^n(\cos n\theta + i\sin n\theta).$$

This is known as **De Moivre's theorem.**
If $z \neq 0 + 0i$, then

$$z^0 = 1 + 0i = 1 \quad \text{and} \quad z^{-n} = \frac{1}{z^n}, \quad n \in N.$$

If $n \in N$ and $z = \rho(\cos\theta + i\sin\theta)$, then

$$\rho^{1/n}\left[\cos\left(\frac{\theta + k\cdot 360°}{n}\right) + i\sin\left(\frac{\theta + k\cdot 360°}{n}\right)\right], \quad k \in J$$

is an ***n*th root of *z*.** For $k = 0$, the nth root

$$\rho^{1/n}\left(\cos\frac{\theta}{n} + i\sin\frac{\theta}{n}\right)$$

is called the **principal root.**

5.6 The components of an ordered pair $(\rho, \theta°)$ or (ρ, θ^R) that locate a point A in the plane are called the **polar coordinates** of A. Each such ordered pair specifies a unique point; however, each point in the plane corresponds to infinitely many ordered pairs of polar coordinates.
Cartesian and polar coordinates are related as follows:

$$x = \rho\cos\theta \qquad \rho = \pm\sqrt{x^2 + y^2},$$

$$y = \rho\sin\theta \qquad \cos\theta = \frac{x}{\rho} = \frac{x}{\pm\sqrt{x^2 + y^2}},$$

$$\sin\theta = \frac{y}{\rho} = \frac{y}{\pm\sqrt{x^2 + y^2}} \quad [x, y) \neq (0, 0)].$$

Review Exercises

A

5.1 *Find real numbers x and y for which each statement is true.*

1. $9x - 6yi = -3 + 4i$ **2.** $-8x - 5i = -2y - 5x^2i$

Write each expression in the form $a + bi$.

3. $(-4 - i) + (-4 + 9i)$

4. $(-3 + i) + (3 - 2i)$

5. $(8 + 6i)(-2 + 5i)$

6. $(5 + 3i)(2 - 4i)$

Write each expression as one of the complex numbers i, -1, $-i$, or 1.

7. i^{13}

8. i^{23}

Write each expression in the form $a + bi$.

9. $\sqrt{-5}(3 - \sqrt{-2})$

10. $(3 - \sqrt{-5})(3 + \sqrt{-5})$

11. Show that $3i$ and $-3i$ satisfy the equation $x^2 + 9 = 0$.

12. Show that $1 + 3i$ and $1 - 3i$ satisfy $x^2 - 2x + 10 = 0$.

5.2 *Express each difference in the form $a + bi$.*

13. $(-7 + 2i) - (8 + 3i)$

14. $(6 - 9i) - (3 - 4i)$

Express each quotient in the form $a + bi$.

15. $\dfrac{8 + 3i}{2 + 4i}$

16. $\dfrac{-7 + i}{6 - 10i}$

Write each expression in the form bi or $a + bi$.

17. $\dfrac{-2}{\sqrt{-6}}$

18. $\dfrac{4}{2 + \sqrt{-5}}$

5.3 *Express each complex number as an ordered pair.*

19. $-4 + i$

20. $2\sqrt{3} - 2i$

Express each complex number in the form $a + bi$.

21. $(9, -6)$

22. $(3, 4)$

Graph each complex number z, its negative $-z$, and its conjugate $\bar{z}$.

23. $-3 + 5i$

24. $(-2, -6)$

Represent each sum graphically and check the results by algebraic methods.

25. $(-4 + 3i) + (8 - 2i)$ **26.** $(-1, -8) + (6, 7)$

Find the absolute value and the principal argument of each complex number.

27. $4 + 2i$ **28.** $(-2, -5)$

5.4 *Represent each complex number in trigonometric form, using the angle with least positive measure.*

29. $1 + i$ **30.** $-3 + 3\sqrt{3}i$

Represent each complex number graphically and write it in the form $a + bi$.

31. $7(\cos 120° + i \sin 120°)$ **32.** $2[\cos(-315°) + i \sin(-315°)]$

Express each product in the form $a + bi$.

33. $3(\cos 200° + i \sin 200°) \cdot 2(\cos 25° + i \sin 25°)$

34. $\sqrt{3}(\cos 50° + i \sin 50°) \cdot \sqrt{12}[\cos(-20°) + i \sin(-20°)]$

Express each quotient in the form $a + bi$.

35. $\dfrac{8(\cos 309° + i \sin 309°)}{9(\cos 129° + i \sin 129°)}$ **36.** $\dfrac{12[\cos(-315°) + i \sin(-315°)]}{3(\cos 270° + i \sin 270°)}$

5.5 *Express each power in the form $a + bi$.*

37. $[3(\cos 15° + i \sin 15°)]^4$ **38.** $(\sqrt{3} - i)^3$

39. $[4(\cos 240° + i \sin 240°)]^{-2}$ **40.** $(4 - 4i)^{-3}$

41. Find the two square roots of $\dfrac{\sqrt{3}}{2} + \dfrac{1}{2}i$.

42. Find the three cube roots of $125(\cos 330° + i \sin 330°)$.

Solve each equation, $x \in C$.

43. $x^4 + 1 = 0$ **44.** $x^5 = -5 - 5\sqrt{3}i$

5.6 *Find four sets of polar coordinates* $(-360° < \theta \leq 360°)$ *for each point.*

45. $(2, 610°)$

46. $(-3, 390°)$

Find the Cartesian coordinates for each point.

47. $\left(3, \frac{\pi^R}{3}\right)$

48. $\left(-2, \frac{3\pi^R}{4}\right)$

Find the pair of polar coordinates (θ *in radians*) *using an angle* α *such that* $0^R < \alpha < 2\pi^R$ *and* $\rho > 0$ *for each point.*

49. $(2, -2)$

50. $(-\sqrt{3}, -1)$

Transform each equation to an equation in polar form.

51. $x = 3$

52. $x^2 + 3y^2 = 5$

Transform each equation to an equation in Cartesian form.

53. $\rho = 5 \sin \theta$

54. $\rho = \dfrac{2}{1 - \cos \theta}$

Graph each equation.

55. $\rho = 1 + \sin \theta$

56. $\rho = 1 - \sin \theta$

57. $\rho = \cos 2\theta$

58. $\rho = 2 \cos 3\theta$

Appendix A

Algebra and Geometry Review

In Appendix A we review the basic ideas of algebra and geometry, using terminology and notation that are needed in the development of various topics in the text.

A.1 Set Notation

Recall that a set is simply a collection of "things." Each "thing" in the set is called a **member** or **element** of the set. The membership can be described by listing the names of the members in braces, { }, or by stating a rule that identifies the members in the set.

Definition A.1 *Two sets are* **equal** *if and only if they have the same members.*

Customarily, the slant bar, /, drawn through symbols for relations, indicates negation. Thus,

$$\{3, 4, 5\} = \{5, 3, 4\} \qquad \text{but} \qquad \{3, 4, 5\} \neq \{4, 5, 6\}.$$

The symbol $\in$ is used to denote membership in a set. For example, we can write

$$3 \in \{3, 4, 5\}$$

to represent the statement "3 is an element of $\{3, 4, 5\}$."

Sets are also designated by means of capital letters, A, B, C, R, etc.

Example If $A = \{2, 3, 4\}$ and $B = \{4, 5, 6\}$, then

$$3 \in A \quad \text{but} \quad 3 \notin B.$$

An unspecified element of a set is usually denoted by a lowercase italic letter such as a, b, c, x, and y or by a lowercase letter from the Greek alphabet such as α, β, γ, and θ. Such a symbol is called a **variable;** a symbol used to denote a specific element is called a **constant.** Variables are used in conjunction with braces in another symbolism that is often useful in discussing sets. This symbolism, illustrated by

$$\{x \mid x \in A \quad \text{and} \quad x \in B\}$$

and read "the set of all x such that x is an element of A and x is an element of B," is called **set-builder notation.** The vertical line, $|$, is read "such that." Set-builder notation names a variable and states conditions on the variable.

Example If $A = \{2, 3, 4\}$ and $B = \{4, 5, 6\}$, then

$$\{x \mid x \in A \quad \textbf{or} \quad x \in B\} = \{2, 3, 4, 5, 6\} \tag{1}$$

because 2, 3, 4, 5, 6 are elements of *one or the other* of the two sets, whereas

$$\{x \mid x \in A \quad \textbf{and} \quad x \in B\} = \{4\} \tag{2}$$

because 4 is an element of *both* sets.

The preceding example is associated with the following two operations on sets.

Definition A.2 *The* **union** *of two sets A and B is the set of all elements that belong to A or to B or to both.*

The symbol $\cup$ is used to denote this operation; $A \cup B$ is read "the union of A and B."

Example If $A = \{2, 3, 4\}$ and $B = \{4, 5, 6\}$, then

$$A \cup B = \{2, 3, 4, 5, 6\}. \tag{3}$$

Notice that each element in $A \cup B$ is listed only once. Since the symbol 4 denotes only one number, it would be redundant to list the name of the element twice and to write $A \cup B = \{2, 3, 4, 4, 5, 6\}$.

A second useful operation on sets is the following.

Definition A.3 *The* **intersection** *of two sets A and B is the set of all elements that belong to both A and B.*

The symbol $\cap$ is used to denote this operation; $A \cap B$ is read "the intersection of A and B."

Example If $A = \{2, 3, 4\}$ and $B = \{4, 5, 6\}$, then

$$A \cap B = \{4\}. \tag{4}$$

Notice from Equations (1) and (3) that the operation of union is associated with the word *or*, and from Equations (2) and (4) that the operation of intersection is associated with the word *and*.

The set that contains no elements is called the **null set** or the **empty set** and is denoted by the symbol $\varnothing$.

Example The intersection of two sets is a set. Therefore, if $C = \{2, 3, 4\}$ and $D = \{5, 6, 7\}$, we can write

$$C \cap D = \{2, 3, 4\} \cap \{5, 6, 7\} = \varnothing.$$

Two sets such as C and D, which do not have any elements in common, are said to be **disjoint.**

A.2 The Set of Real Numbers

Recall from your study of algebra that each real number can be associated with one and only one point on a number line and each point on the line can be associated with one and only one real number. Several examples are shown in Figure A.1. The real number corresponding to a point on the line is called the **coordinate** of the point, and the point is called the **graph** of the number. A number whose graph is to the right of the graph of zero is **positive,** and a number whose graph is to the left is **negative.**

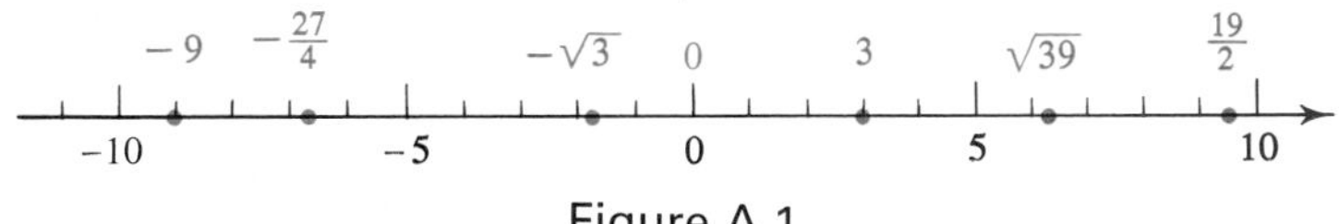

Figure A.1

The following sets of numbers are contained in the set of **real numbers R.** The capital letters shown are the ones that are sometimes associated with the respective sets. These are the symbols that we use in this text.

1. The set N of **natural numbers,** among whose elements are such numbers as 1, 2, 6, 25, and 624.
2. The set J of **integers,** whose elements consist of the natural numbers, their negatives, and zero. Included are such elements as -24, -9, 0, 7, and 314.
3. The set Q of **rational numbers,** whose elements are all those numbers that can be represented in the form $\frac{a}{b}$, or a/b, where a and b are integers and b is not zero. Included are such elements as $-\frac{1}{3}$, 0, $\frac{1}{4}$, and 6.
4. The set H of **irrational numbers,** whose elements are those numbers that are the coordinates of points on the number line that are not the graphs of rational numbers. An irrational number cannot be represented in the form a/b, where a and b are integers.

Figure A.2 summarizes the relationships among these sets of numbers.

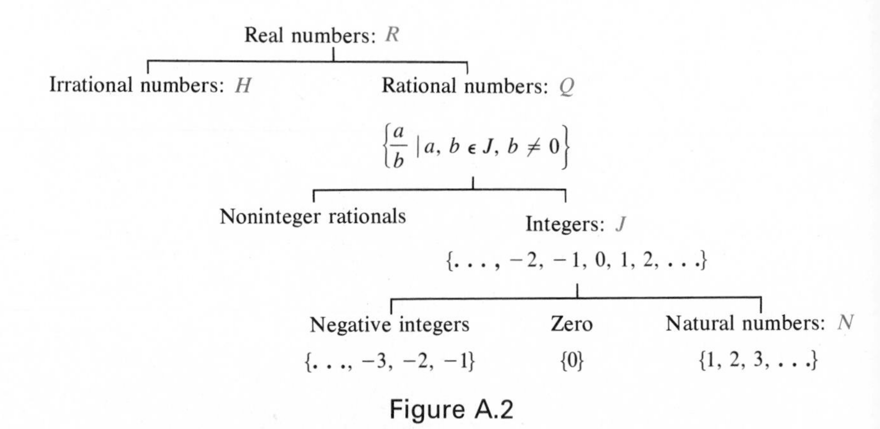

Figure A.2

Radical notation is useful to represent some irrational numbers.

Definition A.4 *For all $a \in R$ where a is positive or zero, $\sqrt{a}$ is the nonnegative number such that*

$$\sqrt{a} \cdot \sqrt{a} = a.$$

Examples

a. $\sqrt{9}$ is the positive number such that $\sqrt{9}\sqrt{9} = 9$.

b. $\sqrt{x}$ is the positive number such that $\sqrt{x}\sqrt{x} = x$.

Frequently, rational numbers are used as approximations for irrational numbers. For example, by using the symbol $\approx$ for "approximately equal to," we write $\sqrt{2} \approx 1.414$, $\sqrt{3} \approx 1.732$, $\sqrt{5} \approx 2.236$, etc. Table I in Appendix C gives rational number approximations for some square roots that are irrational numbers.

If *every* element of a set A is also an element of a set B, then A is said to be a **subset** of B. Thus the sets H and Q can be viewed as subsets of R, the set J as a subset of Q or R, and the set N as a subset of J, Q, or R.

If the elements of a set can be paired with the numerals 1, 2, 3, . . . , n for some natural number n so that each element is paired with exactly one numeral and each numeral is paired with exactly one element, then the set is said to be a **finite set.** A set that is not the empty set and is not finite is called an **infinite set.** Some infinite sets can be designated by using three dots and some of the members listed to establish a pattern as shown in Figure A.2.

Often we wish to consider the nonnegative member of the pair of real numbers a and $-a$, where $a \neq 0$. The special notation $|a|$ is used for this purpose. To extend the notation for all $a \in R$, we define $|0| = 0$.

Definition A.5 *For all* $a \in R$,

if a is positive or zero, then $|a| = a$;

if a is negative, then $|a| = -a$.

$|a|$ *is called the* **absolute value** *of* a.

Examples

a. $|4| = 4$ b. $|-4| = -(-4) = 4$ c. $|0| = 0$

The order of the elements of the set of real numbers can be established by the following.

Definition A.6 *For all* $a, b \in R$, b *is* **less than** a *if for some positive real number* c,

$$b + c = a.$$

For such conditions, a is said to be **greater than** b.

Examples

a. 4 is less than 7 because there is a *positive number* c (equal to 3) such that $4 + c = 7$.

b. -9 is less than -5 because there is a *positive number* c (equal to 4) such that $-9 + c = -5$.

The number line shown in Figure A.3 is very helpful in visualizing whether one real number is less than or greater than a second real number. If the graph of b lies to the *left* of the graph of a, then b is less

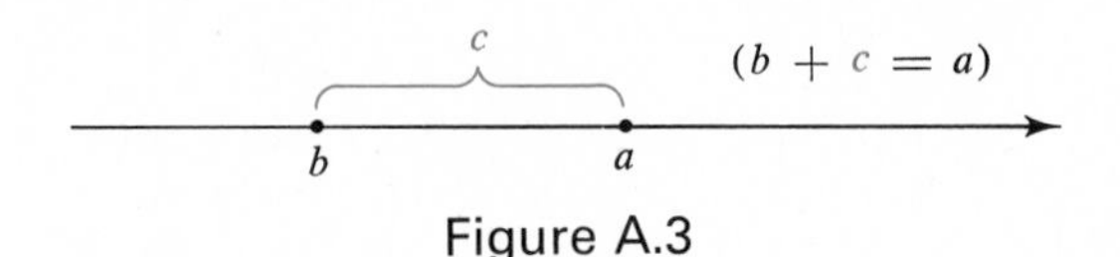

Figure A.3

than a. The following symbols are used in connection with the property of order:

$<$ read "is less than";

$\leq$ read "is less than or equal to";

$>$ read "is greater than";

$\geq$ read "is greater than or equal to."

Observe that $b < a$ and $a > b$ are equivalent statements.

Examples

a. $4 < 7$ and $7 > 4$

b. $-9 < -5$ and $-5 > -9$

c. $x < 7$ is equivalent to $7 > x$

In your study of arithmetic and algebra, you applied the operations of addition and multiplication to elements of the set of real numbers. You also made assumptions, called **axioms,** about these operations and studied the logical consequences of these axioms. For your reference, we include a list of some of these axioms and properties on page 237.

A.3 Equations in Two Variables; Relations and Functions

Recall from your study of algebra that the ordered pair (2, 9) is called a **solution** of the equation $y = x + 7$ because the replacement of x by

For $a, b, c, d \in R$:

Equality Axioms

1. $a = a$	Reflexive law
2. If $a = b$, then $b = a$	Symmetric law
3. If $a = b$ and $b = c$, then $a = c$	Transitive law
4. If $a = b$, then a may be substituted for b or b for a in any expression.	Substitution law

Order Axioms

1. Exactly one of the following is true: $a < b, \quad a = b, \quad \text{or} \quad a > b.$	Trichotomy law
2. If $a < b$ and $b < c$, then $a < c$.	Transitive law
3. If $a, b > 0$, then $a + b > 0 \quad \text{and} \quad a \cdot b > 0$	Closure for positive numbers

Axioms for Operations

1. $a + b \in R$	Closure law for addition
2. $a + b = b + a$	Commutative law of addition
3. $(a + b) + c = a + (b + c)$	Associative law of addition
4. $a + 0 = 0 + a = a$	Identity law of addition
5. $a + (-a) = (-a) + a = 0$	Additive inverse law
6. $a \cdot b \in R$	Closure law for multiplication
7. $a \cdot b = b \cdot a$	Commutative law of multiplication
8. $(a \cdot b) \cdot c = a \cdot (b \cdot c)$	Associative law of multiplication
9. $a \cdot 1 = 1 \cdot a = a$	Identity law of multiplication
10. $a \cdot \frac{1}{a} = \frac{1}{a} \cdot a = 1 \quad (a \neq 0)$	Multiplicative inverse law
11. $a \cdot (b + c) = a \cdot b + a \cdot c$	Distributive law

Some Properties

1. $a \cdot 0 = 0$

2. $-(-a) = a$

3. $-a = -1 \cdot a$

4. If $a = b$, then $a + c = b + c$

5. If $a + c = b + c$, then $a = b$

6. If $a = b$, then $a \cdot c = b \cdot c$

7. If $a \cdot c = b \cdot c$, then $a = b \quad (c \neq 0)$

8. $a - b = a + (-b)$

9. $\frac{a}{b} = \frac{c}{d}$ if and only if $a \cdot d = b \cdot c \; (b, d \neq 0)$

10. $\frac{a}{b} = \frac{a \cdot c}{b \cdot c} \quad (b, c \neq 0)$

11. $\frac{a}{c} + \frac{b}{c} = \frac{a + b}{c} \quad (c \neq 0)$

12. $\frac{a}{b} \cdot \frac{c}{d} = \frac{a \cdot c}{b \cdot d} \quad (b, d \neq 0)$

13. $\frac{a}{b} = a \cdot \frac{1}{b} \quad (b \neq 0)$

14. $\frac{a}{b} \div \frac{c}{d} = \frac{a}{b} \cdot \frac{d}{c} \quad (b, c, d \neq 0)$

15. $\frac{a}{b} = \frac{-a}{-b} = -\frac{a}{-b} = -\frac{-a}{b};$

$$\frac{-a}{b} = \frac{a}{-b} = -\frac{a}{b} = -\frac{-a}{-b} \quad (b \neq 0)$$

16. $ab = 0$ if and only if $a = 0$ or $b = 0$

the first component 2 and the replacement of y by the second component 9 results in a true statement, $9 = 2 + 7$. The ordered pair (2, 9) is said to *satisfy* the equation. The set of all such ordered pairs that satisfy an equation is called the **solution set** of the equation.

The equation $y = x + 7$ does not result in a true statement for all ordered pairs of real numbers. For example, (3, 5) is not a solution because $5 \neq 3 + 7$. Such equations are called **conditional equations.** On the other hand, if an equation such as

$$x + y = y + x$$

results in a true statement for every replacement of x and y, for which both members have meaning, the equation is call an **identity.** The left-hand member and the right-hand member of identities are called **equivalent expressions.**

To show that an equation is not an identity, it is necessary only to find one replacement for the variables that will make the equation false. Such a procedure is referred to as showing a **counterexample.**

Example We can show by counterexample that $x^2 - 3y = 3y - x^2$ is not an identity by substituting some arbitrary values for the variables. If we use 1 for x and 1 for y, the left-hand member of the equation equals $1^2 - 3 \cdot 1$, or -2, and the right-hand member equals $3 \cdot 1 - 1^2$, or 2. Hence,

$$x^2 - 3y \neq 3y - x^2$$

for the ordered pair (1, 1), and the equation is not an identity.

Example We can show that the equation

$$(x - y)^2 + 2xy = x^2 + y^2$$

is an identity by simplifying the left-hand member

$$\begin{aligned}(x - y)^2 + 2xy &= x^2 - 2xy + y^2 + 2xy \\ &= x^2 + y^2.\end{aligned}$$

Thus, the left-hand member and the right-hand member are equivalent and the equation is an identity.

Sets of ordered pairs play an important role in mathematics and are given a special name, as follows:

Definition A.7 *A* **relation** *is a set of ordered pairs.*

The set of first components in a set of ordered pairs is called the **domain** of the relation and the set of second components is called the **range.**

Examples

a. $\{(5, 8), (6, 9), (7, 10)\}$ is a relation with domain $\{5, 6, 7\}$ and range $\{8, 9, 10\}$.

b. $\{(5, 8), (5, 9), (7, 10)\}$ is a relation with domain $\{5, 7\}$ and range $\{8, 9, 10\}$.

The set of ordered pairs exhibited in Example a is a special kind of relation in which each element in the domain is paired with only one element in the range. Such relations are given a special name.

Definition A.8 *A* **function** *is a relation in which each element in the domain is associated with only one element in the range.*

Note that the relation in Example b above is not a function because 5, an element in the domain, is associated with more than one element in the range.

The relations above were exhibited by listing the members in the set of ordered pairs. They can also be described by equations that specify which ordered pairs are in the relation and which are not; those that are in the relation are elements of the solution set. For example,

$$\{(x, y) \mid y = x + 3, \quad x \in \{5, 6, 7\}\}$$

also specifies the function in Example a above:

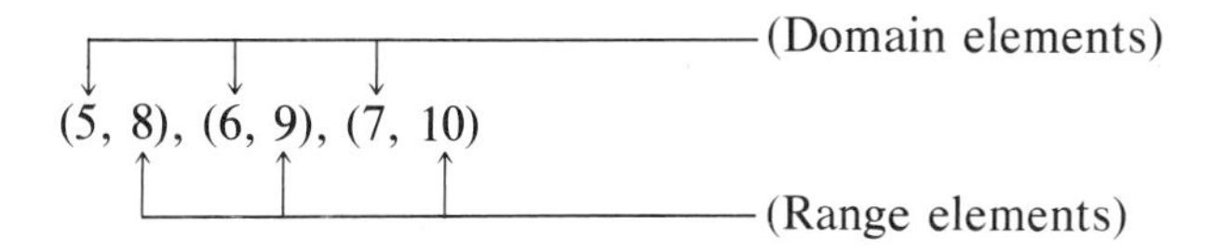

The elements in the range, 8, 9, and 10, are obtained by replacing x in the equation $y = x + 3$ with 5, 6, and 7 in turn and solving for y. This function consists of three ordered pairs only.

Example Given the function

$$\{(x, y) \mid y = x - 3, \quad x \in \{1, 2, 3\}\}:$$

1. The domain is $\{1, 2, 3\}$.
2. Replacing x with each element of the domain in turn, you can find the associated values of y to be -2, -1, and 0, respectively. Thus, the range is $\{-2, -1, 0\}$.
3. The function is $\{(1, -2), (2, -1), (3, 0)\}$.

If the domain of a function is the set of real numbers R, rather than a finite set, the function consists of an *infinite set* of ordered pairs and obviously not all its elements can be listed. *If the domain is not specified, we shall assume that it includes all real numbers for which both members of the defining equation are defined.*

Examples

a. The domain of the function defined by $y = x + 3$ is the set R because $x + 3$, and therefore y, are defined for all real number replacements for x.

b. The domain of the function defined by $y = \dfrac{3}{x - 4}$ is the set R with the restriction that $x \neq 4$, because $x - 4 = 0$ and y is undefined for $x = 4$.

Functions are sometimes designated by names (linear, quadratic, etc.) or by means of a single symbol, generally the symbol f. If the discussion includes a consideration of more than one function, other letters such as g, h, F, P, etc., are used.

The symbol for the function can be used in conjunction with the variable representing an element in the domain to represent the associated element in the range. For example, $f(x)$ (read "f of x" or "the value of f at x") is the element in the range of f associated or paired with the element x in the domain. Thus, we can discuss functions defined by an equation such as

$$y = x - 2$$

$$f(x) = x - 2,$$

$$g(x) = x - 2,$$

etc.

where the symbols $f(x)$, $g(x)$, etc., are playing exactly the same role as y. The variable x represents an element in the domain, and y, $f(x)$, $g(x)$, etc., represent an element in the range.

The notation $f(x)$ is especially useful because, by replacing x with a specific real number a in the domain, the notation $f(a)$ will then denote the paired, or corresponding, element in the range of the function.

Example If $f(x) = 2x - 5$,

$$f(3) = 2(3) - 5 = 6 - 5 = 1,$$
$$f(0) = 2(0) - 5 = 0 - 5 = -5,$$
$$f(-3) = 2(-3) - 5 = -6 - 5 = -11,$$
$$f(a) = 2(a) - 5 = 2a - 5.$$

A.4 Graphs of Relations and Functions

As you probably recall from your study of algebra, an association exists between ordered pairs of real numbers and the points on a geometric plane, sometimes called the **real plane.** If two number lines are drawn perpendicular to each other at their origins as in Figure A.4, they form a rectangular coordinate system. These perpendicular number lines are called **axes** (singular, axis). The horizontal line is usually called the ***x*-axis,** and the vertical line is usually called the ***y*-axis.** The x-axis and the y-axis divide the plane into four regions, called **quadrants,** as shown.

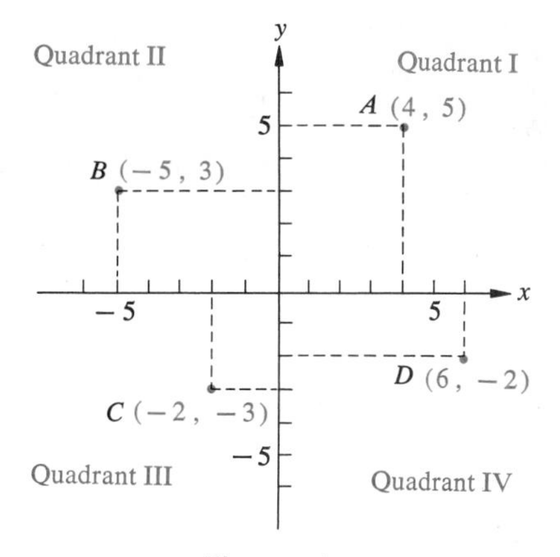

Figure A.4

For a given rectangular coordinate system, any point in the plane can be associated with a unique ordered pair of real numbers. For example, the point A in Quadrant I, in Figure A.4, can be associated with the ordered pair (4, 5), where 4 represents the distance of the point to the right of the y-axis and 5 represents the distance of the point above the

x-axis. The points B, C, and D correspond to the ordered pairs $(-5, 3)$, $(-2, -3)$, and $(6, -2)$, respectively.

The *components* of an ordered pair associated with a point in the plane are called the **coordinates** of the point. The first component is called the **abscissa** of the point; the second component is called the **ordinate** of the point. The *point* is called the **graph** of the ordered pair. An ordered pair is sometimes used as a name for the corresponding point. Thus, we occasionally speak of the point (2, 3) or the point $(4, -3)$ or, in general, the point (x, y).

Recall that in Section A.3, a function was defined in part as a set of ordered pairs. Consequently, a function—or at least a part of a function—can now be displayed as a set of points on the plane. The phrase "part of a function" is used here since the domain or range or both may be the infinite set of real numbers, which cannot be shown on a finite line. For example, consider the function

$$y = x - 4.$$

If arbitrary values are assigned to x, say -3, 0, 3, and 6, and the corresponding values for y are computed, the four ordered pairs

$$(-3, -7), \quad (0, -4), \quad (3, -1), \quad \text{and} \quad (6, 2)$$

are obtained, which are solutions to the equation. These points can be located on a coordinate system as shown in Figure A.5a. These points appear to lie on a straight line, and in fact they do. It can be shown that the coordinates of any point on the line (Figure A.5b) constitute a solu-

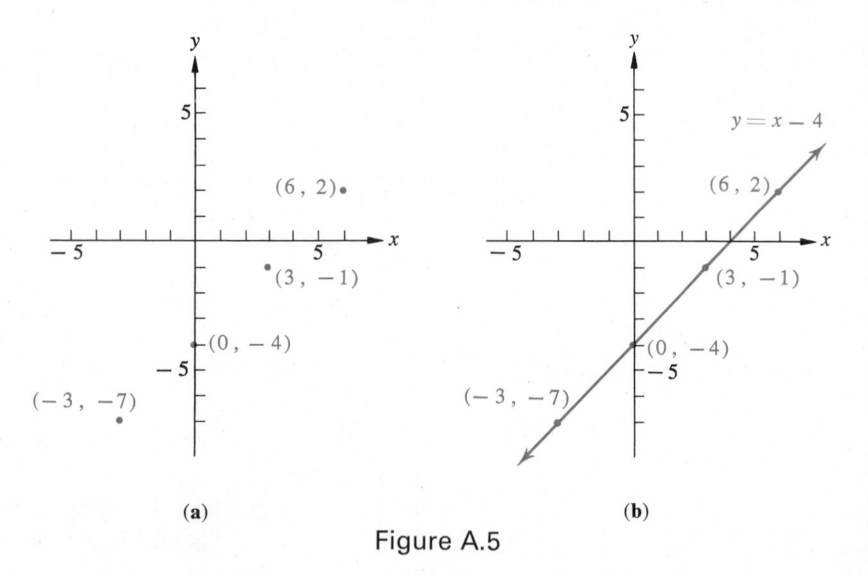

Figure A.5

tion of the first-degree equation in two variables, $y = x - 4$ and that, conversely, every solution of $y = x - 4$ corresponds to a point on this line. The line is referred to as the *graph of the function* $y = x - 4$ or, alternatively, as the *graph of the solution set of the equation* or simply as the *graph of the equation*. Obviously, only part of the line can be displayed, so an arrowhead is placed at each end to indicate that the graph, or line, continues in both directions indefinitely. Furthermore, since the graphs of first-degree equations

$$y = mx + b \qquad \text{or} \qquad ax + by = c$$

are straight lines, it is only necessary to find two ordered pairs in the solution set to sketch the graph. Two pairs that are easy to find are $(0, b)$ and $(a, 0)$, in which the numbers a and b are called the **intercepts.**

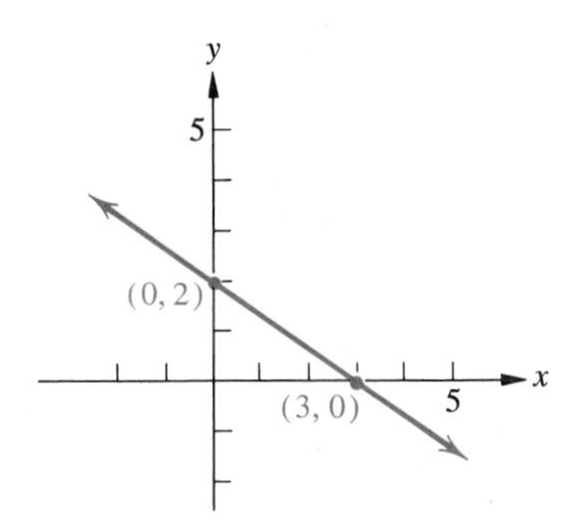

Example The graph of the function $2x + 3y = 6$ can be obtained by first finding the x and y intercepts.

If $y = 0$, then $x = 3$ and $(3, 0)$ is a solution.

If $x = 0$, then $y = 2$ and $(0, 2)$ is a solution.

The x and y intercepts are 3 and 2, respectively. See the graph.

First-degree equations are also called **linear equations,** and the functions defined by these equations are called **linear functions.**

For any two points P_1 and P_2 on a line, the set of points containing P_1 and P_2 and all points lying between P_1 and P_2 is called a **line segment.** We shall designate the line segment by $\overline{P_1P_2}$. A line segment has the important property of length. If we designate the endpoints of a line segment $\overline{P_1P_2}$ with the pairs of coordinates (x_1, y_1) and (x_2, y_2), as in Figure A.6, the length, or measure, of the line segment, which we shall designate by P_1P_2 or some variable such as d, can be determined by an application of the Pythagorean theorem (see Section A.8). From the right triangle shown, we observe that the lengths of the perpendicular sides are $y_2 - y_1$ and $x_2 - x_1$; the square of the distance between P_1 and P_2 is given by

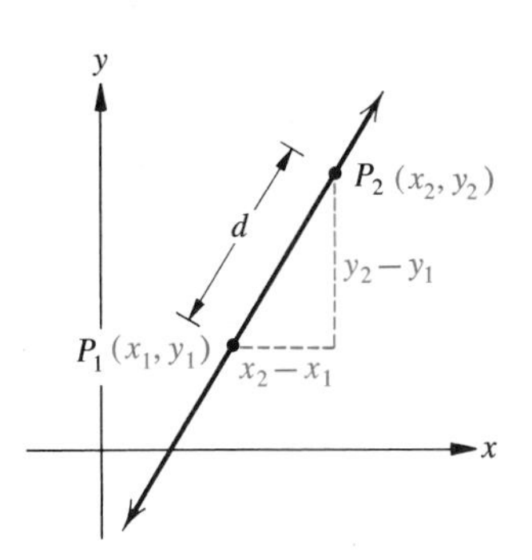

Figure A.6

$$d^2 = (x_2 - x_1)^2 + (y_2 - y_1)^2.$$

By considering only the positive square root of the right-hand member, we have that

$$d = \sqrt{(x_2 - x_1)^2 + (y_2 - y_1)^2}.$$

Example The distance between the pair of points (1, 3) and (5, 6) is given by

$$\begin{aligned} d &= \sqrt{(5-1)^2 + (6-3)^2} \\ &= \sqrt{16+9} = \sqrt{25} = 5. \end{aligned}$$

As with first-degree equations in two variables, solutions of higher-degree equations in two variables are also ordered pairs, which can be found by arbitrarily assigning values to x and finding the associated values of y. For example, in the equation

$$y = x^2 + 2,$$

if $x = -2$, then

$$y = (-2)^2 + 2 = 6,$$

and $(-2, 6)$ is a solution. Similarly, by replacing x with -1, 0, 1, and 2, we obtain $(-1, 3)$, $(0, 2)$, $(1, 3)$, and $(2, 6)$ as additional solutions. Plotting these points on the plane, we have the graph in Figure A.7a. Clearly, these points do not lie on a straight line. By plotting additional solutions of $y = x^2 + 2$, such as

$$\left(-\frac{3}{2}, \frac{17}{4}\right), \quad \left(-\frac{1}{2}, \frac{9}{4}\right), \quad \left(\frac{1}{2}, \frac{9}{4}\right), \quad \text{and} \quad \left(\frac{3}{2}, \frac{17}{4}\right),$$

we have the additional points in Figure A.7b. These points can now be connected in sequence from left to right by a smooth curve as shown. We assume that this curve is a good approximation to the graph of $y = x^2 + 2$.

More generally, the graph of any equation of the form

$$y = ax^2 + bx + c \quad (a, b, c, x \in R,\ a \neq 0),$$

is called a **parabola.** Notice that such an equation defines a function, since with each x, an equation of this form will associate one and only one y. Such functions are called **quadratic functions.**

The graph of

$$x^2 + y^2 = r^2,$$

where r is a real number greater than zero, is a **circle** with center at the origin and radius of length r. However, although this equation defines a relation, it does not define a function.

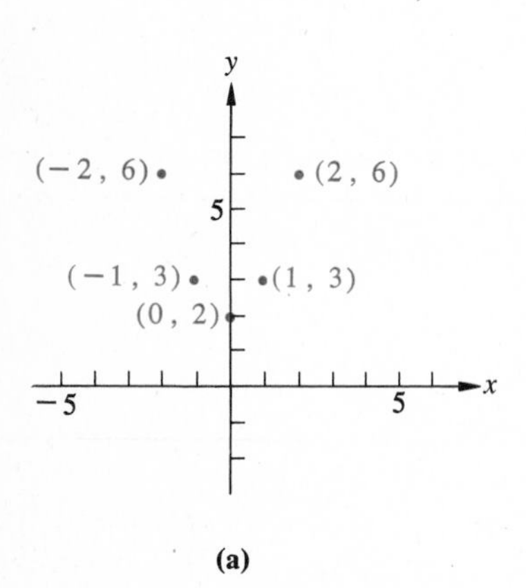

(a)

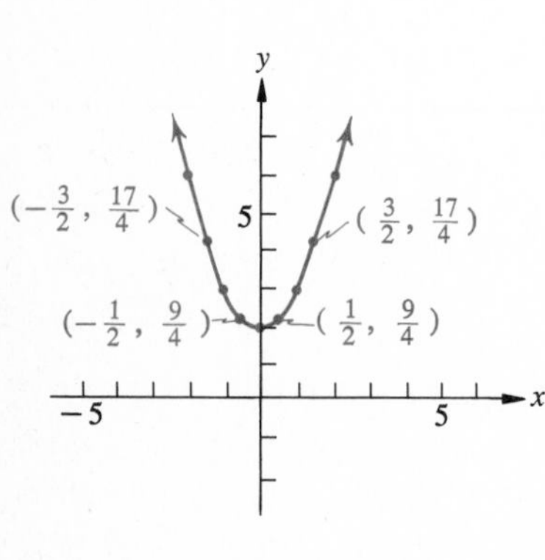

(b)

Figure A.7

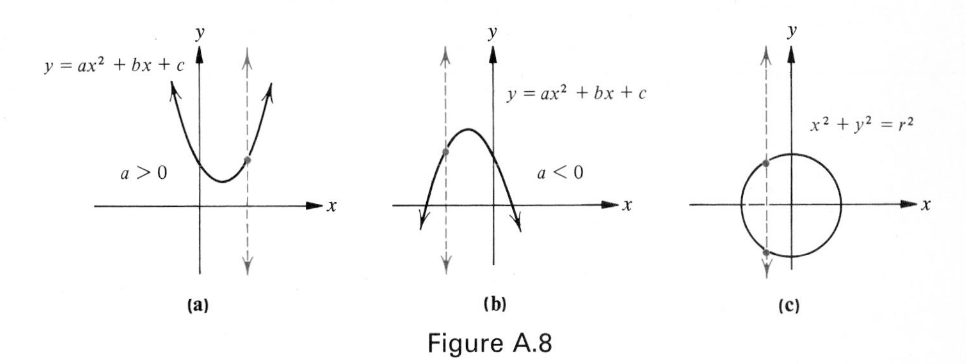

Figure A.8

Whether or not a relation is also a function can be determined by inspecting its graph. For example, consider the graphs of equations $y = ax^2 + bx + c$ and $x^2 + y^2 = r^2$ in Figure A.8. Imagine a vertical line moving across each graph from left to right. Since the line cuts the graph of $y = ax^2 + bx + c$ at *only one point at each position, this equation defines a function;* that is, for each x in this relation, there is only one y. Since the line cuts the graph of $x^2 + y^2 = r^2$ at *more than one point at certain positions, the equation does not define a function,* although it does define a relation. There are ordered pairs in the relation which have the same first components and different second components.

A.5 Zeros of Functions; Equations in One Variable

In any function, the element(s) in the domain that are paired with the element 0 in the range are called **zeros of the function.**

Example The zeros of

$$f = \{(-3, 1), (-2, 0), (-1, -2), (0, -4), (1, 0)\}$$

are -2 and 1 because each of these first components is paired with a second component 0.

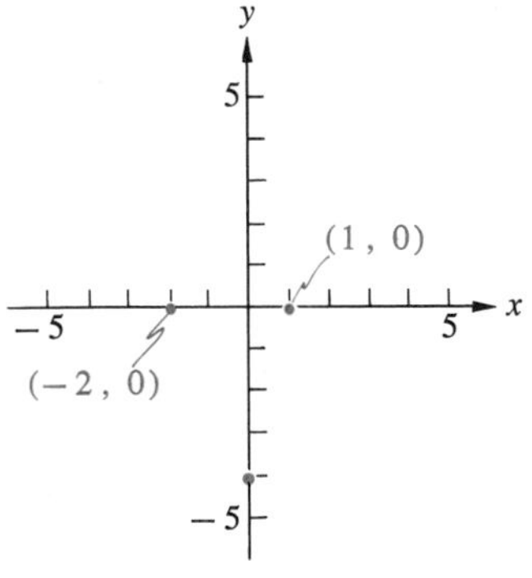

The graph of each zero in the above example lies on the x-axis, as shown in Figure A.9.

Figure A.9

For any function

$$y = f(x),$$

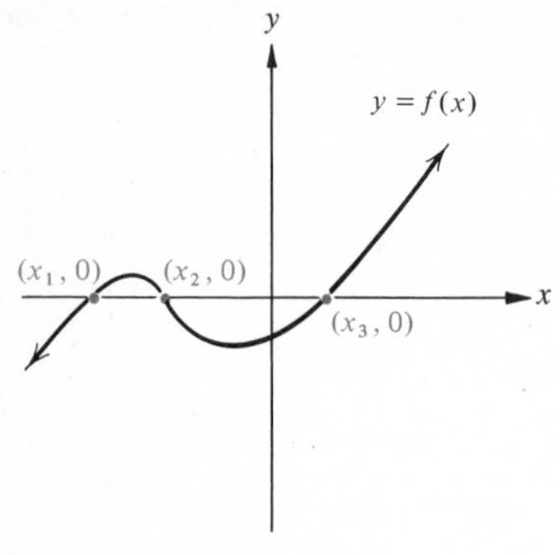

Figure A.10

the zeros of the function are values of x (x_1, x_2, x_3 in Figure A.10) for which

$$y = f(x) = 0.$$

These values of x are the solutions of the equation in one variable, $f(x) = 0$, and the set of such numbers is the solution set of the equation.

Note that the ordered pairs $(x_1, 0)$, $(x_2, 0)$, and $(x_3, 0)$ are the coordinates of the points of intersection of the graph of $y = f(x)$ and the x-axis.

Thus, there are three different names for a single idea:

1. The zeros of $y = f(x)$.
2. The x intercepts of the graph of $y = f(x)$.
3. Solutions of $f(x) = 0$.

In many simple cases, such as

$$x - 2 = 0,$$

the solution set of an equation in one variable is evident by inspection. However, this is not the case for most first-degree equations. Recall from your work in algebra that solutions of first-degree equations in one variable can usually be found by generating **equivalent equations** (a set of equations with the same solution set) until you obtain an equation whose solution set is evident by inspection.

In general, if each member of an equation is multiplied by the same nonzero number, or if the same quantity is added to both members, the result is an equation equivalent to the original equation.

Solutions of quadratic equations in one variable are obtained by various methods. For example if the left-hand member of a quadratic equation of the form

$$ax^2 + bx + c = 0 \quad (a \neq 0)$$

is factorable, the solutions can be obtained from the fact that the left-hand member will equal zero for values of x for which one or both of the factors equals zero. Such a procedure of solving a quadratic equation is called *solution by factoring*.

Example We can solve $x^2 + x = 30$, by first writing the equation equivalently as

$$x^2 + x - 30 = 0, \tag{1}$$

in which form the right-hand member is zero. Factoring the left-hand member yields

$$(x + 6)(x - 5) = 0. \tag{2}$$

The left-hand member equals zero for values of x for which

$$x + 6 = 0 \quad \text{or} \quad x - 5 = 0 \quad \text{or both.}$$

By inspection, the solution set of (2) and, therefore, of (1) is $\{-6, 5\}$.

Quadratic equations of the form

$$x^2 = a \quad (a \geq 0)$$

can be solved by a method called *extraction of roots*. If the equation has a solution, then x must be a square root of a (see Definition A.4). Since each positive real number a has two square roots, the solution set, for $a > 0$, is $\{\sqrt{a}, -\sqrt{a}\}$. If $a = 0$, the solution set is $\{0\}$.

Example The method of extraction of roots can be used to find the solution set of the equation $4x^2 = 24$. Multiplying each member by 1/4 yields

$$x^2 = 6$$

Hence, $x = \sqrt{6}$ or $x = -\sqrt{6}$ and the solution set is $\{\sqrt{6}, -\sqrt{6}\}$.

From the method of extraction of roots applied to the equation

$$(x - a)^2 = b \quad (b \geq 0).$$

we obtain

$$x - a = \sqrt{b} \quad \text{and} \quad x - a = -\sqrt{b},$$

from which the solution set $\{a + \sqrt{b}, a - \sqrt{b}\}$ is obtained. If $b = 0$, the solution set is $\{a\}$. From this it follows that we can find the solution set of any quadratic equation by first rewriting the equation in the form

$$(x - a)^2 = b.$$

This procedure of solving a quadratic equation is called *completing the square*.

Applying this procedure to the general quadratic equation

$$ax^2 + bx + c = 0 \quad (a \neq 0)$$

results in the formula

$$x = \frac{-b \pm \sqrt{b^2 - 4ac}}{2a},$$

called the **quadratic formula,** which gives us the solutions of any quadratic equation in the form $ax^2 + bx + c = 0$ expressed in terms of a, b, and c. The solution set is

$$\left\{\frac{-b + \sqrt{b^2 - 4ac}}{2a}, \frac{-b - \sqrt{b^2 - 4ac}}{2a}\right\}.$$

If $a, b, c \in R$, the elements in the solution set are real numbers if and only if $b^2 - 4ac \geq 0$.

Example The equation $x^2 - 5x = 6$ can be solved by using the quadratic formula as follows. First, rewrite the equation as

$$x^2 - 5x - 6 = 0.$$

Then, replacing a with 1, b with -5, and c with -6 in the quadratic formula yields

$$x = \frac{-(-5) \pm \sqrt{(-5)^2 - 4(1)(-6)}}{2(1)},$$

from which

$$x = \frac{5 + \sqrt{25 + 24}}{2} = 6 \quad \text{or} \quad x = \frac{5 - \sqrt{25 + 24}}{2} = -1,$$

and the solution set is $\{6, -1\}$.

A.6 Inverse of a Function

If the components of each ordered pair in a given relation are interchanged, the resulting relation and the given relation are called **inverses** of each other.

Example $\{(1, 3), (5, 7), (11, 13)\}$ and $\{(3, 1), (7, 5), (13, 11)\}$ are inverse relations.

The inverse of a relation $\mathcal{R}$ is denoted by $\mathcal{R}^{-1}$ (read "$\mathcal{R}$ inverse" or "the inverse of $\mathcal{R}$"). It is evident from the above example that the domain and range of $\mathcal{R}^{-1}$ are the range and domain, respectively, of $\mathcal{R}$. We have used the script $\mathcal{R}$ for a relation so that it should not be confused with the R that represents the set of real numbers. Furthermore, it should be understood that -1 in the inverse notation symbol $\mathcal{R}^{-1}$ is not an exponent.

The inverse of the relation

$$\mathcal{R}\colon y = 3x - 5 \tag{1}$$

can be obtained by replacing x with y and y with x. Hence, the inverse relation of (1) is given by

$$\mathcal{R}^{-1}\colon x = 3y - 5 \tag{2}$$

or, when y is expressed in terms of x in the defining equation,

$$\mathcal{R}^{-1}\colon y = \frac{1}{3}(x + 5). \tag{2'}$$

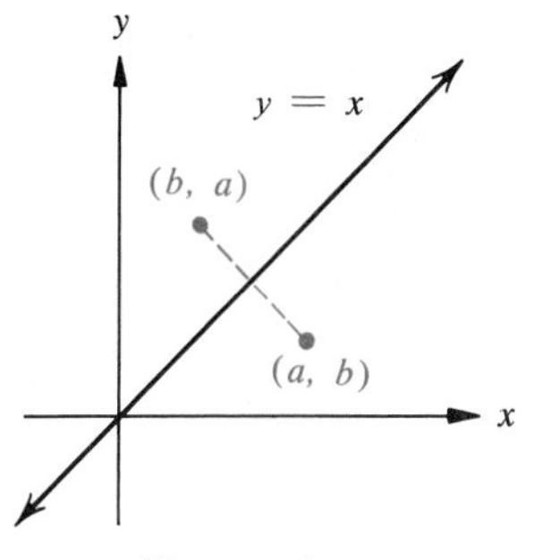

Figure A.11

Notice in the above example that $y = 3x - 5$ and $x = 3y - 5$ from (1) and (2) define inverse relations. The domains and ranges of the relations are interchanged because the variables have been interchanged. However, $x = 3y - 5$ from (2) and $y = \frac{1}{3}(x + 5)$ from (2′) are equivalent equations and define the same relation.

The graphs of inverse relations are always located symmetrically with respect to the graph of the linear equation $y = x$. To see this, notice the graphs of the ordered pairs (a, b) and (b, a) in Figure A.11. They are the same distance from, but on opposite sides of, the line that is the graph of the equation $y = x$. The graph of $y = x$ serves as a reflecting line, or mirror, for these points.

Using the example above, Figure A.12 on page 250 shows the graphs of

$$y = 3x - 5$$

and its inverse,

$$x = 3y - 5 \qquad \text{or} \qquad y = \frac{1}{3}(x + 5),$$

together with the graph of $y = x$.

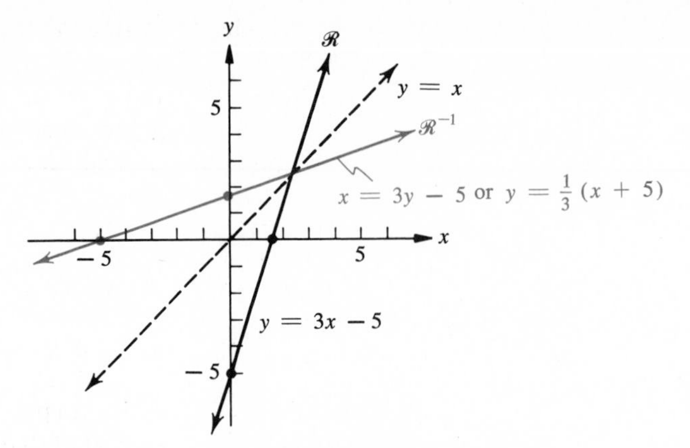

Figure A.12

Because every function is a relation, *every function has an inverse, but the inverse is not always a function.* For example, consider the function

$$y = x^2. \tag{3}$$

Its graph and the graph of its inverse,

$$x = y^2 \qquad \text{or} \qquad y = \pm\sqrt{x}. \tag{4}$$

are shown in Figure A.13. Since for all but one value of x ($x = 0$), the inverse specified by (4) associates two different values of y (one positive and one negative) in the range, the inverse of (3) is not a function. Observe in Figure A.13 that the graphs of these relations are symmetric with respect to the graph of $y = x$.

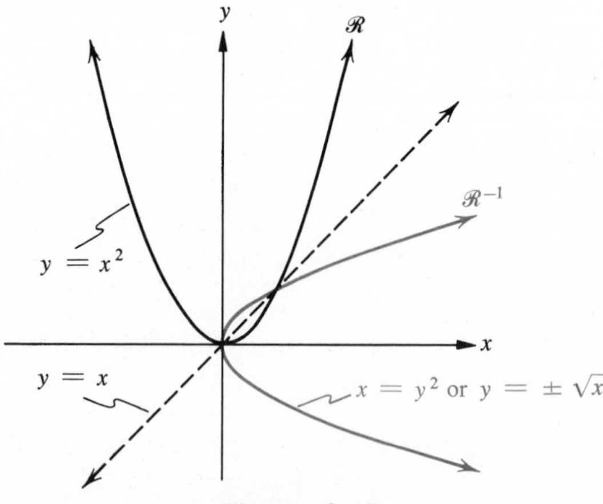

Figure A.13

In order for a function to have an inverse that is also a function, it must be a *one-to-one function;* that is, each element in the domain of the original function must be associated with one and only one element in its range and each element in its range must be associated with one and only one element in its domain.

A.7 Angles and Their Measure

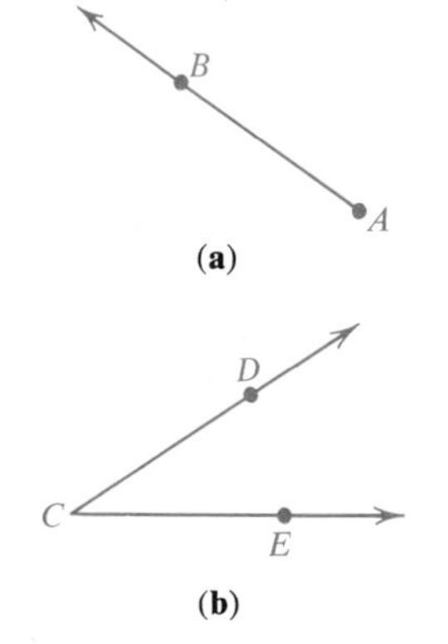

Figure A.14

In geometry, a **ray** is a figure such as that shown in Figure A.14a, where the arrow indicates that the ray extends infinitely far in one direction. The point A (points are usually named with capital letters) in A.14a is the **endpoint** of that ray, and point C is the endpoint of the two rays in A.14b.

A ray is named by naming its endpoint and one other point on the ray. In Figure A.14a the ray is named by $\overrightarrow{AB}$ (read "ray AB"), and in A.14b the rays are named by $\overrightarrow{CD}$ and $\overrightarrow{CE}$.

An **angle** is the union of two rays having a common endpoint called the **vertex** of the angle. The rays are the **sides** of the angle. In Figure A.14b, an angle is formed by $\overrightarrow{CD}$ and $\overrightarrow{CE}$, where $\overrightarrow{CD}$ and $\overrightarrow{CE}$ are the sides, and point C is the vertex.

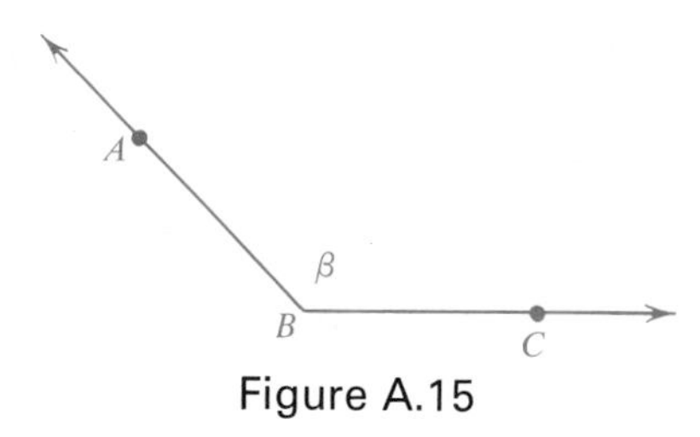

Figure A.15

An angle can be named in several ways. For example, if a point is the vertex of only one angle, as in Figure A.15, then that letter can be used to name the angle; in this case, $\angle B$. Also, an angle can be named using a single letter placed "inside" the angle, such as β (beta), or by using three letters, where the middle letter names the vertex, such as $\angle ABC$ or $\angle CBA$.

If a point is the vertex of more than one angle, as in Figure A.16, each angle can also be named either by using three letters, such as $\angle DEF$, $\angle DEG$, or $\angle GEF$, or by using a single letter "inside" the angle, such as α (alpha), or γ (gamma). However, since three angles have the vertex E, "$\angle E$" cannot be used since it would not clearly name any one of them.

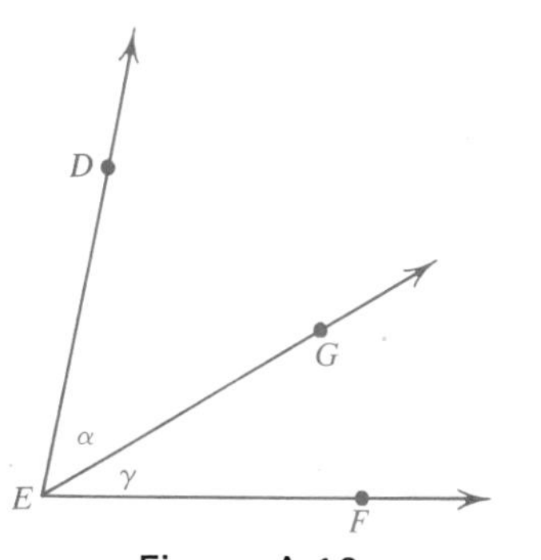

Figure A.16

Angles can be measured by various units of measurement, but in geometry they are commonly measured in **degrees.** Such measures can be assigned to an angle by using a circle whose center is at the vertex of the angle. If a circle is divided into 360 arcs of equal length (Figure A.17 on page 252), the number of arcs intercepted by a central angle α is the measure of the angle in degree units and is designated by $\alpha°$ (the measure of the angle is independent of the radius of the measuring circle). From Figure A.17, we have that $\alpha = 40°$.

Each degree is divided into sixty minutes (60′) and each minute of a degree is divided into sixty seconds (60″). To convert from tenths of degrees to minutes, you simply multiply the number of tenths by 60.

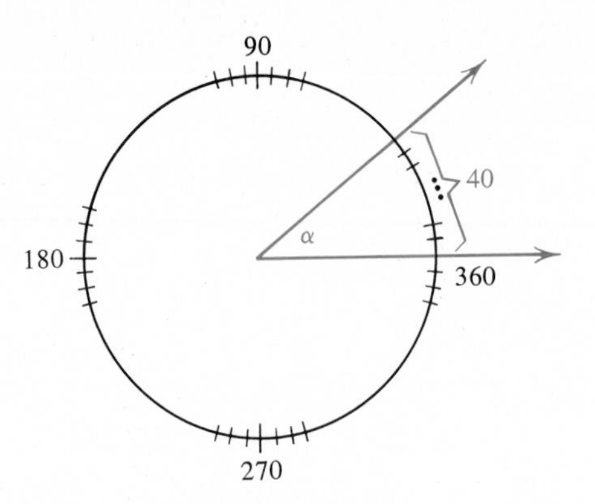

Figure A.17

Examples

a. $0.4° = (0.4 \times 60)' = 24'$

b. $35.2° = 35° + (0.2 \times 60)' = 35°\ 12'$

c. $141.7° = 141° + (0.7 \times 60)' = 141°\ 42'$

Conversely, to convert from minutes to tenths of degrees, simply divide the number of minutes by 60.

Examples

a. $36' = \left(\frac{36}{60}\right)^{\circ} = 0.6°$

b. $75°\ 18' = 75° + \left(\frac{18}{60}\right)^{\circ} = 75.3°$

c. $123°\ 54' = 123° + \left(\frac{54}{60}\right)^{\circ} = 123.9°$

A **protractor,** usually in the shape of a semicircle, as shown in Figure A.18, with 180 equally spaced marks on the semicircle, can be used to measure angles. If an angle is placed with its vertex at the center of the semicircle with one side passing through the 0° mark, the measure of the angle can be read directly.

Examples In Figure A.18,

$$\angle ABC = 50°, \quad \angle ABE = 110°, \quad \text{and} \quad \angle DBE = 70°.$$

Also

$$\angle CBE = 110° - 50° = 60°.$$

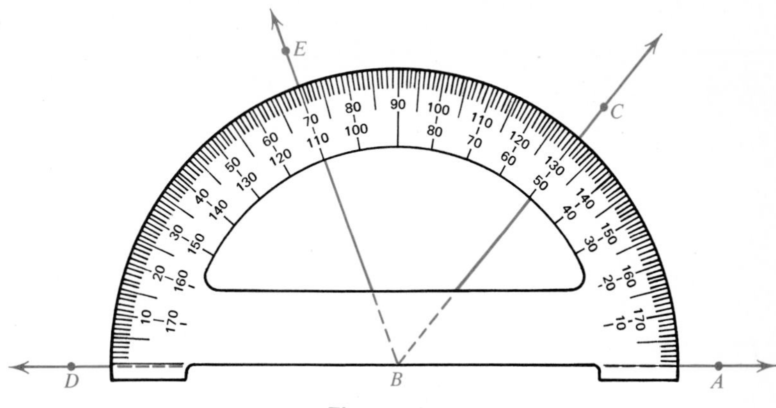

Figure A.18

If the measure of an angle is less than 90°, the angle is an **acute angle;** if more than 90° and less than 180°, it is an **obtuse angle;** if equal to 90°, it is a **right angle.** If two lines, or rays, form a right angle, the lines, or rays, are **perpendicular,** indicated by the symbol $\perp$. Thus $\overrightarrow{DE} \perp \overrightarrow{FG}$ is read "ray *DE* is perpendicular to ray *FG*."

Examples In the figure, α is an acute angle, β is an obtuse angle, and γ is a right angle. Also, $\overrightarrow{AC} \perp \overrightarrow{AB}$.

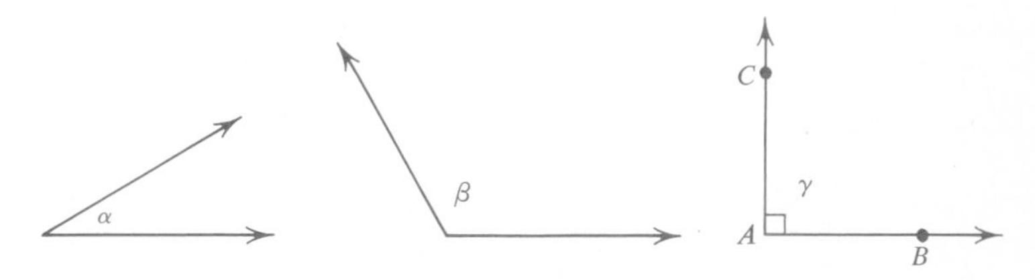

If the sum of the measures of two angles is 90°, the angles are **complementary;** if the sum is 180°, the angles are **supplementary.**

Examples

a. If $\angle D = 32°$, $\angle E = 148°$, and $\angle F = 58°$, then $\angle D$ and $\angle F$ are complementary, and $\angle D$ and $\angle E$ are supplementary.

b. In the figure, angles α and β are complementary, and angles γ and θ (theta) are supplementary.

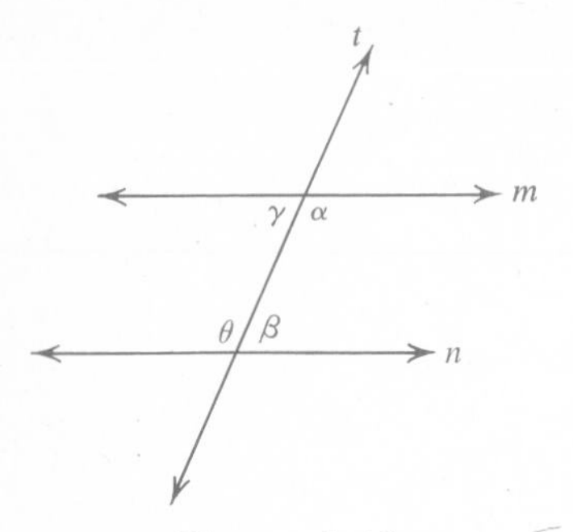

Figure A.19

Frequently, pairs of supplementary angles are introduced through the use of parallel lines. In Figure A.19, in which lines m and n are intersected by a transversal, line t, angles α and β are called interior angles on the same side of the transversal. If lines m and n are parallel, it follows that α and β are supplementary, as are γ and θ.

A.8 Triangles

A **triangle** is a figure formed by three line segments joining three points that are not in the same line. The line segments are the **sides,** and the three points are the **vertices** (plural of **vertex**) of the triangle. The letters assigned to the vertices can be used to name a triangle.

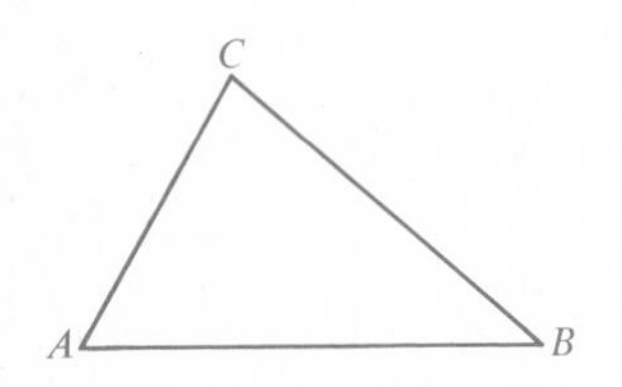

Example In the figure, the points A, B, and C are the vertices of $\triangle ABC$ (read "triangle ABC"), which can also be named by $\triangle ACB$, $\triangle BAC$, $\triangle BCA$, $\triangle CAB$, or $\triangle CBA$. The sides are $\overline{AB}$, $\overline{BC}$, and $\overline{CA}$; their lengths (positive real numbers) can be represented by AB, BC, and CA, respectively.

If each angle of a triangle is acute, the triangle is an **acute triangle;** if one angle is obtuse, it is an **obtuse triangle;** if one angle is a right angle, it is a **right triangle.** In a right triangle, the side opposite the right angle is the **hypotenuse,** and the other sides are the **legs.**

Examples In the figure, $\triangle DEF$ is an acute triangle, $\triangle KGH$ is an obtuse triangle, and $\triangle PQR$ is a right triangle. In $\triangle PQR$, $\overline{PQ}$ and $\overline{PR}$ are the legs, and $\overline{RQ}$ is the hypotenuse.

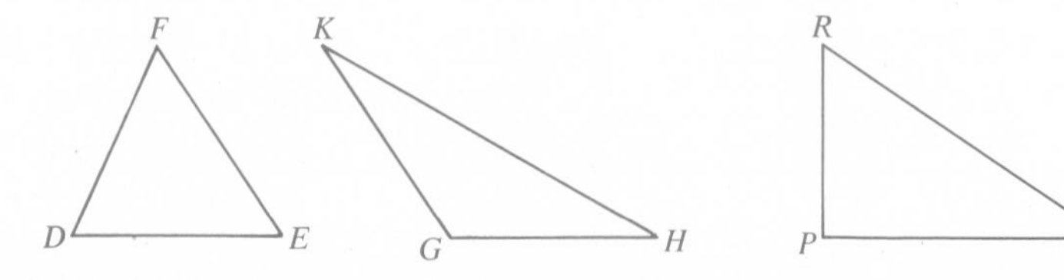

Triangles also have names related to the relative lengths of the sides. If the three sides are of different lengths, the triangle is a **scalene triangle;** if two sides have equal lengths, it is an **isosceles triangle;** if all three sides have equal lengths, it is an **equilateral triangle.**

Examples In the figure, $\triangle ABC$ is scalene, $\triangle DEF$ is isosceles ($FD = FE$), and $\triangle HKL$ is equilateral ($HK = KL = LH$). Note, further, that $\triangle ABC$ is an obtuse triangle, whereas $\triangle DEF$ and $\triangle HKL$ are both acute triangles.

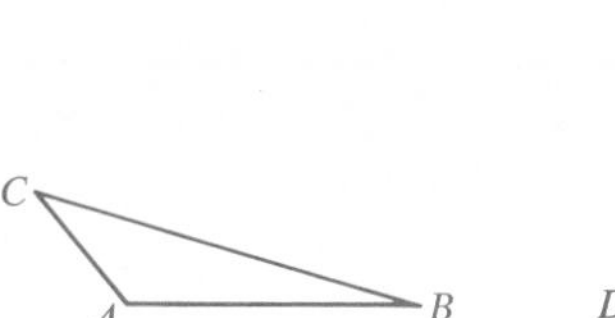

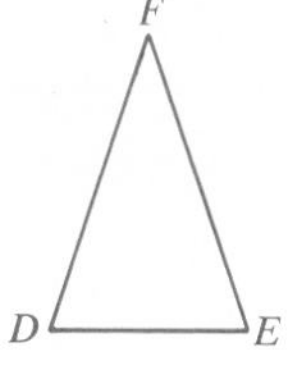

An **altitude** of a triangle is the length of a segment from any vertex of a triangle perpendicular to the opposite side. Sometimes the segment itself is called the altitude.

Examples In the figure, in $\triangle ABC$, h, or $\overline{CD}$, is the altitude to $\overline{AB}$. In $\triangle EFG$, $\overline{GH}$ is the altitude from G, and $\overline{FL}$ is the altitude from F.

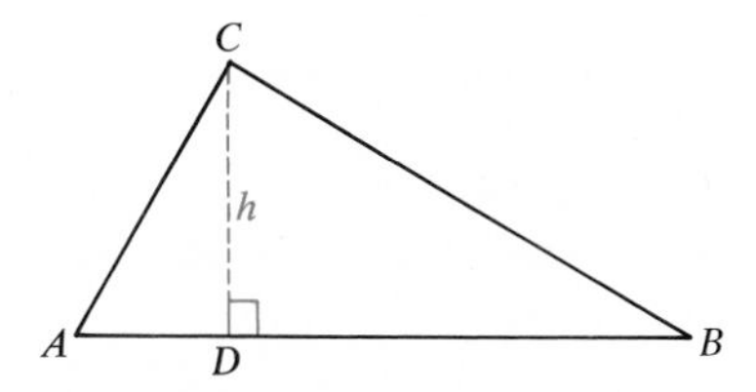

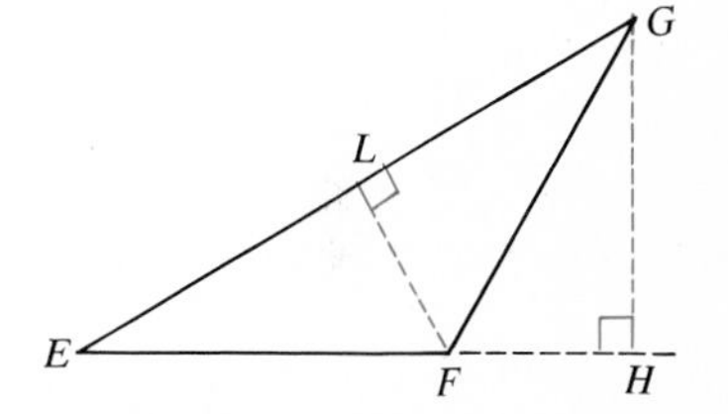

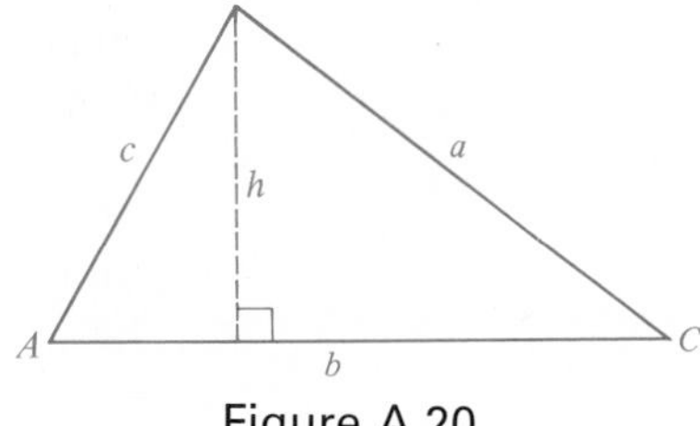

Figure A.20

The following properties hold for any triangle (see Figure A.20).

1. The perimeter is $P = a + b + c$.
2. The area is $\mathscr{A} = \frac{1}{2}bh$.
3. $\angle A + \angle B + \angle C = 180°$.

As in Properties 1 and 2, lowercase letters are often used to indicate the lengths of the sides opposite the angles named by the corresponding capital letters.

Examples

a. The perimeter of $\triangle ABC$ is

$$P = 3 + 5 + 6 = 14.$$

a.

b. If $h = 6$ and $CD = 10$, the area of $\triangle CDE$ is

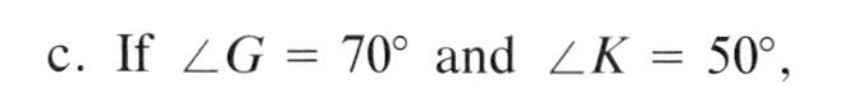

$$\mathscr{A} = \frac{1}{2}(10)(6) = 30.$$

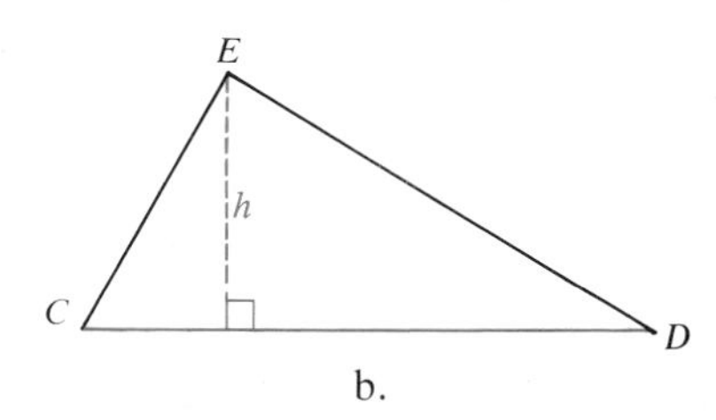

b.

c. If $\angle G = 70°$ and $\angle K = 50°$,

$$70° + 50° + \angle H = 180°,$$

from which we have that $\angle H = 60°$.

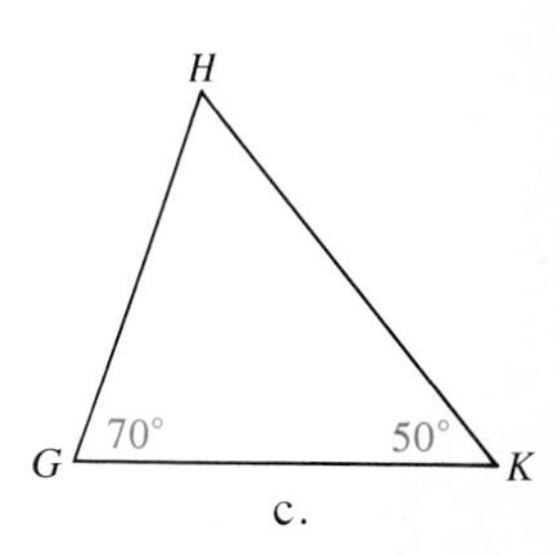

c.

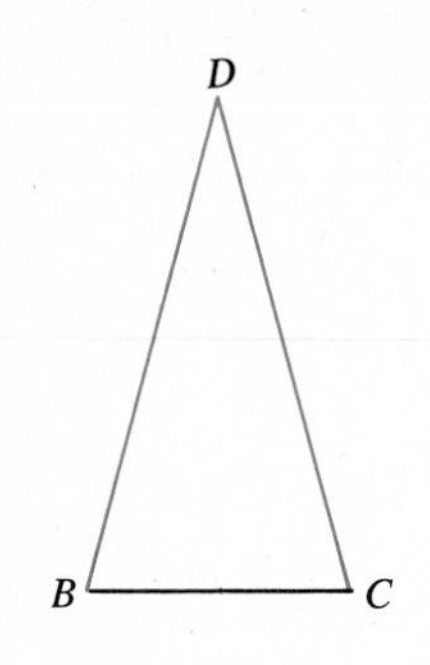

In addition to Properties 1, 2, and 3, the following properties also apply to certain special triangles.

4. In an isosceles triangle, the angles opposite the equal sides are equal.

Example If $BD = CD$, then $\angle B = \angle C$.

5. If a triangle is equilateral, then (see Figure A.21)

$$\angle A = \angle B = \angle C = 60°, \quad h = \frac{s}{2}\sqrt{3}, \quad \text{and} \quad \mathscr{A} = \frac{\sqrt{3}}{4}s^2.$$

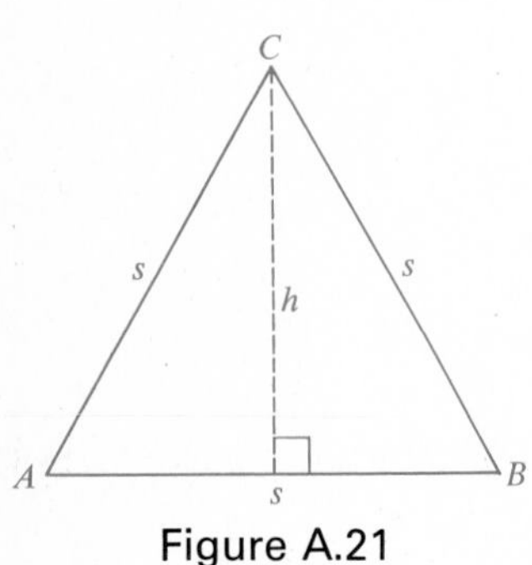

Figure A.21

Example If $s = 12$, then

$$h = \frac{12}{2}\sqrt{3} = 6\sqrt{3} \quad \text{and} \quad \mathscr{A} = \frac{\sqrt{3}}{4}(12)^2 = 36\sqrt{3}.$$

6. In a right triangle ($\angle C = 90°$),
   (i) $a^2 + b^2 = c^2$ (the Pythagorean theorem).
   (ii) If $a = b$, then $\angle A = \angle B = 45°$, and $c = a\sqrt{2}$.
   (iii) If one acute angle is 30°, and the length of the opposite side is a, then the other acute angle is 60°, the length of the hypotenuse is $2a$, and the length of the longer leg is $a\sqrt{3}$.

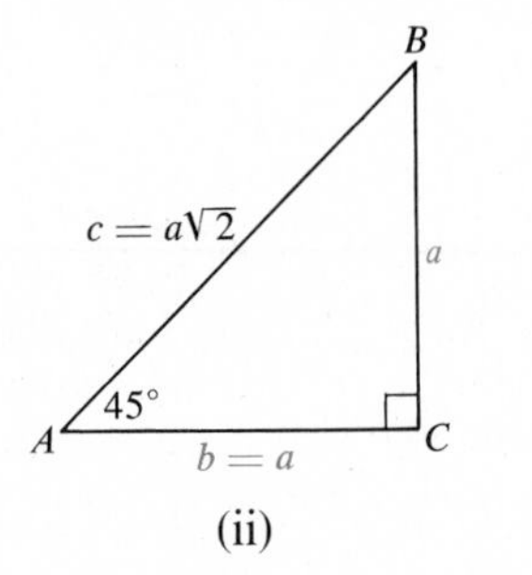

(ii)

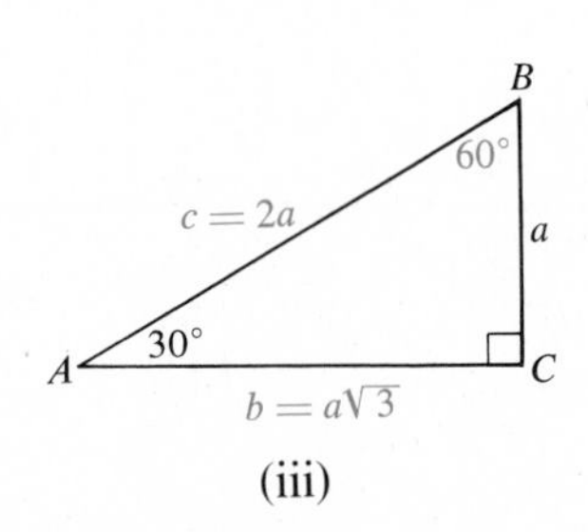

(iii)

Examples Using the measurements of the appropriate triangle:

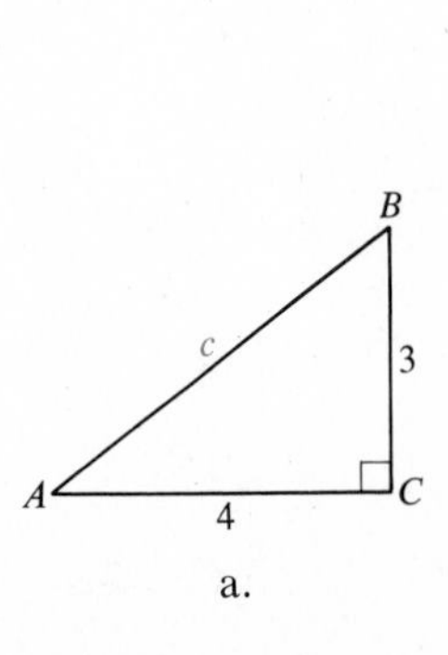

a.

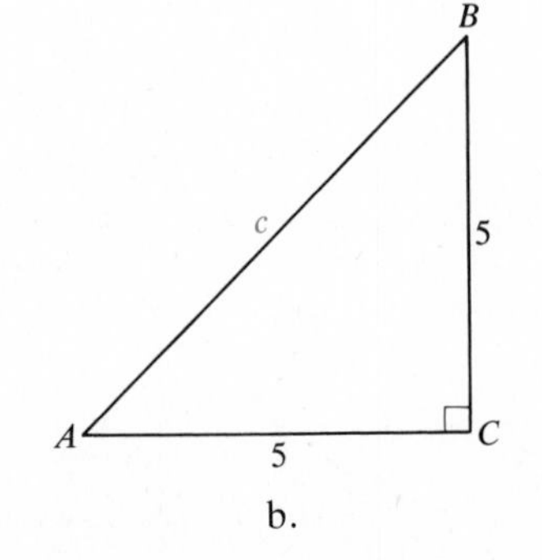

b.

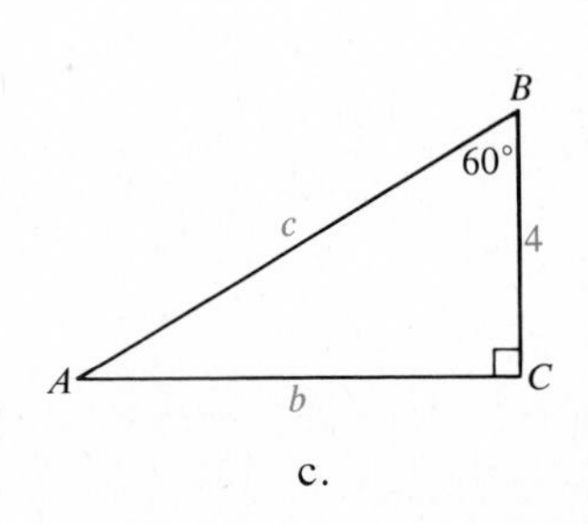

c.

a. $c^2 = 3^2 + 4^2 = 9 + 16 = 25;\ c = 5$

b. $\angle A = \angle B = 45°;\ c = 5\sqrt{2}$

c. $\angle A = 30°;\ c = 2 \cdot 4 = 8;\ b = 4\sqrt{3}$

A.9 Congruent and Similar Triangles

If two triangles are the same size and shape, they are said to be **congruent triangles.** In Figure A.22, the triangles are congruent. This is in-

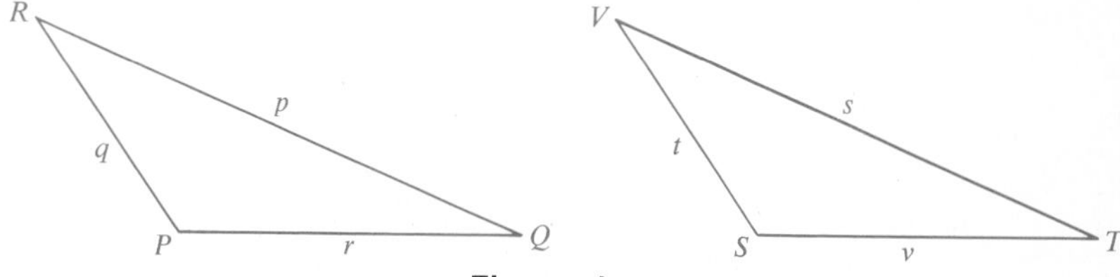

Figure A.22

dicated by use of the symbol ≅ (read "is congruent to"). Thus, in Figure A.22, $\triangle PQR \cong \triangle STV$. From this it follows that

$$\angle P = \angle S, \quad p = s,$$
$$\angle Q = \angle T, \quad q = t,$$
$$\angle R = \angle V, \quad r = v.$$

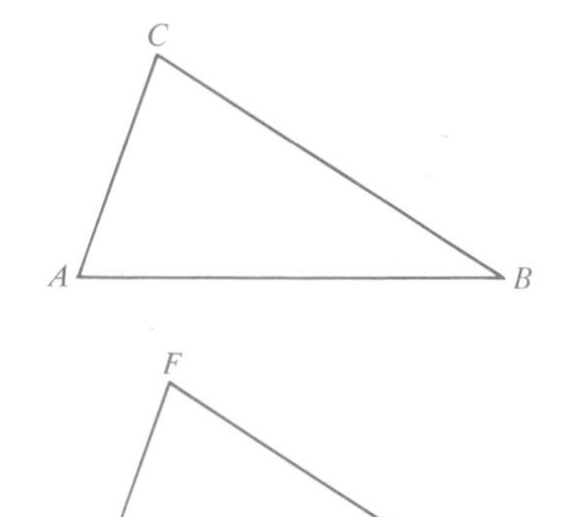

Figure A.23

If two triangles are the same shape, but not necessarily the same size, they are said to be **similar triangles.** In Figure A.23, the triangles are similar. This is written $\triangle ABC \sim \triangle DEF$, where "~" is read "is similar to." In two similar triangles, the corresponding angles are equal—that is, $\angle A = \angle D$, $\angle B = \angle E$, $\angle C = \angle F$; and the corresponding sides are proportional—that is,

$$\frac{AC}{DF} = \frac{AB}{DE} = \frac{BC}{EF}.$$

Example Given that $\triangle RST \sim \triangle ADF$, we can find AD as follows. Since the triangles are similar, we have

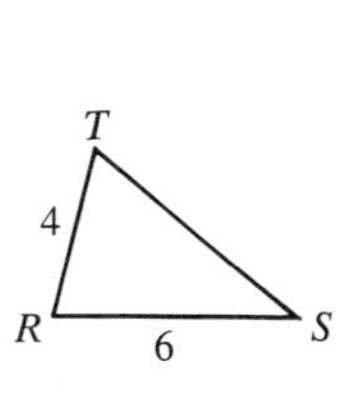

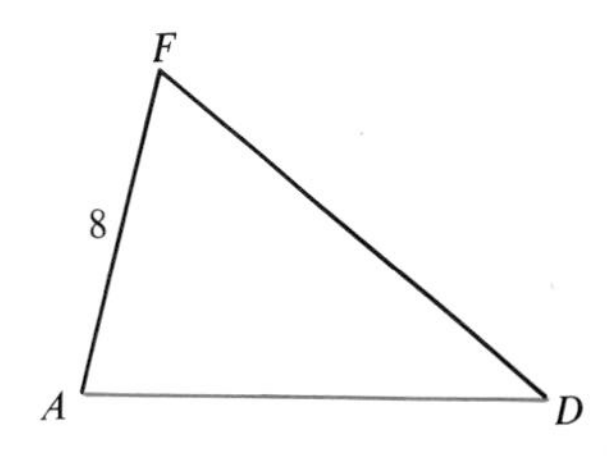

$$\frac{AD}{RS} = \frac{AF}{RT},$$

$$\frac{AD}{6} = \frac{8}{4},$$

from which

$$AD = 12.$$

The following properties can be used to prove that two triangles are similar:

1. two angles of one triangle are equal, respectively, to two angles of the other, or
2. one angle of one triangle is equal to one angle of the other, and the including sides are proportional, or
3. three sides of one triangle are proportional to three sides of the other.

Examples Using the measurements in the figures:

a. $\angle P = \angle M$ and $\angle Q = \angle N$;

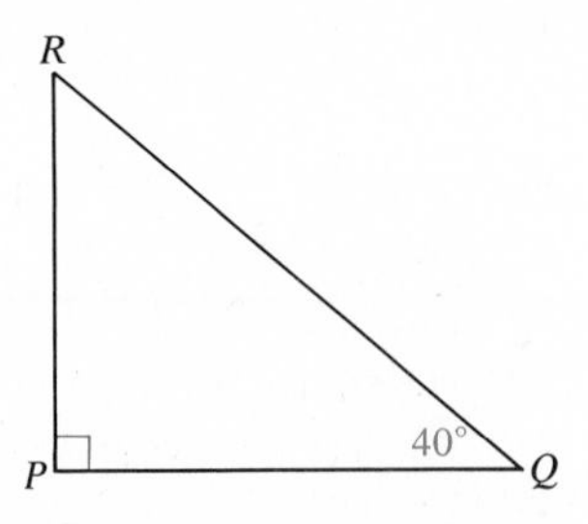

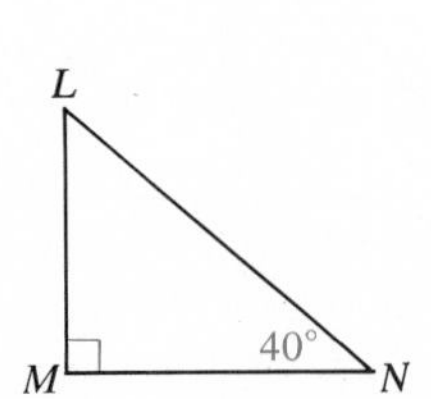

therefore, by Property 1, $\triangle PQR \sim \triangle MNL$.

b. $\angle A = \angle D$ and $\dfrac{AB}{DE} = \dfrac{AC}{DF}$, because $\dfrac{9}{6} = \dfrac{6}{4}$;

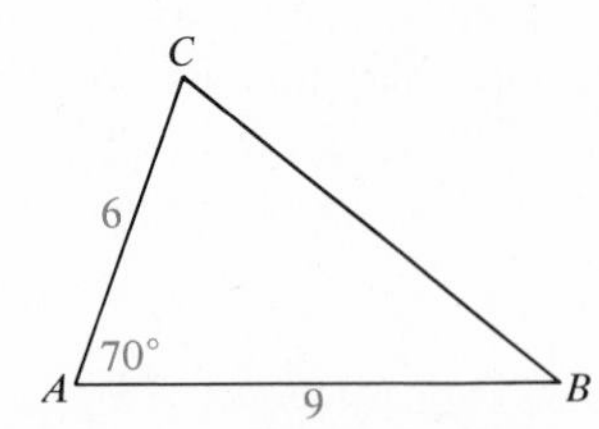

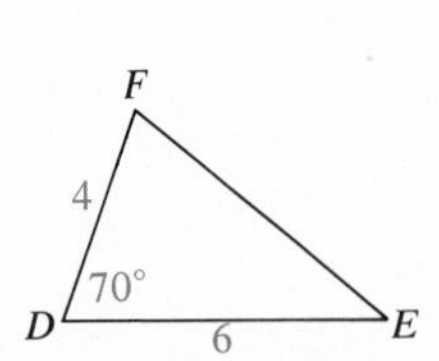

therefore, by Property 2, $\triangle ABC \sim \triangle DEF$.

c. $\frac{GK}{SW} = \frac{KH}{WT} = \frac{GH}{ST}$ because $\frac{8}{4} = \frac{10}{5} = \frac{12}{6}$;

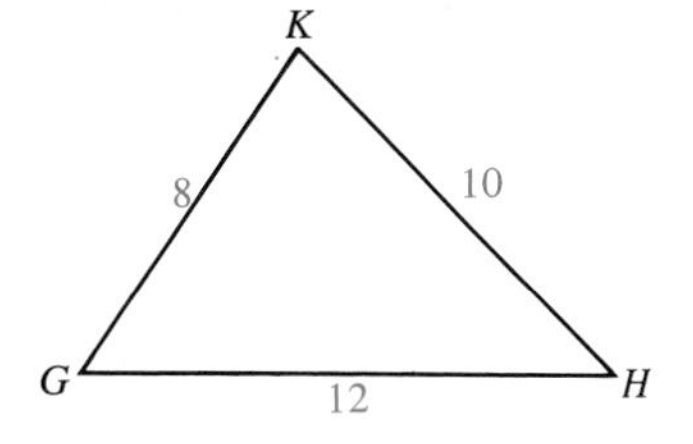

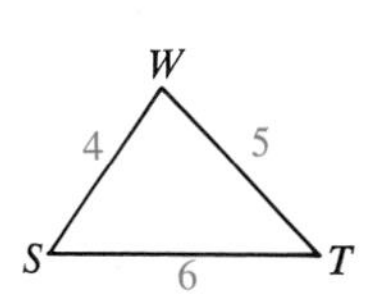

therefore, by Property 3, $\triangle GHK \sim \triangle STW$.

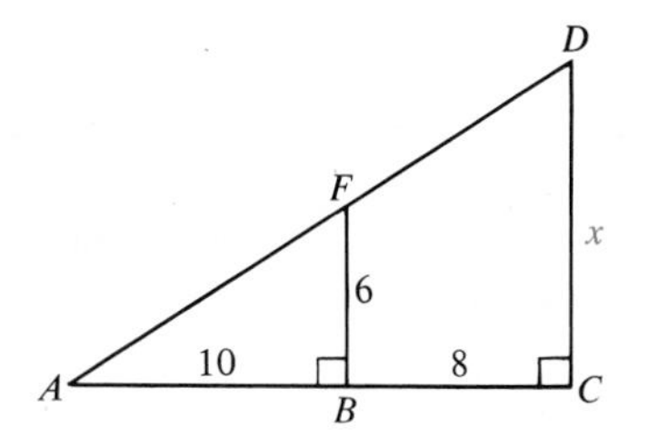

Example To find the length x in the figure, we proceed as follows. $\angle A = \angle A$ and $\angle ABF = \angle ACD$; hence, by Property 1, we have $\triangle ABF \sim \triangle ACD$. Also, $AC = AB + BC = 18$. Because the triangles are similar, the corresponding sides are proportional. Thus,

$$\frac{DC}{FB} = \frac{AC}{AB} \quad \text{and} \quad \frac{x}{6} = \frac{18}{10},$$

from which

$$x = \frac{6 \cdot 18}{10} = \frac{54}{5}.$$

Appendix B

Linear Interpolation

If Table II or Table III in Appendix C is used and we wish to obtain approximations for function values between two entries, we can use a method called **linear interpolation** to approximate such values.

The linear interpolation method assumes that differences in function values are directly proportional to the differences of the measures of the angles over small intervals. In fact, the differences are not directly proportional and we are simply obtaining closer rational number approximations for the function values than could be obtained by reading the nearest value from Table II or Table III.

Example Find an approximation for $\sin 0.714^{R}$ using linear interpolation.

Solution From Table III, we find that

$$\sin 0.71^{R} \approx 0.6518 \qquad \text{and} \qquad \sin 0.72^{R} \approx 0.6594.$$

The following arrangement of the data helps us to set up the direct proportion that we assume.

$$0.010\left\{\begin{array}{l} 0.004\left\{\begin{array}{l}\sin 0.710^{R} \approx 0.6518 \\ \sin 0.714^{R} \approx \ ?\end{array}\right\} d \\ \qquad\ \ \sin 0.720^{R} \approx 0.6594 \end{array}\right\} 0.0076 \quad (d = \text{difference}).$$

We have

$$\frac{0.004}{0.010} \approx \frac{d}{0.0076},$$

$$d \approx \frac{4}{10}(0.0076) \approx 0.0030.$$

Thus,

$$\sin 0.714 \approx 0.6518 + 0.0030 = 0.6548.$$

The next example illustrates a case where the difference d obtained from the proportion is *subtracted* from one of the values obtained from the table.

Example Find an approximation for $\cos 0.393^{R}$.

Solution From Table III,

$$\cos 0.390^{R} \approx 0.9249 \qquad \text{and} \qquad \cos 0.400^{R} \approx 0.9211.$$

By arranging the data for interpolation, we obtain

$$0.010\left\{\begin{array}{l} 0.003\left\{\begin{array}{l}\cos 0.390^{R} \approx 0.9249 \\ \cos 0.393^{R} \approx ?\end{array}\right\} d \\ \qquad\quad \cos 0.400^{R} \approx 0.9211 \end{array}\right\} 0.0038.$$

We have

$$\frac{0.003}{0.010} \approx \frac{d}{0.0038},$$

$$d \approx \frac{3}{10}(0.0038) \approx 0.0011.$$

Thus,

$$\cos 0.393^{R} \approx 0.9249 - 0.0011 = 0.9238.$$

Notice that we *subtracted* the difference, 0.0011, because in this interval *values for* $\cos \alpha$ *decrease as* α *increases*.

Linear interpolation can also be used to find an approximation for the measure of an angle associated with a given element in the range that is not an entry in the table.

Example Find an approximation for an angle α in radian measure such that $\tan \alpha = 0.3369$ and $0^{R} \leq \alpha \leq 1.57^{R}$.

Solution From Table III, we observe that

$$\tan 0.320^{R} \approx 0.3314 \qquad \text{and} \qquad \tan 0.330^{R} \approx 0.3425.$$

(continued)

By arranging the data for interpolation, we obtain

$$0.010\left\{\begin{array}{l} d\left\{\begin{array}{ll} \tan 0.320^{R} \approx 0.3314 \\ \tan \alpha \quad\quad \approx 0.3369 \end{array}\right\} 0.0055 \\ \tan 0.330^{R} \approx 0.3425 \end{array}\right\} 0.0111.$$

We have

$$\frac{d}{0.010} \approx \frac{0.0055}{0.0111},$$

$$d \approx \frac{55}{111}(0.010) \approx 0.005.$$

Thus,

$$\alpha \approx 0.320 + 0.005 = 0.325.$$

In the above examples we used Table III because the measure of the angle was in radians. A similar procedure of linear interpolation is also applicable when the measure of an angle is in degrees. An example is shown in the exercises.

With practice, the computations involved in linear interpolation, as in the examples above, can often be done mentally.

EXERCISE SET B.1

Use linear interpolation to find an approximation for each function value.

Example $\cos 9° 14'$

Solution From Table II,

$$\cos 9° 12' \approx 0.9871 \qquad \text{and} \qquad \cos 9° 18' \approx 0.9869.$$

The data arranged for interpolation appear as

$$6'\left\{\begin{array}{l} 2'\left\{\begin{array}{l} \cos 9° 12' \approx 0.9871 \\ \cos 9° 14' \approx ? \end{array}\right\} d \\ \cos 9° 18' \approx 0.9869 \end{array}\right\} 0.0002.$$

Therefore,

$$\frac{2}{6} \approx \frac{d}{0.0002},$$

$$d \approx \frac{1}{3}(0.0002) \approx 0.0001.$$

Thus, $\cos 9°\, 14' \approx 0.9871 - 0.0001 = 0.9870.$

1. $\sin 3°\, 38'$ **2.** $\cos 10°\, 34'$ **3.** $\cot 60°\, 15'$ **4.** $\csc 45°\, 37'$

5. $\tan 42.32°$ **6.** $\cot 17.06°$ **7.** $\cos 15.62°$ **8.** $\sin 82.48°$

9. $\sin 0.255^{R}$ **10.** $\tan 0.536^{R}$ **11.** $\sec 0.904^{R}$ **12.** $\cot 1.362^{R}$

Use linear interpolation to find an approximation for an acute angle α in degree measure for each of the following.

Example $\sin \alpha = 0.4457$

Solution From Table II,

$$\sin 26.4° \approx 0.4446 \quad \text{and} \quad \sin 26.5° \approx 0.4462.$$

The data arranged for interpolation appears as

$$0.10\left\{\begin{array}{l} d\left\{\begin{array}{l}\sin 26.4° \approx 0.4446 \\ \sin 26.?° \approx 0.4457\end{array}\right\} 0.0011 \\ \sin 26.5° \approx 0.4462\end{array}\right\} 0.0016.$$

Therefore,

$$\frac{d}{0.10} \approx \frac{0.0011}{0.0016},$$

$$d \approx 0.10\left(\frac{11}{16}\right) \approx 0.07.$$

Thus, $\alpha \approx 26.4° + 0.07° = 26.47°.$

13. $\sin \alpha = 0.2745$ **14.** $\cos \alpha = 0.9332$ **15.** $\tan \alpha = 0.4819$

16. $\csc \alpha = 1.9461$ **17.** $\sec \alpha = 1.2664$ **18.** $\cot \alpha = 1.0522$

19. $\sin \alpha = 0.7887$ **20.** $\cos \alpha = 0.4605$

Use linear interpolation to find an approximation for an acute angle α in radian measure for each of the following.

21. $\sin \alpha = 0.3099$ **22.** $\cos \alpha = 0.7944$ **23.** $\tan \alpha = 1.578$

24. $\csc \alpha = 2.901$ **25.** $\sec \alpha = 1.440$ **26.** $\cot \alpha = 0.3838$

27. $\sin \alpha = 0.8243$ **28.** $\cos \alpha = 0.7325$

Appendix C

Tables

TABLE I Powers, roots, and reciprocals

N	N^2	$\sqrt{N}$	$\sqrt{10N}$	$1/N$	N	N^2	$\sqrt{N}$	$\sqrt{10N}$	$1/N$
1	1	1.0 00	3.1 62	1.00 00	51	2 601	7.1 41	22.5 83	.019 61
2	4	1.4 14	4.4 72	.500 00	52	2 704	7.2 11	22.8 04	.019 23
3	9	1.7 32	5.4 77	.333 33	53	2 809	7.2 80	23.0 22	.018 87
4	16	2.0 00	6.3 25	.250 00	54	2 916	7.3 48	23.2 38	.018 52
5	25	2.2 36	7.0 71	.200 00	55	3 025	7.4 16	23.4 52	.018 18
6	36	2.4 49	7.7 46	.166 67	56	3 136	7.4 83	23.6 64	.017 86
7	49	2.6 46	8.3 67	.142 86	57	3 249	7.5 50	23.8 75	.017 54
8	64	2.8 28	8.9 44	.125 00	58	3 364	7.6 16	24.0 83	.017 24
9	81	3.0 00	9.4 87	.111 11	59	3 481	7.6 81	24.2 90	.016 95
10	100	3.1 62	10.0 00	.100 00	**60**	3 600	7.7 46	24.4 95	.016 67
11	121	3.3 17	10.4 88	.090 91	61	3 721	7.8 10	24.6 98	.016 39
12	144	3.4 64	10.9 54	.083 33	62	3 844	7.8 74	24.9 00	.016 13
13	169	3.6 06	11.4 02	.076 92	63	3 969	7.9 37	25.1 00	.015 87
14	196	3.7 42	11.8 32	.071 43	64	4 096	8.0 00	25.2 98	.015 62
15	225	3.8 73	12.2 47	.066 67	65	4 225	8.0 62	25.4 95	.015 38
16	256	4.0 00	12.6 49	.062 50	66	4 356	8.1 24	25.6 90	.015 15
17	289	4.1 23	13.0 38	.058 82	67	4 489	8.1 85	25.8 84	.014 93
18	324	4.2 43	13.4 16	.055 56	68	4 624	8.2 46	26.0 77	.014 71
19	361	4.3 59	13.7 84	.052 63	69	4 761	8.3 07	26.2 68	.014 49
20	400	4.4 72	14.1 42	.050 00	**70**	4 900	8.3 67	26.4 58	.014 29
21	441	4.5 83	14.4 91	.047 62	71	5 041	8.4 26	26.6 46	.014 08
22	484	4.6 90	14.8 32	.045 45	72	5 184	8.4 85	26.8 33	.013 89
23	529	4.7 96	15.1 66	.043 48	73	5 329	8.5 44	27.0 19	.013 70
24	576	4.8 99	15.4 92	.041 67	74	5 476	8.6 02	27.2 03	.013 51
25	625	5.0 00	15.8 11	.040 00	75	5 625	8.6 60	27.3 86	.013 33
26	676	5.0 99	16.1 25	.038 46	76	5 776	8.7 18	27.5 68	.013 16
27	729	5.1 96	16.4 32	.037 04	77	5 929	8.7 75	27.7 49	.012 99
28	784	5.2 92	16.7 33	.035 71	78	6 084	8.8 32	27.9 28	.012 82
29	841	5.3 85	17.0 29	.034 48	79	6 241	8.8 88	28.1 07	.012 66
30	900	5.4 77	17.3 21	.033 33	**80**	6 400	8.9 44	28.2 84	.012 50
31	961	5.5 68	17.6 07	.032 26	81	6 561	9.0 00	28.4 60	.012 35
32	1 024	5.6 57	17.8 89	.031 25	82	6 724	9.0 55	28.6 36	.012 20
33	1 089	5.7 45	18.1 66	.030 30	83	6 889	9.1 10	28.8 10	.012 05
34	1 156	5.8 31	18.4 39	.029 41	84	7 056	9.1 65	28.9 83	.011 90
35	1 225	5.9 16	18.7 08	.028 57	85	7 225	9.2 20	29.1 55	.011 76
36	1 296	6.0 00	18.9 74	.027 78	86	7 396	9.2 74	29.3 26	.011 63
37	1 369	6.0 83	19.2 35	.027 03	87	7 569	9.3 27	29.4 96	.011 49
38	1 444	6.1 64	19.4 94	.026 32	88	7 744	9.3 81	29.6 65	.011 36
39	1 521	6.2 45	19.7 48	.025 64	89	7 921	9.4 34	29.8 33	.011 24
40	1 600	6.3 25	20.0 00	.025 00	**90**	8 100	9.4 87	30.0 00	.011 11
41	1 681	6.4 03	20.2 48	.024 39	91	8 281	9.5 39	30.1 66	.010 99
42	1 764	6.4 81	20.4 94	.023 81	92	8 464	9.5 92	30.3 32	.010 87
43	1 849	6.5 57	20.7 36	.023 26	93	8 649	9.6 44	30.4 96	.010 75
44	1 936	6.6 33	20.9 76	.022 73	94	8 836	9.6 95	30.6 59	.010 64
45	2 025	6.7 08	21.2 13	.022 22	95	9 025	9.7 47	30.8 22	.010 53
46	2 116	6.7 82	21.4 48	.021 74	96	9 216	9.7 98	30.9 84	.010 42
47	2 209	6.8 56	21.6 79	.021 28	97	9 409	9.8 49	31.1 45	.010 31
48	2 304	6.9 28	21.9 09	.020 83	98	9 604	9.8 99	31.3 05	.010 20
49	2 401	7.0 00	22.1 36	.020 41	99	9 801	9.9 50	31.4 64	.010 10
50	2 500	7.0 71	22.3 61	.020 00	**100**	10 000	10.0 00	31.6 23	.010 00
N	N^2	$\sqrt{N}$	$\sqrt{10N}$	$1/N$	N	N^2	$\sqrt{N}$	$\sqrt{10N}$	$1/N$

TABLE II Values of trigonometric functions

θ deg	deg	min	sin θ	cos θ	tan θ	csc θ	sec θ	cot θ			
0.0	0	0	0.0000	1.0000	0.0000	No value	1.0000	No value	90	0	90.0
0.1	0	6	0.0017	1.0000	0.0017	572.96	1.0000	572.96	89	54	89.9
0.2	0	12	0.0035	1.0000	0.0035	286.48	1.0000	286.48	89	48	89.8
0.3	0	18	0.0052	1.0000	0.0052	190.99	1.0000	190.98	89	42	89.7
0.4	0	24	0.0070	1.0000	0.0070	143.24	1.0000	143.24	89	36	89.6
0.5	0	30	0.0087	1.0000	0.0087	114.59	1.0000	114.59	89	30	89.5
0.6	0	36	0.0105	0.9999	0.0105	95.495	1.0001	95.490	89	24	89.4
0.7	0	42	0.0122	0.9999	0.0122	81.853	1.0001	81.847	89	18	89.3
0.8	0	48	0.0140	0.9999	0.0140	71.622	1.0001	71.615	89	12	89.2
0.9	0	54	0.0157	0.9999	0.0157	63.665	1.0001	63.657	89	6	89.1
1.0	1	0	0.0175	0.9998	0.0175	57.299	1.0002	57.290	89	0	89.0
1.1	1	6	0.0192	0.9998	0.0192	52.090	1.0002	52.081	88	54	88.9
1.2	1	12	0.0209	0.9998	0.0209	47.750	1.0002	47.740	88	48	88.8
1.3	1	18	0.0227	0.9997	0.0227	44.077	1.0003	44.066	88	42	88.7
1.4	1	24	0.0244	0.9997	0.0244	40.930	1.0003	40.917	88	36	88.6
1.5	1	30	0.0262	0.9997	0.0262	38.202	1.0003	38.188	88	30	88.5
1.6	1	36	0.0279	0.9996	0.0279	35.815	1.0004	35.801	88	24	88.4
1.7	1	42	0.0297	0.9996	0.0297	33.708	1.0004	33.694	88	18	88.3
1.8	1	48	0.0314	0.9995	0.0314	31.836	1.0005	31.821	88	12	88.2
1.9	1	54	0.0332	0.9995	0.0332	30.161	1.0005	30.145	88	6	88.1
2.0	2	0	0.0349	0.9994	0.0349	28.654	1.0006	28.636	88	0	88.0
2.1	2	6	0.0366	0.9993	0.0367	27.290	1.0007	27.271	87	54	87.9
2.2	2	12	0.0384	0.9993	0.0384	26.050	1.0007	26.031	87	48	87.8
2.3	2	18	0.0401	0.9992	0.0402	24.918	1.0008	24.898	87	42	87.7
2.4	2	24	0.0419	0.9991	0.0419	23.880	1.0009	23.859	87	36	87.6
2.5	2	30	0.0436	0.9990	0.0437	22.926	1.0010	22.904	87	30	87.5
2.6	2	36	0.0454	0.9990	0.0454	22.044	1.0010	22.022	87	24	87.4
2.7	2	42	0.0471	0.9989	0.0472	21.229	1.0011	21.205	87	18	87.3
2.8	2	48	0.0488	0.9988	0.0489	20.471	1.0012	20.446	87	12	87.2
2.9	2	54	0.0506	0.9987	0.0507	19.766	1.0013	19.740	87	6	87.1
3.0	3	0	0.0523	0.9986	0.0524	19.107	1.0014	19.081	87	0	87.0
3.1	3	6	0.0541	0.9985	0.0542	18.492	1.0015	18.464	86	54	86.9
3.2	3	12	0.0558	0.9984	0.0559	17.914	1.0016	17.886	86	48	86.8
3.3	3	18	0.0576	0.9983	0.0577	17.372	1.0017	17.343	86	42	86.7
3.4	3	24	0.0593	0.9982	0.0594	16.862	1.0018	16.832	86	36	86.6
3.5	3	30	0.0610	0.9981	0.0612	16.380	1.0019	16.350	86	30	86.5
3.6	3	36	0.0628	0.9980	0.0629	15.926	1.0020	15.895	86	24	86.4
3.7	3	42	0.0645	0.9979	0.0647	15.496	1.0021	15.464	86	18	86.3
3.8	3	48	0.0663	0.9978	0.0664	15.089	1.0022	15.056	86	12	86.2
3.9	3	54	0.0680	0.9977	0.0682	14.703	1.0023	14.669	86	6	86.1
4.0	4	0	0.0698	0.9976	0.0699	14.336	1.0024	14.301	86	0	86.0
4.1	4	6	0.0715	0.9974	0.0717	13.987	1.0026	13.951	85	54	85.9
4.2	4	12	0.0732	0.9973	0.0734	13.654	1.0027	13.617	85	48	85.8
4.3	4	18	0.0750	0.9972	0.0752	13.337	1.0028	13.300	85	42	85.7
4.4	4	24	0.0767	0.9971	0.0769	13.035	1.0030	12.996	85	36	85.6
4.5	4	30	0.0785	0.9969	0.0787	12.746	1.0031	12.706	85	30	85.5
4.6	4	36	0.0802	0.9968	0.0805	12.469	1.0032	12.429	85	24	85.4
4.7	4	42	0.0819	0.9966	0.0822	12.204	1.0034	12.163	85	18	85.3
4.8	4	48	0.0837	0.9965	0.0840	11.951	1.0035	11.909	85	12	85.2
4.9	4	54	0.0854	0.9963	0.0857	11.707	1.0037	11.665	85	6	85.1
			cos θ	sin θ	cot θ	sec θ	csc θ	tan θ	deg	min	θ deg

TABLE II (*continued*)

θ deg	deg	min	sin θ	cos θ	tan θ	csc θ	sec θ	cot θ			
5.0	5	0	0.0872	0.9962	0.0875	11.474	1.0038	11.430	85	0	85.0
5.1	5	6	0.0889	0.9960	0.0892	11.249	1.0040	11.205	84	54	84.9
5.2	5	12	0.0906	0.9959	0.0910	11.034	1.0041	10.988	84	48	84.8
5.3	5	18	0.0924	0.9957	0.0928	10.826	1.0043	10.780	84	42	84.7
5.4	5	24	0.0941	0.9956	0.0945	10.626	1.0045	10.579	84	36	84.6
5.5	5	30	0.0958	0.9954	0.0963	10.433	1.0046	10.385	84	30	84.5
5.6	5	36	0.0976	0.9952	0.0981	10.248	1.0048	10.199	84	24	84.4
5.7	5	42	0.0993	0.9951	0.0998	10.069	1.0050	10.019	84	18	84.3
5.8	5	48	0.1011	0.9949	0.1016	9.8955	1.0051	9.8448	84	12	84.2
5.9	5	54	0.1028	0.9947	0.1033	9.7283	1.0053	9.6768	84	6	84.1
6.0	6	0	0.1045	0.9945	0.1051	9.5668	1.0055	9.5144	84	0	84.0
6.1	6	6	0.1063	0.9943	0.1069	9.4105	1.0057	9.3573	83	54	83.9
6.2	6	12	0.1080	0.9942	0.1086	9.2593	1.0059	9.2052	83	48	83.8
6.3	6	18	0.1097	0.9940	0.1104	9.1129	1.0061	9.0579	83	42	83.7
6.4	6	24	0.1115	0.9938	0.1122	8.9711	1.0063	8.9152	83	36	83.6
6.5	6	30	0.1132	0.9936	0.1139	8.8337	1.0065	8.7769	83	30	83.5
6.6	6	36	0.1149	0.9934	0.1157	8.7004	1.0067	8.6428	83	24	83.4
6.7	6	42	0.1167	0.9932	0.1175	8.5711	1.0069	8.5126	83	18	83.3
6.8	6	48	0.1184	0.9930	0.1192	8.4457	1.0071	8.3863	83	12	83.2
6.9	6	54	0.1201	0.9928	0.1210	8.3238	1.0073	8.2636	83	6	83.1
7.0	7	0	0.1219	0.9925	0.1228	8.2055	1.0075	8.1444	83	0	83.0
7.1	7	6	0.1236	0.9923	0.1246	8.0905	1.0077	8.0285	82	54	82.9
7.2	7	12	0.1253	0.9921	0.1263	7.9787	1.0079	7.9158	82	48	82.8
7.3	7	18	0.1271	0.9919	0.1281	7.8700	1.0082	7.8062	82	42	82.7
7.4	7	24	0.1288	0.9917	0.1299	7.7642	1.0084	7.6996	82	36	82.6
7.5	7	30	0.1305	0.9914	0.1317	7.6613	1.0086	7.5958	82	30	82.5
7.6	7	36	0.1323	0.9912	0.1334	7.5611	1.0089	7.4947	82	24	82.4
7.7	7	42	0.1340	0.9910	0.1352	7.4635	1.0091	7.3962	82	18	82.3
7.8	7	48	0.1357	0.9907	0.1370	7.3684	1.0093	7.3002	82	12	82.2
7.9	7	54	0.1374	0.9905	0.1388	7.2757	1.0096	7.2066	82	6	82.1
8.0	8	0	0.1392	0.9903	0.1405	7.1853	1.0098	7.1154	82	0	82.0
8.1	8	6	0.1409	0.9900	0.1423	7.0972	1.0101	7.0264	81	54	81.9
8.2	8	12	0.1426	0.9898	0.1441	7.0112	1.0103	6.9395	81	48	81.8
8.3	8	18	0.1444	0.9895	0.1459	6.9273	1.0106	6.8548	81	42	81.7
8.4	8	24	0.1461	0.9893	0.1477	6.8454	1.0108	6.7720	81	36	81.6
8.5	8	30	0.1478	0.9890	0.1495	6.7655	1.0111	6.6912	81	30	81.5
8.6	8	36	0.1495	0.9888	0.1512	6.6874	1.0114	6.6122	81	24	81.4
8.7	8	42	0.1513	0.9885	0.1530	6.6111	1.0116	6.5350	81	18	81.3
8.8	8	48	0.1530	0.9882	0.1548	6.5366	1.0119	6.4596	81	12	81.2
8.9	8	54	0.1547	0.9880	0.1566	6.4637	1.0122	6.3859	81	6	81.1
9.0	9	0	0.1564	0.9877	0.1584	6.3925	1.0125	6.3138	81	0	81.0
9.1	9	6	0.1582	0.9874	0.1602	6.3228	1.0127	6.2432	80	54	80.9
9.2	9	12	0.1599	0.9871	0.1620	6.2547	1.0130	6.1742	80	48	80.8
9.3	9	18	0.1616	0.9869	0.1638	6.1880	1.0133	6.1066	80	42	80.7
9.4	9	24	0.1633	0.9866	0.1655	6.1227	1.0136	6.0405	80	36	80.6
9.5	9	30	0.1650	0.9863	0.1673	6.0589	1.0139	5.9758	80	30	80.5
9.6	9	36	0.1668	0.9860	0.1691	5.9963	1.0142	5.9124	80	24	80.4
9.7	9	42	0.1685	0.9857	0.1709	5.9351	1.0145	5.8502	80	18	80.3
9.8	9	48	0.1702	0.9854	0.1727	5.8751	1.0148	5.7894	80	12	80.2
9.9	9	54	0.1719	0.9851	0.1745	5.8164	1.0151	5.7297	80	6	80.1
			cos θ	sin θ	cot θ	sec θ	csc θ	tan θ	deg	min	θ deg

II

TABLE II (*continued*)

θ deg	**deg**	**min**	**sin θ**	**cos θ**	**tan θ**	**csc θ**	**sec θ**	**cot θ**			
10.0	10	0	0.1736	0.9848	0.1763	5.7588	1.0154	5.6713	80	0	80.0
10.1	10	6	0.1754	0.9845	0.1781	5.7023	1.0157	5.6140	79	54	79.9
10.2	10	12	0.1771	0.9842	0.1799	5.6470	1.0161	5.5578	79	48	79.8
10.3	10	18	0.1788	0.9839	0.1817	5.5928	1.0164	5.5027	79	42	79.7
10.4	10	24	0.1805	0.9836	0.1835	5.5396	1.0167	5.4486	79	36	79.6
10.5	10	30	0.1822	0.9833	0.1853	5.4874	1.0170	5.3955	79	30	79.5
10.6	10	36	0.1840	0.9829	0.1871	5.4362	1.0174	5.3435	79	24	79.4
10.7	10	42	0.1857	0.9826	0.1890	5.3860	1.0177	5.2924	79	18	79.3
10.8	10	48	0.1874	0.9823	0.1908	5.3367	1.0180	5.2422	79	12	79.2
10.9	10	54	0.1891	0.9820	0.1926	5.2883	1.0184	5.1929	79	6	79.1
11.0	11	0	0.1908	0.9816	0.1944	5.2408	1.0187	5.1446	79	0	79.0
11.1	11	6	0.1925	0.9813	0.1962	5.1942	1.0191	5.0970	78	54	78.9
11.2	11	12	0.1942	9.9810	0.1980	5.1484	1.0194	5.0504	78	48	78.8
11.3	11	18	0.1959	0.9806	0.1998	5.1034	1.0198	5.0045	78	42	78.7
11.4	11	24	0.1977	0.9803	0.2016	5.0593	1.0201	4.9595	78	36	78.6
11.5	11	30	0.1994	0.9799	0.2035	5.0159	1.0205	4.9152	78	30	78.5
11.6	11	36	0.2011	0.9796	0.2053	4.9732	1.0209	4.8716	78	24	78.4
11.7	11	42	0.2028	0.9792	0.2071	4.9313	1.0212	4.8288	78	18	78.3
11.8	11	48	0.2045	0.9789	0.2089	4.8901	1.0216	4.7867	78	12	78.2
11.9	11	54	0.2062	0.9785	0.2107	4.8496	1.0220	4.7453	78	6	78.1
12.0	12	0	0.2079	0.9781	0.2126	4.8097	1.0223	4.7046	78	0	78.0
12.1	12	6	0.2096	0.9778	0.2144	4.7706	1.0227	4.6646	77	54	77.9
12.2	12	12	0.2113	0.9774	0.2162	4.7321	1.0231	4.6252	77	48	77.8
12.3	12	18	0.2130	0.9770	0.2180	4.6942	1.0235	4.5864	77	42	77.7
12.4	12	24	0.2147	0.9767	0.2199	4.6569	1.0239	4.5483	77	36	77.6
12.5	12	30	0.2164	0.9763	0.2217	4.6202	1.0243	4.5107	77	30	77.5
12.6	12	36	0.2181	0.9759	0.2235	4.5841	1.0247	4.4737	77	24	77.4
12.7	12	42	0.2198	0.9755	0.2254	4.5486	1.0251	4.4374	77	18	77.3
12.8	12	48	0.2215	0.9751	0.2272	4.5137	1.0255	4.4015	77	12	77.2
12.9	12	54	0.2232	0.9748	0.2290	4.4793	1.0259	4.3662	77	6	77.1
13.0	13	0	0.2250	0.9744	0.2309	4.4454	1.0263	4.3315	77	0	77.0
13.1	13	6	0.2267	0.9740	0.2327	4.4121	1.0267	4.2972	76	54	76.9
13.2	13	12	0.2284	0.9736	0.2345	4.3792	1.0271	4.2635	76	48	76.8
13.3	13	18	0.2300	0.9732	0.2364	4.3469	1.0276	4.2303	76	42	76.7
13.4	13	24	0.2317	0.9728	0.2382	4.3150	1.0280	4.1976	76	36	76.6
13.5	13	30	0.2334	0.9724	0.2401	4.2837	1.0284	4.1653	76	30	76.5
13.6	13	36	0.2351	0.9720	0.2419	4.2528	1.0288	4.1335	76	24	76.4
13.7	13	42	0.2368	0.9715	0.2438	4.2223	1.0293	4.1022	76	18	76.3
13.8	13	48	0.2385	0.9711	0.2456	4.1923	1.0297	4.0713	76	12	76.2
13.9	13	54	0.2402	0.9707	0.2475	4.1627	1.0302	4.0408	76	6	76.1
14.0	14	0	0.2419	0.9703	0.2493	4.1336	1.0306	4.0108	76	0	76.0
14.1	14	6	0.2436	0.9699	0.2512	4.1048	1.0311	3.9812	75	54	75.9
14.2	14	12	0.2453	0.9694	0.2530	4.0765	1.0315	3.9520	75	48	75.8
14.3	14	18	0.2470	0.9690	0.2549	4.0486	1.0320	3.9232	75	42	75.7
14.4	14	24	0.2487	0.9686	0.2568	4.0211	1.0324	3.8947	75	36	75.6
14.5	14	30	0.2504	0.9681	0.2586	3.9939	1.0329	3.8667	75	30	75.5
14.6	14	36	0.2521	0.9677	0.2605	3.9672	1.0334	3.8391	75	24	75.4
14.7	14	42	0.2538	0.9673	0.2623	3.9408	1.0338	3.8118	75	18	75.3
14.8	14	48	0.2554	0.9668	0.2642	3.9147	1.0343	3.7849	75	12	75.2
14.9	14	54	0.2571	0.9664	0.2661	3.8890	1.0348	3.7583	75	6	75.1
			cos θ	**sin θ**	**cot θ**	**sec θ**	**csc θ**	**tan θ**	**deg**	**min**	**θ deg**

TABLE II (*continued*)

θ deg	deg min	sin θ	cos θ	tan θ	csc θ	sec θ	cot θ		
15.0	15 0	0.2588	0.9659	0.2679	3.8637	1.0353	3.7321	75 0	75.0
15.1	15 6	0.2605	0.9655	0.2698	3.8387	1.0358	3.7062	74 54	74.9
15.2	15 12	0.2622	0.9650	0.2717	3.8140	1.0363	3.6806	74 48	74.8
15.3	15 18	0.2639	0.9646	0.2736	3.7897	1.0367	3.6554	74 42	74.7
15.4	15 24	0.2656	0.9641	0.2754	3.7657	1.0372	3.6305	74 36	74.6
15.5	15 30	0.2672	0.9636	0.2773	3.7420	1.0377	3.6059	74 30	74.5
15.6	15 36	0.2689	0.9632	0.2792	3.7186	1.0382	3.5816	74 24	74.4
15.7	15 42	0.2706	0.9627	0.2811	3.6955	1.0388	3.5576	74 18	74.3
15.8	15 48	0.2723	0.9622	0.2830	3.6727	1.0393	3.5339	74 12	74.2
15.9	15 54	0.2740	0.9617	0.2849	3.6502	1.0398	3.5105	74 6	74.1
16.0	16 0	0.2756	0.9613	0.2867	3.6280	1.0403	3.4874	74 0	74.0
16.1	16 6	0.2773	0.9608	0.2886	3.6060	1.0408	3.4646	73 54	73.9
16.2	16 12	0.2790	0.9603	0.2905	3.5843	1.0413	3.4420	73 48	73.8
16.3	16 18	0.2807	0.9598	0.2924	3.5629	1.0419	3.4197	73 42	73.7
16.4	16 24	0.2823	0.9593	0.2943	3.5418	1.0424	3.3977	73 36	73.6
16.5	16 30	0.2840	0.9588	0.2962	3.5209	1.0429	3.3759	73 30	73.5
16.6	16 36	0.2857	0.9583	0.2981	3.5003	1.0435	3.3544	73 24	74.4
16.7	16 42	0.2874	0.9578	0.3000	3.4800	1.0440	3.3332	73 18	73.3
16.8	16 48	0.2890	0.9573	0.3019	3.4598	1.0446	3.3122	73 12	73.2
16.9	16 54	0.2907	0.9568	0.3038	3.4399	1.0451	3.2914	73 6	73.1
17.0	17 0	0.2924	0.9563	0.3057	3.4203	1.0457	3.2709	73 0	73.0
17.1	17 6	0.2940	0.9558	0.3076	3.4009	1.0463	3.2506	72 54	72.9
17.2	17 12	0.2957	0.9553	0.3096	3.3817	1.0468	3.2305	72 48	72.8
17.3	17 18	0.2974	0.9548	0.3115	3.3628	1.0474	3.2106	72 42	72.7
17.4	17 24	0.2990	0.9542	0.3134	3.3440	1.0480	3.1910	72 36	72.6
17.5	17 30	0.3007	0.9537	0.3153	3.3255	1.0485	3.1716	72 30	72.5
17.6	17 36	0.3024	0.9532	0.3172	3.3072	1.0491	3.1524	72 24	72.4
17.7	17 42	0.3040	0.9527	0.3191	3.2891	1.0497	3.1334	72 18	72.3
17.8	17 48	0.3057	0.9521	0.3211	3.2712	1.0503	3.1146	72 12	72.2
17.9	17 54	0.3074	0.9516	0.3230	3.2536	1.0509	3.0961	72 6	72.1
18.0	18 0	0.3090	0.9511	0.3249	3.2361	1.0515	3.0777	72 0	72.0
18.1	18 6	0.3107	0.9505	0.3268	3.2188	1.0521	3.0595	71 54	71.9
18.2	18 12	0.3123	0.9500	0.3288	3.2017	1.0527	3.0415	71 48	71.8
18.3	18 18	0.3140	0.9494	0.3307	3.1848	1.0533	3.0237	71 42	71.7
18.4	18 24	0.3156	0.9489	0.3327	3.1681	1.0539	3.0061	71 36	71.6
18.5	18 30	0.3173	0.9483	0.3346	3.1515	1.0545	2.9887	71 30	71.5
18.6	18 36	0.3190	0.9478	0.3365	3.1352	1.0551	2.9714	71 24	71.4
18.7	18 42	0.3206	0.9472	0.3385	3.1190	1.0557	2.9544	71 18	71.3
18.8	18 48	0.3223	0.9466	0.3404	3.1030	1.0564	2.9375	71 12	71.2
18.9	18 54	0.3239	0.9461	0.3424	3.0872	1.0570	2.9208	71 6	71.1
19.0	19 0	0.3256	0.9455	0.3443	2.0716	1.0576	2.9042	71 0	71.0
19.1	19 6	0.3272	0.9449	0.3463	3.0561	1.0583	2.8878	70 54	70.9
19.2	19 12	0.3289	0.9444	0.3482	3.0407	1.0589	2.8716	70 48	70.8
19.3	19 18	0.3305	0.9438	0.3502	3.0256	1.0595	2.8556	70 42	70.7
19.4	19 24	0.3322	0.9432	0.3522	3.0106	1.0602	2.8397	70 36	70.6
19.5	19 30	0.3338	0.9426	0.3541	2.9957	1.0608	2.8239	70 30	70.5
19.6	19 36	0.3355	0.9421	0.3561	2.9811	1.0615	2.8083	70 24	70.4
19.7	19 42	0.3371	0.9415	0.3581	2.9665	1.0622	2.7929	70 18	70.3
19.8	19 48	0.3387	0.9409	0.3600	2.9521	1.0628	2.7776	70 12	70.2
19.9	19 54	0.3404	0.9403	0.3620	2.9379	1.0635	2.7625	70 6	70.1
		cos θ	sin θ	cot θ	sec θ	csc θ	tan θ	deg min	θ deg

TABLE II (*continued*)

θ deg	deg	min	sin θ	cos θ	tan θ	csc θ	sec θ	cot θ			
20.0	20	0	0.3420	0.9397	0.3640	2.9238	1.0642	2.7475	70	0	70.0
20.1	20	6	0.3437	0.9391	0.3659	2.9099	1.0649	2.7326	69	54	69.9
20.2	20	12	0.3453	0.9385	0.3679	2.8960	1.0655	2.7179	69	48	69.8
20.3	20	18	0.3469	0.9379	0.3699	2.8824	1.0662	2.7034	69	42	69.7
20.4	20	24	0.3486	0.9373	0.3719	2.8688	1.0669	2.6889	69	36	69.6
20.5	20	30	0.3502	0.9367	0.3739	2.8555	1.0676	2.6746	69	30	69.5
20.6	20	36	0.3518	0.9361	0.3759	2.8422	1.0683	2.6605	69	24	69.4
20.7	20	42	0.3535	0.9354	0.3779	2.8291	1.0690	2.6464	69	18	69.3
20.8	20	48	0.3551	0.9348	0.3799	2.8161	1.0697	2.6325	69	12	69.2
20.9	20	54	0.3567	0.9342	0.3819	2.8032	1.0704	2.6187	69	6	69.1
21.0	21	0	0.3584	0.9336	0.3839	2.7904	1.0711	2.6051	69	0	69.0
21.1	21	6	0.3600	0.9330	0.3859	2.7778	1.0719	2.5916	68	54	68.9
21.2	21	12	0.3616	0.9323	0.3879	2.7653	1.0726	2.5782	68	48	68.8
21.3	21	18	0.3633	0.9317	0.3899	2.7529	1.0733	2.5649	68	42	68.7
21.4	21	24	0.3649	0.9311	0.3919	2.7407	1.0740	2.5517	68	36	68.6
21.5	21	30	0.3665	0.9304	0.3939	2.7285	1.0748	2.5386	68	30	68.5
21.6	21	36	0.3681	0.9298	0.3959	2.7165	1.0755	2.5257	68	24	68.4
21.7	21	42	0.3697	0.9291	0.3979	2.7046	1.0763	2.5129	68	18	68.3
21.8	21	48	0.3714	0.9285	0.4000	2.6927	1.0770	2.5002	68	12	68.2
21.9	21	54	0.3730	0.9278	0.4020	2.6811	1.0778	2.4876	68	6	68.1
22.0	22	0	0.3746	0.9272	0.4040	2.6695	1.0785	2.4751	68	0	68.0
22.1	22	6	0.3762	0.9265	0.4061	2.6580	1.0793	2.4627	67	54	67.9
22.2	22	12	0.3778	0.9259	0.4081	2.6466	1.0801	2.4504	67	48	67.8
22.3	22	18	0.3795	0.9252	0.4101	2.6354	1.0808	2.4383	67	42	67.7
22.4	22	24	0.3811	0.9245	0.4122	2.6242	1.0816	2.4262	67	36	67.6
22.5	22	30	0.3827	0.9239	0.4142	2.6131	1.0824	2.4142	67	30	67.5
22.6	22	36	0.3843	0.9232	0.4163	2.6022	1.0832	2.4023	67	24	67.4
22.7	22	42	0.3859	0.9225	0.4183	2.5913	1.0840	2.3906	67	18	67.3
22.8	22	48	0.3875	0.9219	0.4204	2.5805	1.0848	2.3789	67	12	67.2
22.9	22	54	0.3891	0.9212	0.4224	2.5699	1.0856	2.3673	67	6	67.1
23.0	23	0	0.3907	0.9205	0.4245	2.5593	1.0864	2.3559	67	0	67.0
23.1	23	6	0.3923	0.9198	0.4265	2.5488	1.0872	2.3445	66	54	66.9
23.3	23	12	0.3939	0.9191	0.4286	2.5384	1.0880	2.3332	66	48	66.8
23.3	23	18	0.3955	0.9184	0.4307	2.5282	1.0888	2.3220	66	42	66.7
23.4	23	24	0.3971	0.9178	0.4327	2.5180	1.0896	2.3109	66	36	66.6
23.5	23	30	0.3987	0.9171	0.4348	2.5078	1.0904	2.2998	66	30	66.5
23.6	23	36	0.4003	0.9164	0.4369	2.4978	1.0913	2.2889	66	24	66.4
23.7	23	42	0.4019	0.9157	0.4390	2.4879	1.0921	2.2781	66	18	66.3
23.8	23	48	0.4035	0.9150	0.4411	2.4780	1.0929	2.2673	66	12	66.2
23.9	23	54	0.4051	0.9143	0.4431	2.4683	1.0938	2.2566	66	6	66.1
24.0	24	0	0.4067	0.9135	0.4452	2.4586	1.0946	2.2460	66	0	66.0
24.1	24	6	0.4083	0.9128	0.4473	2.4490	1.0955	2.2355	65	54	65.9
24.2	24	12	0.4099	0.9121	0.4494	2.4395	1.0963	2.2251	65	48	65.8
24.3	24	18	0.4115	0.9114	0.4515	2.4301	1.0972	2.2148	65	42	65.7
24.4	24	24	0.4131	0.9107	0.4536	2.4207	1.0981	2.2045	65	36	65.6
24.5	24	30	0.4147	0.9100	0.4557	2.4114	1.0989	2.1943	65	30	65.5
24.6	24	36	0.4163	0.9092	0.4578	2.4022	1.0998	2.1842	65	24	65.4
24.7	24	42	0.4179	0.9085	0.4599	2.3931	1.1007	2.1742	65	18	65.3
24.8	24	48	0.4195	0.9078	0.4621	2.3841	1.1016	2.1642	65	12	65.2
24.9	24	54	0.4210	0.9070	0.4642	2.3751	1.1025	2.1543	65	6	65.1
			cos θ	sin θ	cot θ	sec θ	csc θ	tan θ	deg	min	θ deg

TABLE II (*continued*)

θ deg	deg	min	sin θ	cos θ	tan θ	csc θ	sec θ	cot θ			
25.0	25	0	0.4226	0.9063	0.4663	2.3662	1.1034	2.1445	65	0	65.0
25.1	25	6	0.4242	0.9056	0.4684	2.3574	1.1043	2.1348	64	54	64.9
25.2	25	12	0.4258	0.9048	0.4706	2.3486	1.1052	2.1251	64	48	64.8
25.3	25	18	0.4274	0.9041	0.4727	2.3400	1.1061	2.1155	64	42	64.7
25.4	25	24	0.4289	0.9033	0.4748	2.3314	1.1070	2.1060	64	36	64.6
25.5	25	30	0.4305	0.9026	0.4770	2.3228	1.1079	2.0965	64	30	64.5
25.6	25	36	0.4321	0.9018	0.4791	2.3144	1.1089	2.0872	64	24	64.4
25.7	25	42	0.4337	0.9011	0.4813	2.3060	1.1098	2.0778	64	18	64.3
25.8	25	48	0.4352	0.9003	0.4834	2.2976	1.1107	2.0686	64	12	64.2
25.9	25	54	0.4368	0.8996	0.4856	2.2894	1.1117	2.0594	64	6	64.1
26.0	26	0	0.4384	0.8988	0.4877	2.2812	1.1126	2.0503	64	0	64.0
26.1	26	6	0.4399	0.8980	0.4899	2.2730	1.1136	2.0413	63	54	63.9
26.2	26	12	0.4415	0.8973	0.4921	2.2650	1.1145	2.0323	63	48	63.8
26.3	26	18	0.4431	0.8965	0.4942	2.2570	1.1155	2.0233	63	42	63.7
26.4	26	24	0.4446	0.8957	0.4964	2.2490	1.1164	2.0145	63	36	63.6
26.5	26	30	0.4462	0.8949	0.4986	2.2412	1.1174	2.0057	63	30	63.5
26.6	26	36	0.4478	0.8942	0.5008	2.2333	1.1184	1.9970	63	24	63.4
26.7	26	42	0.4493	0.8934	0.5029	2.2256	1.1194	1.9883	63	18	63.3
26.8	26	48	0.4509	0.8926	0.5051	2.2179	1.1203	1.9797	63	12	63.2
26.9	26	54	0.4524	0.8918	0.5073	2.2103	1.1213	1.9711	63	6	63.1
27.0	27	0	0.4540	0.8910	0.5095	2.2027	1.1223	1.9626	63	0	63.0
27.1	27	6	0.4555	0.8902	0.5117	2.1952	1.1233	1.9542	62	54	62.9
27.2	27	12	0.4571	0.8894	0.5139	2.1877	1.1243	1.9458	62	48	62.8
27.3	27	18	0.4586	0.8886	0.5161	2.1803	1.1253	1.9375	62	42	62.7
27.4	27	24	0.4602	0.8878	0.5184	2.1730	1.1264	1.9292	62	36	62.6
27.5	27	30	0.4617	0.8870	0.5206	2.1657	1.1274	1.9210	62	30	62.5
27.6	27	36	0.4633	0.8862	0.5228	2.1584	1.1284	1.9128	62	24	62.4
27.7	27	42	0.4648	0.8854	0.5250	2.1513	1.1294	1.9047	62	18	62.3
27.8	27	48	0.4664	0.8846	0.5272	2.1441	1.1305	1.8967	62	12	62.2
27.9	27	54	0.4679	0.8838	0.5295	2.1371	1.1315	1.8887	62	6	62.1
28.0	28	0	0.4695	0.8829	0.5317	2.1301	1.1326	1,8807	62	0	62.0
28.1	28	6	0.4710	0.8821	0.5339	2.1231	1.1336	1.8728	61	54	61.9
28.2	28	12	0.4726	0.8813	0.5362	2.1162	1.1347	1.8650	61	48	61.8
28.3	28	18	0.4741	0.8805	0.5384	2.1093	1.1357	1.8572	61	42	61.7
28.4	28	24	0.4756	0.8796	0.5407	2.1025	1.1368	1.8495	61	36	61.6
28.5	28	30	0.4772	0.8788	0.5430	2.0957	1.1379	1.8418	61	30	61.5
28.6	28	36	0.4787	0.8780	0.5452	2.0890	1.1390	1.8341	61	24	61.4
28.7	28	42	0.4802	0.8771	0.5475	2.0824	1.1401	1.8265	61	18	61.3
28.8	28	48	0.4818	0.8763	0.5498	2.0758	1.1412	1.8190	61	12	61.2
28.9	28	54	0.4833	0.8755	0.5520	2.0692	1.1423	1.8115	61	6	61.1
29.0	29	0	0.4848	0.8746	0.5543	2.0627	1.1434	1.8040	61	0	61.0
29.1	29	6	0.4863	0.8738	0.5566	2.0562	1.1445	1.7966	60	54	60.9
29.2	29	12	0.4879	0.8729	0.5589	2.0598	1.1456	1.7893	60	48	60.8
29.3	29	18	0.4894	0.8721	0.5612	2.0434	1.1467	1.7820	60	42	60.7
29.4	29	24	0.4909	0.8712	0.5635	2.0371	1.1478	1.7747	60	36	60.6
29.5	29	30	0.4924	0.8704	0.5658	2.0308	1.1490	1.7675	60	30	60.5
29.6	29	36	0.4939	0.8695	0.5681	2.0245	1.1501	1.7603	60	24	60.4
29.7	29	42	0.4955	0.8686	0.5704	2.0183	1.1512	1.7532	60	18	60.3
29.8	29	48	0.4970	0.8678	0.5727	2.0122	1.1524	1.7461	60	12	60.2
29.9	29	54	0.4985	0.8669	0.5750	2.0061	1.1535	1.7391	60	6	60.1
			cos θ	sin θ	cot θ	sec θ	csc θ	tan θ	deg	min	θ deg

TABLE II *(continued)*

θ deg	deg	min	sin θ	cos θ	tan θ	csc θ	sec θ	cot θ			
30.0	30	0	0.5000	0.8660	0.5774	2.0000	1.1547	1.7321	60	0	60.0
30.1	30	6	0.5015	0.8652	0.5797	1.9940	1.1559	1.7251	59	54	59.9
30.2	30	12	0.5030	0.8643	0.5820	1.9880	1.1570	1.7182	59	48	59.8
30.3	30	18	0.5045	0.8634	0.5844	1.9821	1.1582	1.7113	59	42	59.7
30.4	30	24	0.5060	0.8625	0.5867	1.9762	1.1594	1.7045	59	36	59.6
30.5	30	30	0.5075	0.8616	0.5890	1.9703	1.1606	1.6977	59	30	59.5
30.6	30	36	0.5090	0.8607	0.5914	1.9645	1.1618	1.6909	59	24	59.4
30.7	30	42	0.5105	0.8599	0.5938	1.9587	1.1630	1.6842	59	18	59.3
30.8	30	48	0.5120	0.8590	0.5961	1.9530	1.1642	1.6775	59	12	59.2
30.9	30	54	0.5135	0.8581	0.5985	1.9473	1.1654	1.6709	59	6	59.1
31.0	31	0	0.5150	0.8572	0.6009	1.9416	1.1666	1.6643	59	0	59.0
31.1	31	6	0.5165	0.8563	0.6032	1.9360	1.1679	1.6577	58	54	58.9
31.2	31	12	0.5180	0.8554	0.6056	1.9304	1.1691	1.6512	58	48	58.8
31.3	31	18	0.5195	0.8545	0.6080	1.9249	1.1703	1.6447	58	42	58.7
31.4	31	24	0.5210	0.8536	0.6104	1.9194	1.1716	1.6383	58	36	58.6
31.5	31	30	0.5225	0.8526	0.6128	1.9139	1.1728	1.6319	58	30	58.5
31.6	31	36	0.5240	0.8517	0.6152	1.9084	1.1741	1.6255	58	24	58.4
31.7	31	42	0.5255	0.8508	0.6176	1.9031	1.1753	1.6191	58	18	58.3
31.8	31	48	0.5270	0.8499	0.6200	1.8977	1.1766	1.6128	58	12	58.2
31.9	31	54	0.5284	0.8490	0.6224	1.8924	1.1779	1.6066	58	6	58.1
32.0	32	0	0.5299	0.8480	0.6249	1.8871	1.1792	1.6003	58	0	58.0
32.1	32	6	0.5314	0.8471	0.6273	1.8818	1.1805	1.5941	57	54	57.9
32.2	32	12	0.5329	0.8462	0.6297	1.8766	1.1818	1.5880	57	48	57.8
32.3	32	18	0.5344	0.8453	0.6322	1.8714	1.1831	1.5818	57	42	57.7
32.4	32	24	0.5358	0.8443	0.6346	1.8663	1.1844	1.5757	57	36	57.6
32.5	32	30	0.5373	0.8434	0.6371	1.8612	1.1857	1.5697	57	30	57.5
32.6	32	36	0.5388	0.8425	0.6395	1.8561	1.1870	1.5637	57	24	57.4
32.7	32	42	0.5402	0.8415	0.6420	1.8510	1.1883	1.5577	57	18	57.3
32.8	32	48	0.5417	0.8406	0.6445	1.8460	1.1897	1.5517	57	12	57.2
32.9	32	54	0.5432	0.8396	0.6469	1.8410	1.1910	1.5458	57	6	57.1
33.0	33	0	0.5446	0.8387	0.6494	1.8361	1.1924	1.5399	57	0	57.0
33.1	33	6	0.5461	0.8377	0.6519	1.8312	1.1937	1.5340	56	54	56.9
33.2	33	12	0.5476	0.8368	0.6544	1.8263	1.1951	1.5282	56	48	56.8
33.3	33	18	0.5490	0.8358	0.6569	1.8214	1.1964	1.5224	56	42	56.7
33.4	33	24	0.5505	0.8348	0.6594	1.8166	1.1978	1.5166	56	36	56.6
33.5	33	30	0.5519	0.8339	0.6619	1.8118	1.1992	1.5108	56	30	56.5
33.6	33	36	0.5534	0.8329	0.6644	1.8070	1.2006	1.5051	56	24	56.4
33.7	33	42	0.5548	0.8320	0.6669	1.8023	1.2020	1.4994	56	18	56.3
33.8	33	48	0.5563	0.8310	0.6694	1.7976	1.2034	1.4938	56	12	56.2
33.9	33	54	0.5577	0.8300	0.6720	1.7929	1.2048	1.4882	56	6	56.1
34.0	34	0	0.5592	0.8290	0.6745	1.7883	1.2062	1.4826	56	0	56.0
34.1	34	6	0.5606	0.8281	0.6771	1.7837	1.2076	1.4770	55	54	55.9
34.2	34	12	0.5621	0.8271	0.6796	1.7791	1.2091	1.4715	55	48	55.8
34.3	34	18	0.5635	0.8261	0.6822	1.7745	1.2105	1.4659	55	42	55.7
34.4	34	24	0.5650	0.8251	0.6847	1.7700	1.2120	1.4605	55	36	55.6
34.5	34	30	0.5664	0.8241	0.6873	1.7655	1.2134	1.4550	55	30	55.5
34.6	34	36	0.5678	0.8231	0.6899	1.7610	1.2149	1.4496	55	24	55.4
34.7	34	42	0.5693	0.8221	0.6924	1.7566	1.2163	1.4442	55	18	55.3
34.8	34	48	0.5707	0.8211	0.6950	1.7522	1.2178	1.4388	55	12	55.2
34.9	34	54	0.5721	0.8202	0.6976	1.7478	1.2193	1.4335	55	6	55.1
			cos θ	sin θ	cot θ	sec θ	csc θ	tan θ	deg	min	θ deg

TABLE II (*continued*)

θ deg	deg	min	sin θ	cos θ	tan θ	csc θ	sec θ	cot θ			
35.0	35	0	0.5736	0.8192	0.7002	1.7434	1.2208	1.4281	55	0	55.0
35.1	35	6	0.5750	0.8181	0.7028	1.7391	1.2223	1.4229	54	54	54.9
35.2	35	12	0.5764	0.8171	0.7054	1.7348	1.2238	1.4176	54	48	54.8
35.3	35	18	0.5779	0.8161	0.7080	1.7305	1.2253	1.4124	54	42	54.7
35.4	35	24	0.5793	0.8151	0.7107	1.7263	1.2268	1.4071	54	36	54.6
35.5	35	30	0.5807	0.8141	0.7133	1.7221	1.2283	1.4019	54	30	54.5
35.6	35	36	0.5821	0.8131	0.7159	1.7179	1.2299	1.3968	54	24	54.4
35.7	35	42	0.5835	0.8121	0.7186	1.7137	1.2314	1.3916	54	18	54.3
35.8	35	48	0.5850	0.8111	0.7212	1.7095	1.2329	1.3865	54	12	54.2
35.9	35	54	0.5864	0.8100	0.7239	1.7054	1.2345	1.3814	54	6	54.1
36.0	36	0	0.5878	0.8090	0.7265	1.7013	1.2361	1.3764	54	0	54.0
36.1	36	6	0.5892	0.8080	0.7292	1.6972	1.2376	1.3713	53	54	53.9
36.2	36	12	0.5906	0.8070	0.7319	1.6932	1.2392	1.3663	53	48	53.8
36.3	36	18	0.5920	0.8059	0.7346	1.6892	1.2408	1.3613	53	42	53.7
36.4	36	24	0.5934	0.8049	0.7373	1.6852	1.2424	1.3564	53	36	53.6
36.5	36	30	0.5948	0.8039	0.7400	1.6812	1.2440	1.3514	53	30	53.5
36.6	36	36	0.5962	0.8028	0.7427	1.6772	1.2456	1.3465	53	24	53.4
36.7	36	42	0.5976	0.8018	0.7454	1.6733	1.2472	1.3416	53	18	53.3
36.8	36	48	0.5990	0.8007	0.7481	1.6694	1.2489	1.3367	53	12	53.2
36.9	36	54	0.6004	0.7997	0.7508	1.6655	1.2505	1.3319	53	6	53.1
37.0	37	0	0.6018	0.7986	0.7536	1.6616	1.2521	1.3270	53	0	53.0
37.1	37	6	0.6032	0.7976	0.7563	1.6578	1.2538	1.3222	52	54	52.9
37.2	37	12	0.6046	0.7965	0.7590	1.6540	1.2554	1.3175	52	48	52.8
37.3	37	18	0.6060	0.7955	0.7618	1.6502	1.2571	1.3127	52	42	52.7
37.4	37	24	0.6074	0.7944	0.7646	1.6464	1.2588	1.3079	52	36	52.6
37.5	37	30	0.6088	0.7934	0.7673	1.6427	1.2605	1.3032	52	30	52.5
37.6	37	36	0.6101	0.7923	0.7701	1.6390	1.2622	1.2985	52	24	52.4
37.7	37	42	0.6115	0.7912	0.7729	1.6353	1.2639	1.2938	52	18	52.3
37.8	37	48	0.6129	0.7902	0.7757	1.6316	1.2656	1.2892	52	12	52.2
37.9	37	54	0.6143	0.7891	0.7785	1.6279	1.2673	1.2846	52	6	52.1
38.0	38	0	0.6157	0.7880	0.7813	1.6243	1.2690	1.2799	52	0	52.0
38.1	38	6	0.6170	0.7869	0.7841	1.6207	1.2708	1.2753	51	54	51.9
38.2	38	12	0.6184	0.7859	0.7869	1.6171	1.2725	1.2708	51	48	51.8
38.3	38	18	0.6198	0.7848	0.7898	1.6135	1.2742	1.2662	51	42	51.7
38.4	38	24	0.6211	0.7837	0.7926	1.6099	1.2760	1.2617	51	36	51.6
38.5	38	30	0.6225	0.7826	0.7954	1.6064	1.2778	1.2572	51	30	51.5
38.6	38	36	0.6239	0.7815	0.7983	1.6029	1.2796	1.2527	51	24	51.4
38.7	38	42	0.6252	0.7804	0.8012	1.5994	1.2813	1.2482	51	18	51.3
38.8	38	48	0.6266	0.7793	0.8040	1.5959	1.2831	1.2437	51	12	51.2
38.9	38	54	0.6280	0.7782	0.8069	1.5925	1.2849	1.2393	51	6	51.1
39.0	39	0	0.6293	0.7771	0.8098	1.5890	1.2868	1.2349	51	0	51.0
39.1	39	6	0.6307	0.7760	0.8127	1.5856	1.2886	1.2305	50	54	50.9
39.2	39	12	0.6320	0.7749	0.8156	1.5822	1.2904	1.2261	50	48	50.8
39.3	39	18	0.6334	0.7738	0.8185	1.5788	1.2923	1.2218	50	42	50.7
39.4	39	24	0.6347	0.7727	0.8214	1.5755	1.2941	1.2174	50	36	50.6
39.5	39	30	0.6361	0.7716	0.8243	1.5721	1.2960	1.2131	50	30	50.5
39.6	39	36	0.6374	0.7705	0.8273	1.5688	1.2978	1.2088	50	24	50.4
39.7	39	42	0.6388	0.7694	0.8302	1.5655	1.2997	1.2045	50	18	50.3
39.8	39	48	0.6401	0.7683	0.8332	1.5622	1.3016	1.2002	50	12	50.2
39.9	39	54	0.6414	0.7672	0.8361	1.5590	1.3035	1.1960	50	6	50.1
			cos θ	sin θ	cot θ	sec θ	csc θ	tan θ	deg	min	θ deg

II

TABLE II *(continued)*

θ deg	deg min	sin θ	cos θ	tan θ	csc θ	sec θ	cot θ		
40.0	40 0	0.6428	0.7660	0.8391	1.5557	1.3054	1.1918	50 0	50.0
40.1	40 6	0.6441	0.7649	0.8421	1.5525	1.3073	1.1875	49 54	49.9
40.2	40 12	0.6455	0.7638	0.8451	1.5493	1.3092	1.1833	49 48	49.8
40.3	40 18	0.6468	0.7627	0.8481	1.5461	1.3112	1.1792	49 42	49.7
40.4	40 24	0.6481	0.7615	0.8511	1.5429	1.3131	1.1750	49 36	49.6
40.5	40 30	0.6494	0.7604	0.8541	1.5398	1.3151	1.1708	49 30	49.5
40.6	40 36	0.6508	0.7593	0.8571	1.5366	1.3171	1.1667	49 24	49.4
40.7	40 42	0.6521	0.7581	0.8601	1.5335	1.3190	1.1626	49 18	49.3
40.8	40 48	0.6534	0.7570	0.8632	1.5304	1.3210	1.1585	49 12	49.2
40.9	40 54	0.6547	0.7559	0.8662	1.5273	1.3230	1.1544	49 6	49.1
41.0	41 0	0.6561	0.7547	0.8693	1.5243	1.3250	1.1504	49 0	49.0
41.1	41 6	0.6574	0.7536	0.8724	1.5212	1.3270	1.1463	48 54	48.9
41.2	41 12	0.6587	0.7524	0.8754	1.5182	1.3291	1.1423	48 48	48.8
41.3	41 18	0.6600	0.7513	0.8785	1.5151	1.3311	1.1383	48 42	48.7
41.4	41 24	0.6613	0.7501	0.8816	1.5121	1.3331	1.1343	48 36	48.6
41.5	41 30	0.6626	0.7490	0.8847	1.5092	1.3352	1.1303	48 30	48.5
41.6	41 36	0.6639	0.7478	0.8878	1.5062	1.3373	1.1263	48 24	48.4
41.7	41 42	0.6652	0.7466	0.8910	1.5032	1.3393	1.1224	48 18	48.3
41.8	41 48	0.6665	0.7455	0.8941	1.5003	1.3414	1.1184	48 12	48.2
41.9	41 54	0.6678	0.7443	0.8972	1.4974	1.3435	1.1145	48 6	48.1
42.0	42 0	0.6691	0.7431	0.9004	1.4945	1.3456	1.1106	48 0	48.0
42.1	42 6	0.6704	0.7420	0.9036	1.4916	1.3478	1.1067	47 54	47.9
42.2	42 12	0.6717	0.7408	0.9067	1.4887	1.3499	1.1028	47 48	47.8
42.3	42 18	0.6730	0.7396	0.9099	1.4859	1.3520	1.0990	47 42	47.7
42.4	42 24	0.6743	0.7385	0.9131	1.4830	1.3542	1.0951	47 36	47.6
42.5	42 30	0.6756	0.7373	0.9163	1.4802	1.3563	1.0913	47 30	47.5
42.6	42 36	0.6769	0.7361	0.9195	1.4774	1.3585	1.0875	47 24	47.4
42.7	42 42	0.6782	0.7349	0.9228	1.4746	1.3607	1.0837	47 18	47.3
42.8	42 48	0.6794	0.7337	0.9260	1.4718	1.3629	1.0799	47 12	47.2
42.9	42 54	0.6807	0.7325	0.9293	1.4690	1.3651	1.0761	47 6	47.1
43.0	43 0	0.6820	0.7314	0.9325	1.4663	1.3673	1.0724	47 0	47.0
43.1	43 6	0.6833	0.7302	0.9358	1.4635	1.3696	1.0686	46 54	46.9
43.2	43 12	0.6845	0.7290	0.9391	1.4608	1.3718	1.0649	46 48	46.8
43.3	43 18	0.6858	0.7278	0.9424	1.4581	1.3741	1.0612	46 42	46.7
43.4	43 24	0.6871	0.7266	0.9457	1.4554	1.3763	1.0575	46 36	46.6
43.5	43 30	0.6884	0.7254	0.9490	1.4527	1.3786	1.0538	46 30	46.5
43.6	43 36	0.6896	0.7242	0.9523	1.4501	1.3809	1.0501	46 24	46.4
43.7	43 42	0.6909	0.7230	0.9556	1.4474	1.3832	1.0464	46 18	46.3
43.8	43 48	0.6921	0.7218	0.9590	1.4448	1.3855	1.0428	46 12	46.2
43.9	43 54	0.6934	0.7206	0.9623	1.4422	1.3878	1.0392	46 6	46.1
44.0	44 0	0.6947	0.7193	0.9657	1.4396	1.3902	1.0355	46 0	46.0
44.1	44 6	0.6959	0.7181	0.9691	1.4370	1.3925	1.0319	45 54	45.9
44.2	44 12	0.6972	0.7169	0.9725	1.4344	1.3949	1.0283	45 48	45.8
44.3	44 18	0.6984	0.7157	0.9759	1.4318	1.3972	1.0247	45 42	45.7
44.4	44 24	0.6997	0.7145	0.9793	1.4293	1.3996	1.0212	45 36	45.6
44.5	44 30	0.7009	0.7133	0.9827	1.4267	1.4020	1.0176	45 30	45.5
44.6	44 36	0.7022	0.7120	0.9861	1.4242	1.4044	1.0141	45 24	45.4
44.7	44 42	0.7034	0.7108	0.9896	1.4217	1.4069	1.0105	45 18	45.3
44.8	44 48	0.7046	0.7096	0.9930	1.4192	1.4093	1.0070	45 12	45.2
44.9	44 54	0.7059	0.7083	0.9965	1.4167	1.4118	1.0035	45 6	45.1
45.0	45 0	0.7071	0.7071	1.0000	1.4142	1.4142	1.0000	45 0	45.0
		cos θ	sin θ	cot θ	sec θ	csc θ	tan θ	deg min	θ deg

TABLE III Values of trigonometric functions (radians)

θ radians or Real Number x	θ degrees	sin θ or sin x	cos θ or cos x	tan θ or tan x	csc θ or csc x	sec θ or sec x	cot θ or cot x
0.00	0° 00′	0.0000	1.000	0.0000	No value	1.000	No value
.01	0° 34′	.0100	1.000	.0100	100.0	1.000	100.0
.02	1° 09′	.0200	0.9998	.0200	50.00	1.000	49.99
.03	1° 43′	.0300	0.9996	.0300	33.34	1.000	33.32
.04	2° 18′	.0400	0.9992	.0400	25.01	1.001	24.99
0.05	2° 52′	0.0500	0.9988	0.0500	20.01	1.001	19.98
.06	3° 26′	.0600	.9982	.0601	16.68	1.002	16.65
.07	4° 01′	.0699	.9976	.0701	14.30	1.002	14.26
.08	4° 35′	.0799	.9968	.0802	12.51	1.003	12.47
.09	5° 09′	.0899	.9960	.0902	11.13	1.004	11.08
0.10	5° 44′	0.0998	0.9950	0.1003	10.02	1.005	9.967
.11	6° 18′	.1098	.9940	.1104	9.109	1.006	9.054
.12	6° 53′	.1197	.9928	.1206	8.353	1.007	8.293
.13	7° 27′	.1296	.9916	.1307	7.714	1.009	7.649
.14	8° 01′	.1395	.9902	.1409	7.166	1.010	7.096
0.15	8° 36′	0.1494	0.9888	0.1511	6.692	1.011	6.617
.16	9° 10′	.1593	.9872	.1614	6.277	1.013	6.197
.17	9° 44′	.1692	.9856	.1717	5.911	1.015	5.826
.18	10° 19′	.1790	.9838	.1820	5.586	1.016	5.495
.19	10° 53′	.1889	.9820	.1923	5.295	1.018	5.200
0.20	11° 28′	0.1987	0.9801	0.2027	5.033	1.020	4.933
.21	12° 02′	.2085	.9780	.2131	4.797	1.022	4.692
.22	12° 36′	.2182	.9759	.2236	4.582	1.025	4.472
.23	13° 11′	.2280	.9737	.2341	4.386	1.027	4.271
.24	13° 45′	.2377	.9713	.2447	4.207	1.030	4.086
0.25	14° 19′	0.2474	0.9689	0.2553	4.042	1.032	3.916
.26	14° 54′	.2571	.9664	.2660	3.890	1.035	3.759
.27	15° 28′	.2667	.9638	.2768	3.749	1.038	3.613
.28	16° 03′	.2764	.9611	.2876	3.619	1.041	3.478
.29	16° 37′	.2860	.9582	.2984	3.497	1.044	3.351
0.30	17° 11′	0.2955	0.9553	0.3093	3.384	1.047	3.233
.31	17° 46′	.3051	.9523	.3203	3.278	1.050	3.122
.32	18° 20′	.3146	.9492	.3314	3.179	1.053	3.018
.33	18° 54′	.3240	.9460	.3425	3.086	1.057	2.920
.34	19° 29′	.3335	.9428	.3537	2.999	1.061	1.827
0.35	20° 03′	0.3429	0.9394	0.3650	2.916	1.065	2.740
.36	20° 38′	.3523	.9359	.3764	2.839	1.068	2.657
.37	21° 12′	.3616	.9323	.3879	2.765	1.073	2.578
.38	21° 46′	.3709	.9287	.3994	2.696	1.077	2.504
.39	22° 21′	.3802	.9249	.4111	2.630	1.081	2.433
0.40	22° 55′	0.3894	0.9211	0.4228	2.568	1.086	2.365
.41	23° 29′	.3986	.9171	.4346	2.509	1.090	2.301
.42	24° 04′	.4078	.9131	.4466	2.452	1.095	2.239
.43	24° 38′	.4169	.9090	.4586	2.399	1.100	2.180
.44	25° 13′	.4259	.9048	.4708	2.348	1.105	2.124
0.45	25° 47′	0.4350	0.9004	0.4831	2.299	1.111	2.070

TABLE III *(continued)*

θ radians or Real Number x	θ degrees	$\sin\theta$ or $\sin x$	$\cos\theta$ or $\cos x$	$\tan\theta$ or $\tan x$	$\csc\theta$ or $\csc x$	$\sec\theta$ or $\sec x$	$\cot\theta$ or $\cot x$
0.45	25° 47′	0.4350	0.9004	0.4831	2.299	1.111	2.070
.46	26° 21′	.4439	.8961	.4954	2.253	1.116	2.018
.47	26° 56′	.4529	.8916	.5080	2.208	1.122	1.969
.48	27° 30′	.4618	.8870	.5206	2.166	1.127	1.921
.49	28° 04′	.4706	.8823	.5334	2.125	1.133	1.875
0.50	28° 39′	0.4794	0.8776	0.5463	2.086	1.139	1.830
.51	29° 13′	.4882	.8727	.5594	2.048	1.146	1.788
.52	29° 48′	.4969	.8678	.5726	2.013	1.152	1.747
.53	30° 22′	.5055	.8628	.5859	1.978	1.159	1.707
.54	30° 56′	.5141	.8577	.5994	1.945	1.166	1.668
0.55	31° 31′	0.5227	0.8525	0.6131	1.913	1.173	1.631
.56	32° 05′	.5312	.8473	.6269	1.883	1.180	1.595
.57	32° 40′	.5396	.8419	.6410	1.853	1.188	1.560
.58	33° 14′	.5480	.8365	.6552	1.825	1.196	1.526
.59	33° 48′	.5564	.8309	.6696	1.797	1.203	1.494
0.60	34° 23′	0.5646	0.8253	0.6841	1.771	1.212	1.462
.61	34° 57′	.5729	.8196	.6989	1.746	1.220	1.431
.62	35° 31′	.5810	.8139	.7139	1.721	1.229	1.401
.63	36° 06′	.5891	.8080	.7291	1.697	1.238	1.372
.64	36° 40′	.5972	.8021	.7445	1.674	1.247	1.343
0.65	37° 15′	0.6052	0.7961	0.7602	1.652	1.256	1.315
.66	37° 49′	.6131	.7900	.7761	1.631	1.266	1.288
.67	38° 23′	.6210	.7838	.7923	1.610	1.276	1.262
.68	38° 58′	.6288	.7776	.8087	1.590	1.286	1.237
.69	39° 32′	.6365	.7712	.8253	1.571	1.297	1.212
0.70	40° 06′	0.6442	0.7648	0.8423	1.552	1.307	1.187
.71	40° 41′	.6518	.7584	.8595	1.534	1.319	1.163
.72	41° 15′	.6594	.7518	.8771	1.517	1.330	1.140
.73	41° 50′	.6669	.7452	.8949	1.500	1.342	1.117
.74	42° 24′	.6743	.7385	.9131	1.483	1.354	1.095
0.75	42° 58′	0.6816	0.7317	0.9316	1.467	1.367	1.073
.76	43° 33′	.6889	.7248	.9505	1.452	1.380	1.052
.77	44° 07′	.6961	.7179	.9697	1.436	1.393	1.031
.78	44° 41′	.7033	.7109	.9893	1.422	1.407	1.011
.79	45° 16′	.7104	.7038	1.009	1.408	1.421	0.9908
0.80	45° 50′	0.7174	0.6967	1.030	1.394	1.435	0.9712
.81	46° 25′	.7243	.6895	1.050	1.381	1.450	.9520
.82	46° 59′	.7311	.6822	1.072	1.368	1.466	.9331
.83	47° 33′	.7379	.6749	1.093	1.355	1.482	.9146
.84	48° 08′	.7446	.6675	1.116	1.343	1.498	.8964
0.85	48° 42′	0.7513	0.6600	1.138	1.331	1.515	0.8785
.86	49° 16′	.7578	.6524	1.162	1.320	1.533	.8609
.87	49° 51′	.7643	.6448	1.185	1.308	1.551	.8437
.88	50° 25′	.7707	.6372	1.210	1.297	1.569	.8267
.89	51° 00′	.7771	.6294	1.235	1.287	1.589	.8100
0.90	51° 34′	0.7833	0.6216	1.260	1.277	1.609	0.7936
.91	52° 08′	.7895	.6137	1.286	1.267	1.629	.7774
.92	52° 43′	.7956	.6058	1.313	1.257	1.651	.7615
.93	53° 17′	.8016	.5978	1.341	1.247	1.673	.7458
.94	53° 51′	.8076	.5898	1.369	1.238	1.696	.7303
0.95	54° 26′	0.8134	0.5817	1.398	1.229	1.719	0.7151

TABLE III *(continued)*

θ radians or Real Number *x*	*θ* degrees	sin *θ* or sin *x*	cos *θ* or cos *x*	tan *θ* or tan *x*	csc *θ* or csc *x*	sec *θ* or sec *x*	cot *θ* or cot *x*
0.95	54° 26′	0.8134	0.5817	1.398	1.229	1.719	0.7151
.96	55° 00′	.8192	.5735	1.428	1.221	1.744	.7001
.97	55° 35′	.8249	.5653	1.459	1.212	1.769	.6853
.98	56° 09′	.8305	.5570	1.491	1.204	1.795	.6707
.99	56° 43′	.8360	.5487	1.524	1.196	1.823	.6563
1.00	57° 18′	0.8415	0.5403	1.557	1.188	1.851	0.6421
1.01	57° 52′	.8468	.5319	1.592	1.181	1.880	.6281
1.02	58° 27′	.8521	.5234	1.628	1.174	1.911	.6142
1.03	59° 01′	.8573	.5148	1.665	1.166	1.942	.6005
1.04	59° 35′	.8624	.5062	1.704	1.160	1.975	.5870
1.05	60° 10′	0.8674	0.4976	1.743	1.153	2.010	0.5736
1.06	60° 44′	.8724	.4889	1.784	1.146	2.046	.5604
1.07	61° 18′	.8772	.4801	1.827	1.140	2.083	.5473
1.08	61° 53′	.8820	.4713	1.871	1.134	2.122	.5344
1.09	62° 27′	.8866	.4625	1.917	1.128	2.162	.5216
1.10	63° 02′	0.8912	0.4536	1.965	1.122	2.205	0.5090
1.11	63° 36′	.8957	.4447	2.014	1.116	2.249	.4964
1.12	64° 10′	.9001	.4357	2.066	1.111	2.295	.4840
1.13	64° 45′	.9044	.4267	2.120	1.106	2.344	.4718
1.14	65° 19′	.9086	.4176	2.176	1.101	2.395	.4596
1.15	65° 53′	0.9128	0.4085	2.234	1.096	2.448	0.4475
1.16	66° 28′	.9168	.3993	2.296	1.091	2.504	.4356
1.17	67° 02′	.9208	.3902	2.360	1.086	2.563	.4237
1.18	67° 37′	.9246	.3809	2.427	1.082	2.625	.4120
1.19	68° 11′	.9284	.3717	2.498	1.077	2.691	.4003
1.20	68° 45′	0.9320	0.3624	2.572	1.073	2.760	0.3888
1.21	69° 20′	.9356	.3530	2.650	1.069	2.833	.3773
1.22	69° 54′	.9391	.3436	2.733	1.065	2.910	.3659
1.23	70° 28′	.9425	.3342	2.820	1.061	2.992	.3546
1.24	71° 03′	.9458	.3248	2.912	1.057	3.079	.3434
1.25	71° 37′	0.9490	0.3153	3.010	1.054	3.171	0.3323
1.26	72° 12′	.9521	.3058	3.113	1.050	3.270	.3212
1.27	72° 46′	.9551	.2963	3.224	1.047	3.375	.3102
1.28	73° 20′	.9580	.2867	3.341	1.044	3.488	.2993
1.29	73° 55′	.9608	.2771	3.467	1.041	3.609	.2884
1.30	74° 29′	0.9636	0.2675	3.602	1.038	3.738	0.2776
1.31	75° 03′	.9662	.2579	3.747	1.035	3.878	.2669
1.32	75° 38′	.9687	.2482	3.903	1.032	4.029	.2562
1.33	76° 12′	.9711	.2385	4.072	1.030	4.193	.2456
1.34	76° 47′	.9735	.2288	4.256	1.027	4.372	.2350
1.35	77° 21′	0.9757	0.2190	4.455	1.025	4.566	0.2245
1.36	77° 55′	.9779	.2092	4.673	1.023	4.779	.2140
1.37	78° 30′	.9799	.1994	4.913	1.021	5.014	.2035
1.38	79° 04′	.9819	.1896	5.177	1.018	5.273	.1931
1.39	79° 38′	.9837	.1798	5.471	1.017	5.561	.1828
1.40	80° 13′	0.9854	0.1700	5.798	1.015	5.883	0.1725
1.41	80° 47′	.9871	.1601	6.165	1.013	6.246	.1622
1.42	81° 22′	.9887	.1502	6.581	1.011	6.657	.1519
1.43	81° 56′	.9901	.1403	7.055	1.010	7.126	.1417
1.44	82° 30′	.9915	.1304	7.602	1.009	7.667	.1315
1.45	83° 05′	0.9927	0.1205	8.238	1.007	8.299	0.1214

TABLE III (continued)

θ radians or Real Number x	θ degrees	sin θ or sin x	cos θ or cos x	tan θ or tan x	csc θ or csc x	sec θ or sec x	cot θ or cot x
1.45	83° 05′	0.9927	0.1205	8.238	1.007	8.299	0.1214
1.46	83° 39′	.9939	.1106	8.989	1.006	9.044	.1113
1.47	84° 13′	.9949	.1006	9.887	1.005	9.938	.1001
1.48	84° 48′	.9959	.0907	10.98	1.004	11.03	.0910
1.49	85° 22′	.9967	.0807	12.35	1.003	12.39	.0810
1.50	85° 57′	0.9975	0.0707	14.10	1.003	14.14	0.0709
1.51	86° 31′	.9982	.0608	16.43	1.002	16.46	.0609
1.52	87° 05′	.9987	.0508	19.67	1.001	19.69	.0508
1.53	87° 40′	.9992	.0408	24.50	1.001	24.52	.0408
1.54	88° 14′	.9995	.0308	32.46	1.000	32.48	.0308
1.55	88° 49′	0.9998	0.0208	48.08	1.000	48.09	0.0208
1.56	89° 23′	.9999	.0108	92.62	1.000	92.63	.0108
1.57	89° 57′	1.000	.0008	1256	1.000	1256	.0008

Answers to Odd-Numbered Exercises

1

Exercise Set 1.1

1.

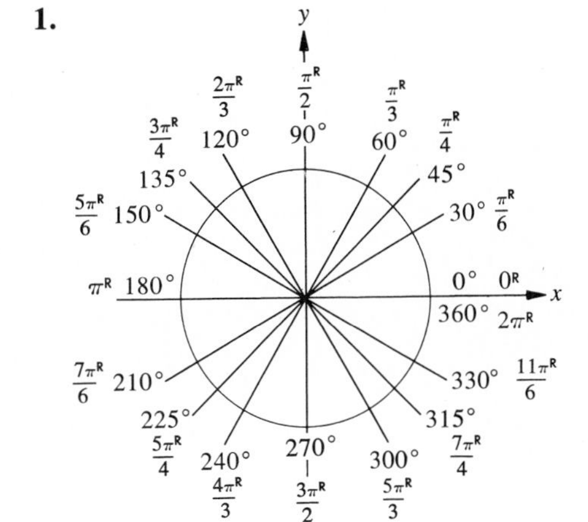

3. 60.0°; Quad. I

5. 840.0°; Quad. II

7. $-288.0°$; Quad.I

9. $-405.0°$; Quad. IV

11. 10.3°; Quad. I

13. 295.7°; Quad. IV

15. $-15.5°$; Quad. IV

17. $-231.5°$; Quad. II

19. 0.52^R; Quad. I

21. 1.31^R; Quad. I

23. 2.50^R; Quad. II

25. 7.01^R; Quad. I

27. -0.65^R; Quad. IV

29. -1.40^R; Quad. IV

31. -4.19^R; Quad. II

33. -8.90^R; Quad. III

35. 0.44^{R}

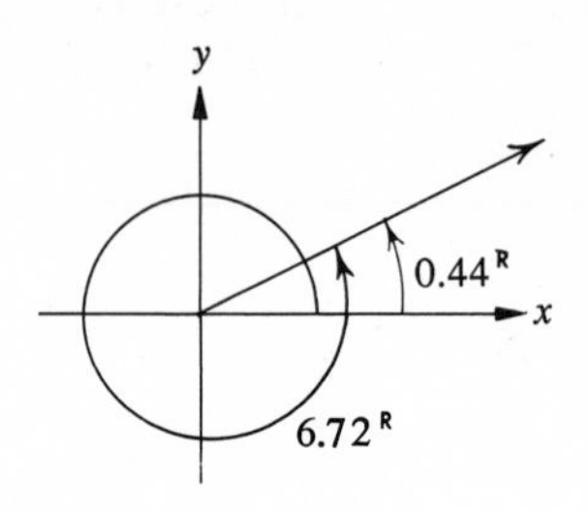

37. $19°$

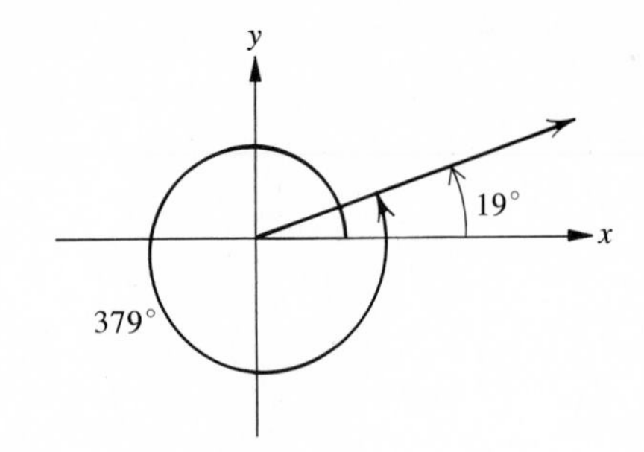

39. 1.45^{R}

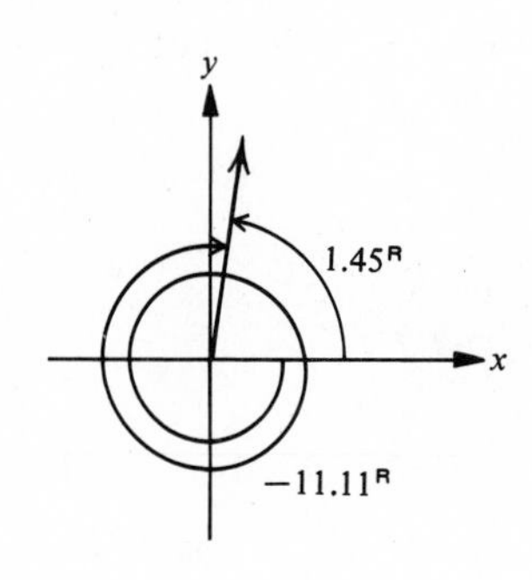

41. $288°$

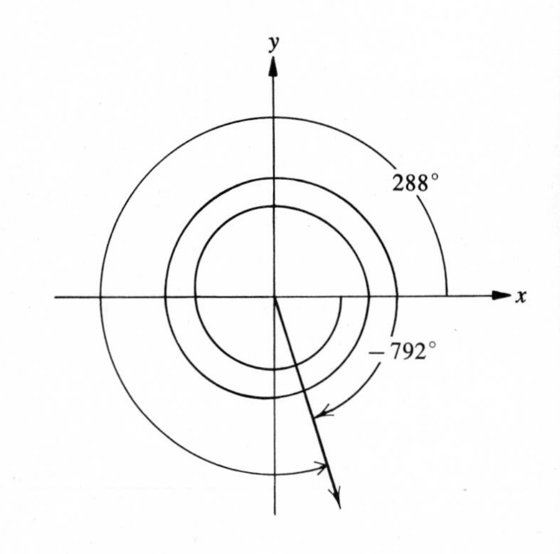

43. $(21 + k \cdot 360)°,\ k \in J$

45. $(242 + k \cdot 360)°,\ k \in J$

47. $(\pi/6 + k \cdot 2\pi)^{\mathsf{R}},\ k \in J$

49. $(5\pi/3 + k \cdot 2\pi)^{\mathsf{R}},\ k \in J$

Exercise Set 1.2

1. $\sin \alpha = 3/5$, $\cos \alpha = 4/5$,
$\tan \alpha = 3/4$, $\csc \alpha = 5/3$,
$\sec \alpha = 5/4$, $\cot \alpha = 4/3$.

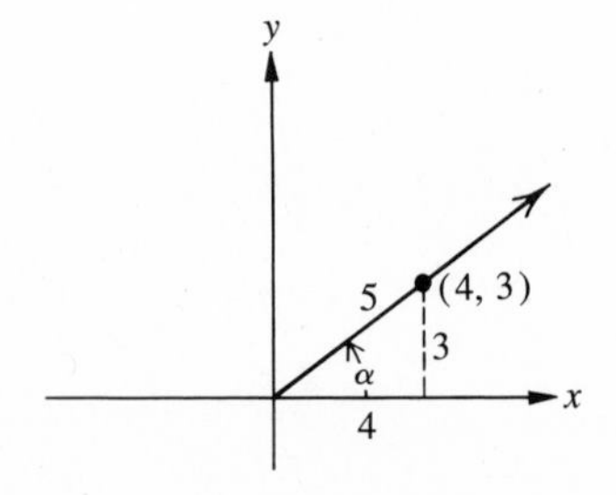

3. $\sin\alpha = -15/17$, $\cos\alpha = 8/17$,
$\tan\alpha = -15/8$, $\csc\alpha = -17/15$,
$\sec\alpha = 17/8$, $\cot\alpha = -8/15$.

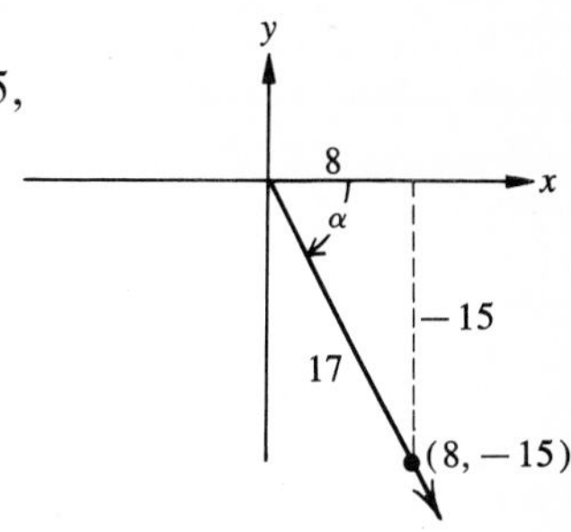

5. $\sin\alpha = 3/\sqrt{10}$, $\cos\alpha = -1/\sqrt{10}$,
$\tan\alpha = -3$, $\csc\alpha = \sqrt{10}/3$,
$\sec\alpha = -\sqrt{10}$, $\cot\alpha = -1/3$.

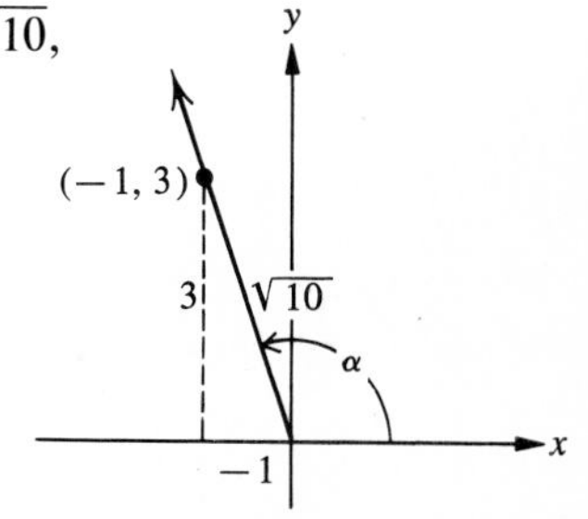

7. $\sin\alpha = 1/\sqrt{2}$, $\cos\alpha = -1/\sqrt{2}$,
$\tan\alpha = -1$, $\csc\alpha = \sqrt{2}$,
$\sec\alpha = -\sqrt{2}$, $\cot\alpha = -1$.

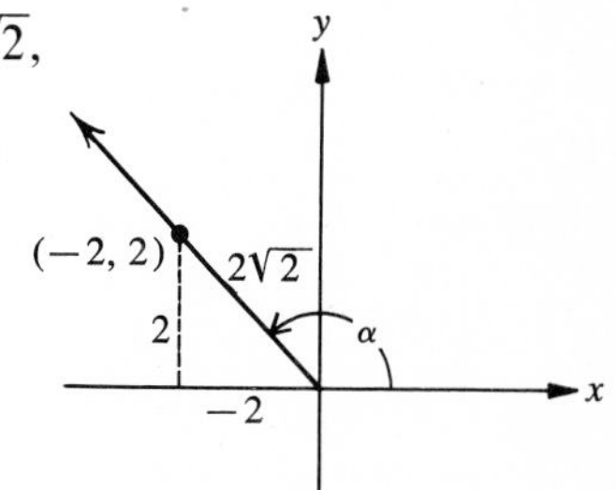

9. $\sin\alpha = -\sqrt{3}/2$, $\cos\alpha = -1/2$,
$\tan\alpha = \sqrt{3}$, $\csc\alpha = -2/\sqrt{3}$,
$\sec\alpha = -2$, $\cot\alpha = 1/\sqrt{3}$.

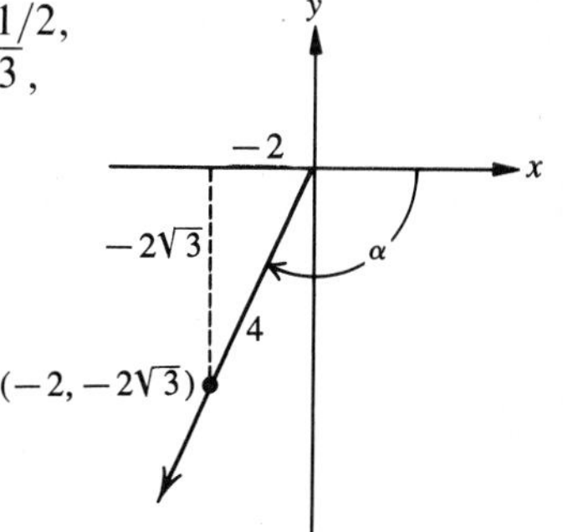

11. $\sin \alpha = -1$, $\cos \alpha = 0$,
tan α is not defined, $\csc \alpha = -1$,
sec α is not defined, $\cot \alpha = 0$.

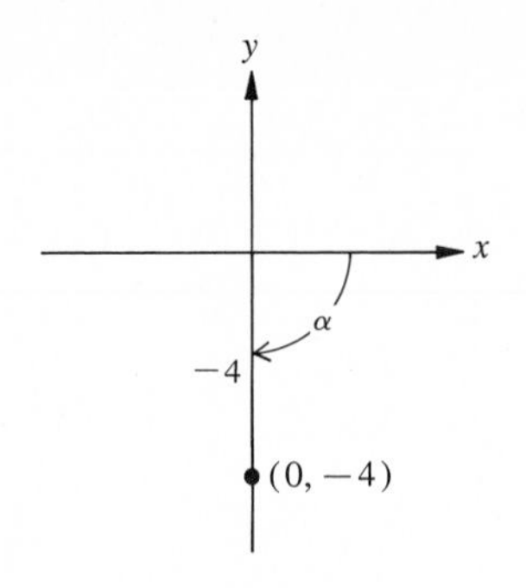

13. Quadrant IV
15. Quadrant I
17. Quadrant III
19. Quadrant II

21. $\sin \alpha = 4/5$, $\cos \alpha = 3/5$, $\tan \alpha = 4/3$, $\csc \alpha = 5/4$, $\sec \alpha = 5/3$, $\cot \alpha = 3/4$

23. $\sin \alpha = 2/5$, $\cos \alpha = \sqrt{21}/5$, $\tan \alpha = 2/\sqrt{21}$, $\csc \alpha = 5/2$, $\sec \alpha = 5/\sqrt{21}$, $\cot \alpha = \sqrt{21}/2$

25. $\sin \alpha = 2\sqrt{10}/7$, $\cos \alpha = 3/7$, $\tan \alpha = 2\sqrt{10}/3$, $\csc \alpha = 7/2\sqrt{10}$, $\sec \alpha = 7/3$, $\cot \alpha = 3/2\sqrt{10}$

27. $\sin \alpha = \sqrt{3}/2$, $\cos \alpha = 1/2$, $\tan \alpha = \sqrt{3}$, $\csc \alpha = 2/\sqrt{3}$, $\sec \alpha = 2$, $\cot \alpha = 1/\sqrt{3}$

29. $\sin \alpha = 1/4$, $\cos \alpha = \sqrt{15}/4$, $\tan \alpha = 1/\sqrt{15}$, $\csc \alpha = 4$, $\sec \alpha = 4/\sqrt{15}$, $\cot \alpha = \sqrt{15}$

31. $\sin \alpha = 1/\sqrt{2}$, $\cos \alpha = 1/\sqrt{2}$, $\tan \alpha = 1$, $\csc \alpha = \sqrt{2}$, $\sec \alpha = \sqrt{2}$, $\cot \alpha = 1$

33. $\sin \alpha = 5/13$, $\cos \alpha = -12/13$, $\tan \alpha = -5/12$, $\csc \alpha = 13/5$, $\sec \alpha = -13/12$, $\cot \alpha = -12/5$

35. $\sin \alpha = -2/3$, $\cos \alpha = -\sqrt{5}/3$, $\tan \alpha = 2/\sqrt{5}$, $\csc \alpha = -3/2$, $\sec \alpha = -3/\sqrt{5}$, $\cot \alpha = \sqrt{5}/2$

37. $\sin \alpha = -3/5$, $\cos \alpha = 4/5$, $\tan \alpha = -3/4$, $\csc \alpha = -5/3$, $\sec \alpha = 5/4$, $\cot \alpha = -4/3$

39. $\sin \alpha = 1/\sqrt{5}$, $\cos \alpha = -2/\sqrt{5}$, $\tan \alpha = -1/2$, $\csc \alpha = \sqrt{5}$, $\sec \alpha = -\sqrt{5}/2$, $\cot \alpha = -2$

41. $\sin \alpha = -3/\sqrt{13}$, $\cos \alpha = 2/\sqrt{13}$, $\tan \alpha = -3/2$, $\csc \alpha = -\sqrt{13}/3$, $\sec \alpha = \sqrt{13}/2$, $\cot \alpha = -2/3$

43. Domain: $\{\alpha \mid \alpha \in A\}$;
range: $\{\sin \alpha \mid -1 \leq \sin \alpha \leq 1\}$

45. Domain: $\{\alpha \mid \alpha \in A,\ \alpha \neq (90 + k \cdot 180)^\circ,\ k \in J\}$;
range: $\{\tan \alpha \mid \tan \alpha \in R\}$

47. Domain: $\{\alpha \mid \alpha \in A,\ \alpha \neq (90 + k \cdot 180)^\circ,\ k \in J\}$;
range: $\{\sec \alpha \mid \sec \alpha \leq -1 \text{ or } \sec \alpha \geq 1\}$

49. In the figure, corresponding sides in similar triangles OAB and $OA'B'$ are proportional. Thus,

$$\frac{x_1}{y_1} = \frac{x_2}{y_2},$$

$$\frac{x_1}{r_1} = \frac{x_2}{r_2},$$

and

$$\frac{y_1}{r_1} = \frac{y_2}{r_2}.$$

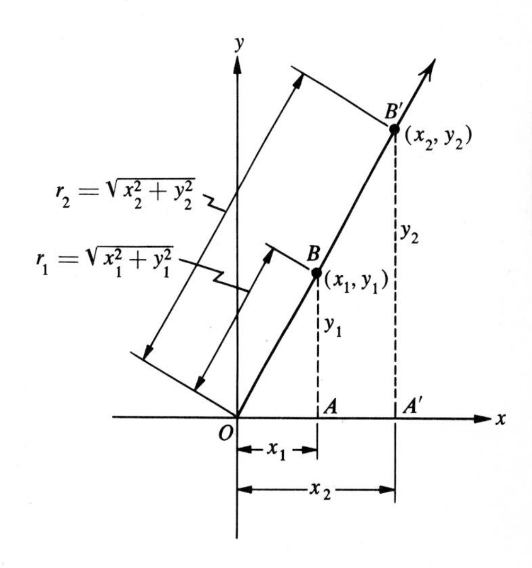

Exercise Set 1.3

1. $\sqrt{3}/2$ **3.** $1/\sqrt{2}$ **5.** $\sqrt{3}/2$ **7.** $1/\sqrt{3}$
9. $\sqrt{2}$ **11.** $2/\sqrt{3}$ **13.** 1 **15.** 0
17. Undefined **19.** -1 **21.** 0 **23.** Undefined
25. 3/4 **27.** 1/2 **29.** -3 **31.** 1/12
33. 3/2 **35.** 0 **37.** $3\sqrt{3}/2\sqrt{2}$ **39.** 2
41. Undefined **43.** 1 **45.** $-1/2$ **47.** Undefined

49. $\dfrac{1}{\sin\alpha}$ **51.** $\dfrac{1}{\sin\alpha\cos\alpha}$ **53.** $\cos^2\alpha$

55. $\dfrac{\cos^3\alpha}{\sin^2\alpha}$ **57.** $\dfrac{1}{\cos\alpha}$ **59.** $\dfrac{\cos\alpha}{\sin\alpha}$

61. Undefined for $\alpha = (90 + k \cdot 180)°$, $k \in J$; equal to zero for $\alpha = (0 + k \cdot 180)°$, $k \in J$

63. Undefined for $\alpha = (90 + k \cdot 180)°$, $k \in J$; no secant function values are equal to zero

65. $\cos\alpha = \sqrt{1 - y^2}$, $\tan\alpha = y/\sqrt{1 - y^2}$, $\csc\alpha = 1/y$, $\sec\alpha = 1/\sqrt{1 - y^2}$, $\cot\alpha = \sqrt{1 - y^2}/y$

67. $1 + \tan^2\alpha = 1 + \dfrac{\sin^2\alpha}{\cos^2\alpha} = \dfrac{\cos^2\alpha + \sin^2\alpha}{\cos^2\alpha} = \dfrac{1}{\cos^2\alpha} = \sec^2\alpha$

Exercise Set 1.4

1. 0.2823 **3.** 2.6325 **5.** 0.9466
7. 0.1086 **9.** 3.9232 **11.** 4.6569

13. 0.7895 **15.** 0.4466 **17.** 1.238
19. 3.122 **21.** $\alpha = \text{Sin}^{-1}\, 0.2169$ **23.** $\alpha = \text{Cot}^{-1}\, 2.0941$
25. $\alpha = \text{Sec}^{-1}\, 2.2013$ **27.** $\cot \alpha = 2.1431$ **29.** $\sec \alpha = 1.4214$
31. $\sin \alpha = 0.4112$ **33.** 21.2° **35.** 26.0°
37. 55.5° **39.** 35.2° **41.** 0.27^R
43. 0.49^R **45.** 1.31^R **47.** 0.83^R
49. 20.7° **51.** 63.7° **53.** 1
55. $1/\sqrt{2}$ **57.** $1/\sqrt{2}$ **59.** 0.1296
61. 0.6372 **63.** 1.009 **65.** x
67. 0.0100 **69.** 0.8415 **71.** 0.5646

Exercise Set 1.5

Where required, the answer was obtained by first finding a reference angle, using 3.14 *for* π.

1. 28° **3.** 32° 40′

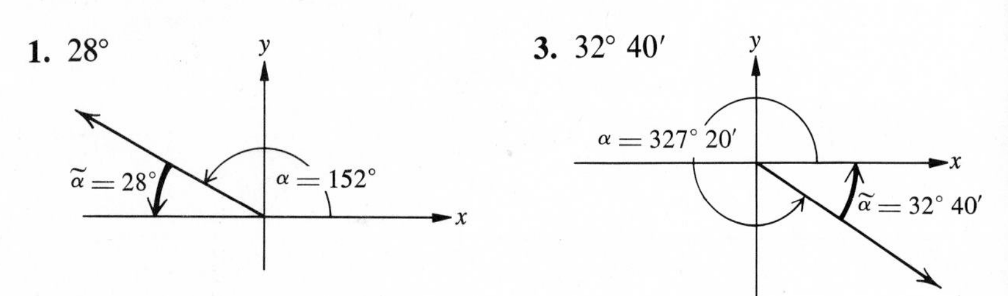

5. 82° 12′ **7.** $\dfrac{\pi^R}{6}$

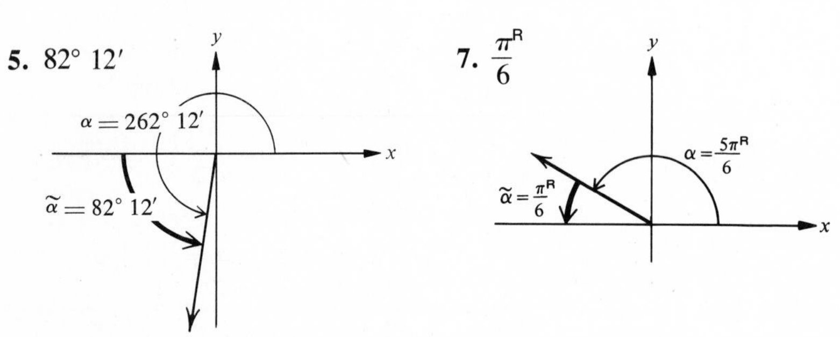

9. $\dfrac{\pi^R}{4}$ **11.** $\dfrac{\pi^R}{3}$

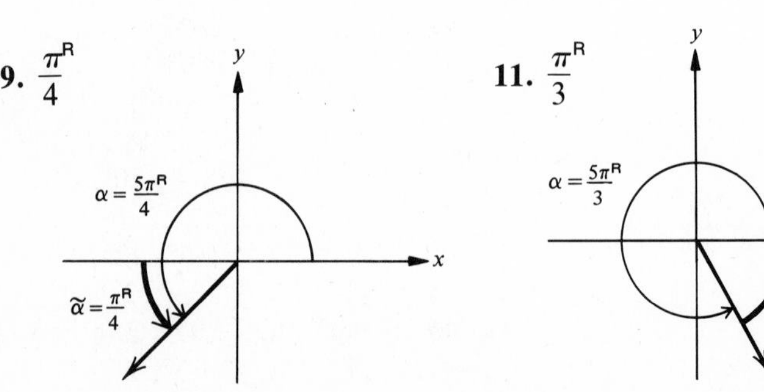

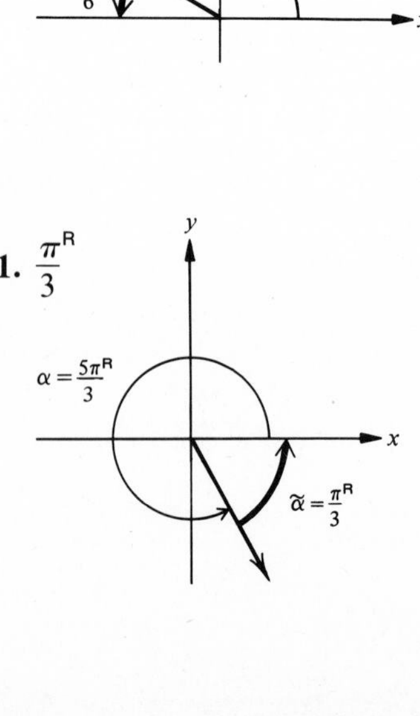

13. 1.18^R

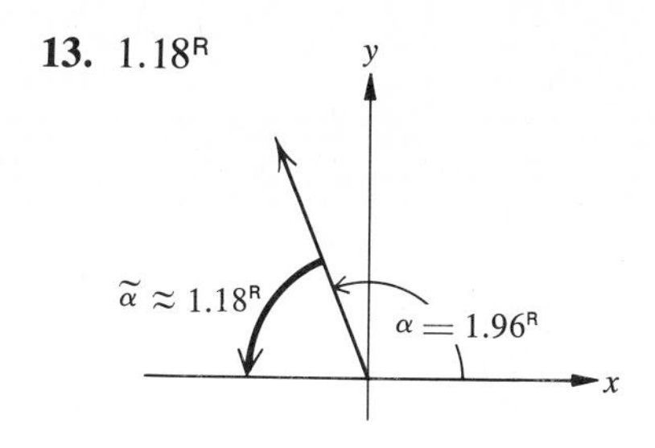

15. 0.13^R

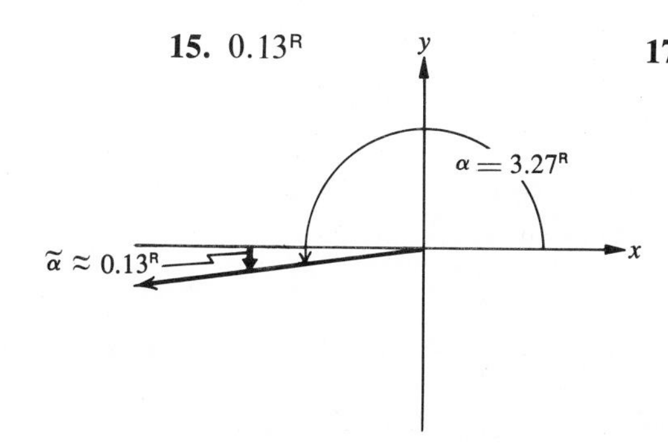

17. 0.33^R

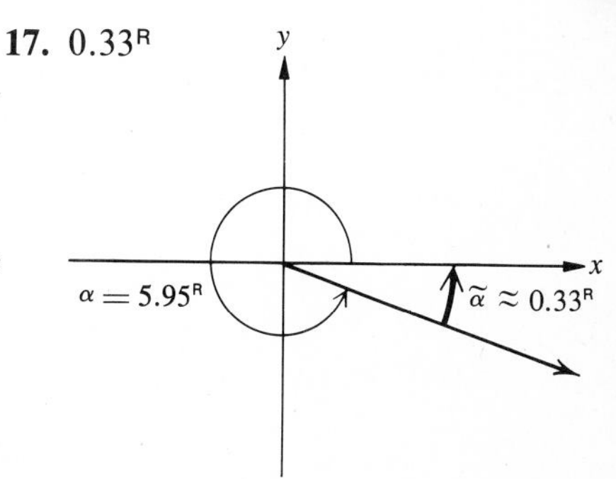

19. $-1/2$ **21.** $1/2$ **23.** 1 **25.** $2/\sqrt{3}$
27. 1 **29.** $-2/\sqrt{3}$ **31.** $-1/\sqrt{2}$ **33.** $-\sqrt{3}$
35. $-2/\sqrt{3}$ **37.** 0.6428 **39.** 8.6428 **41.** -3.1716
43. -0.2622 **45.** 0.0752 **47.** -1.6390 **49.** -1.704
51. -0.0508 **53.** 6.246 **55.** -0.9780 **57.** 0.9086
59. -0.9902

Exercise Set 1.6

1a. 30° **b.** {30°, 330°} **3a.** 150° **b.** {150°, 330°}
5a. 30° **b.** {30°, 210°} **7a.** 45° **b.** {45°, 315°}
9a. 210° **b.** {210°, 330°} **11a.** 210° **b.** {210°, 330°}
13a. 135° **b.** {135°, 225°} **15a.** 45° **b.** {45°, 135°}
17a. 0° **b.** {0°, 180°} **19a.** 150° **b.** {150°, 210°}
21a. 0° **b.** {0°, 180°} **23a.** 0° **b.** {0°, 180°}
25a. 14° **b.** {14°, 166°} **27a.** 111° **b.** {111°, 291°}
29a. 14° 18′ **b.** {14° 18′, 345° 42′} **31a.** 155.3° **b.** {155.3°, 204.7°}
33a. 25.4° **b.** {25.4°, 205.4°} **35a.** 231.4° **b.** {231.4°, 308.6°}
37a. 213.2° **b.** {213.2°, 326.8°} **39a.** 66.7° **b.** {66.7°, 293.3°}
41a. 45.6° **b.** {45.6°, 225.6°} **43a.** 0.70^R **b.** {0.70^R, 2.44^R}
45a. 0.23^R **b.** {0.23^R, 3.37^R} **47a.** 0.98^R **b.** {0.98^R, 5.30^R}
49a. 2.89^R **b.** {2.89^R, 3.39^R} **51a.** 3.26^R **b.** {3.26^R, 6.16^R}
53a. 3.63^R **b.** {3.63^R, 5.79^R}

Exercise Set 1.7

Where required, the answer was obtained by first finding a reference angle, using 3.14 for π.

1. $\sqrt{3}$ **3.** $1/\sqrt{2}$ **5.** $\sqrt{3}$ **7.** -1
9. $\sqrt{2}$ **11.** 0 **13.** $-1/\sqrt{2}$ **15.** $\sqrt{3}$
17. $\sqrt{3}$ **19.** 0.3453 **21.** -0.9839 **23.** -1.0205
25. 0.6524 **27.** -0.8253 **29.** -1.880 **31.** $-1/2$
33. $-1/\sqrt{3}$ **35.** $\sqrt{3}/2$ **37.** $-1/\sqrt{2}$ **39.** 0.3090
41. -2.7929 **43.** -0.8704 **45.** -2.234 **47.** 0.9713
49. $\{\alpha \mid \alpha \approx 16.2° + k \cdot 360° \text{ or } \alpha \approx 163.8° + k \cdot 360°\}$
51. $\{\alpha \mid \alpha \approx 142.0° + k \cdot 180°\}$
53. $\{\alpha \mid \alpha \approx 231.3° + k \cdot 360° \text{ or } 308.7° + k \cdot 360°\}$
55. $\{\alpha \mid \alpha \approx 16.8° + k \cdot 180°\}$

Review Exercises, Chapter 1

Where required, the answer was obtained by first finding a reference angle, using 3.14 for π.

1. 72.0° **2.** 138.1° **3.** 1.31^R
4. 4.48^R **5.** 218° **6.** 0.95^R
7. $(131 + k \cdot 360)°$, $k \in J$ **8.** $(7\pi/4 + k \cdot 2\pi)^R$, $k \in J$
9. $\sin \alpha = 3/5$, $\cos \alpha = -4/5$, $\tan \alpha = -3/4$, $\csc \alpha = 5/3$, $\sec \alpha = -5/4$, $\cot \alpha = -4/3$
10. $\sin \alpha = 2/\sqrt{5}$, $\cos \alpha = -1/\sqrt{5}$, $\tan \alpha = -2$, $\csc \alpha = \sqrt{5}/2$, $\sec \alpha = -\sqrt{5}$, $\cot \alpha = -1/2$
11. $\sin \alpha = -1/2$, $\cos \alpha = \sqrt{3}/2$, $\tan \alpha = -1/\sqrt{3}$, $\csc \alpha = -2$, $\sec \alpha = 2/\sqrt{3}$, $\cot \alpha = -\sqrt{3}$
12. $\sin \alpha = -15/17$, $\cos \alpha = -8/17$, $\tan \alpha = 15/8$, $\csc \alpha = -17/15$, $\sec \alpha = -17/8$, $\cot \alpha = 8/15$
13. Quadrant III **14.** Quadrant III
15. $\sin \alpha = 4/5$, $\cos \alpha = 3/5$, $\tan \alpha = 4/3$, $\csc \alpha = 5/4$, $\sec \alpha = 5/3$, $\cot \alpha = 3/4$
16. $\sin \alpha = 4/\sqrt{65}$, $\cos \alpha = 7/\sqrt{65}$, $\tan \alpha = 4/7$, $\csc \alpha = \sqrt{65}/4$, $\sec \alpha = \sqrt{65}/7$, $\cot \alpha = 7/4$
17. $\sin \alpha \approx -1/3$, $\cos \alpha = -2\sqrt{2}/3$, $\tan \alpha = 1/2\sqrt{2}$, $\csc \alpha = -3$, $\sec \alpha = -3/2\sqrt{2}$, $\cot \alpha = 2\sqrt{2}$
18. $\sin \alpha = \sqrt{21}/5$, $\cos \alpha = -2/5$, $\tan \alpha = -\sqrt{21}/2$, $\csc \alpha = 5/\sqrt{21}$, $\sec \alpha = -5/2$, $\cot \alpha = -2/\sqrt{21}$
19. $2/\sqrt{3}$ **20.** $\sqrt{2}$ **21.** Undefined
22. 0 **23.** 4/3 **24.** -1
25. $\sqrt{2}/6$ **26.** 13/2 **27.** $\sin \alpha \cos \alpha$

28. $\dfrac{1}{\sin\alpha\cos\alpha}$ **29.** $\dfrac{\cos\alpha}{\sin\alpha}$ **30.** $\dfrac{\sin\alpha}{\cos\alpha}$

31. 5.3955 **32.** 1.3972 **33.** 2.999

34. 0.7452 **35.** $\alpha = \text{Tan}^{-1}\ 1.4321$ **36.** $\cos\alpha = 0.4232$

37. 55.9° **38.** 52.7° **39.** 0.22^{R}

40. 0.69^{R}

41. 51° 20′

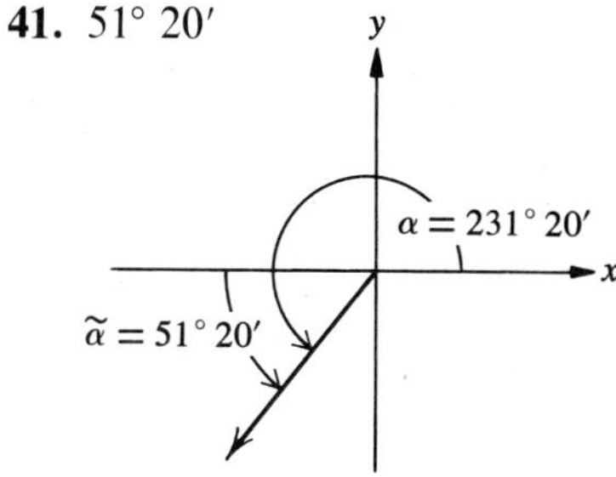

42. 58° 17′

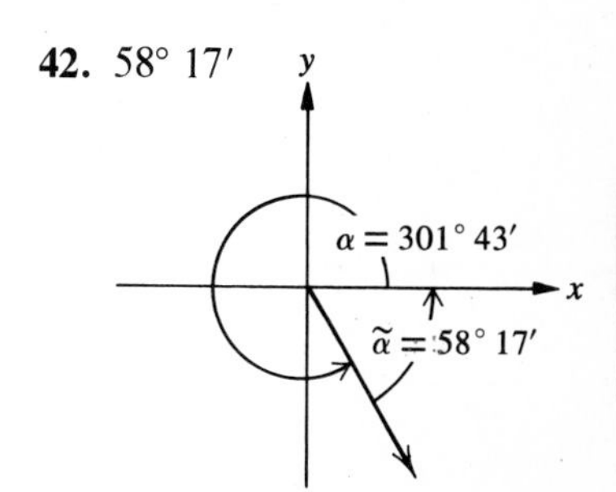

43. 0.69^{R}

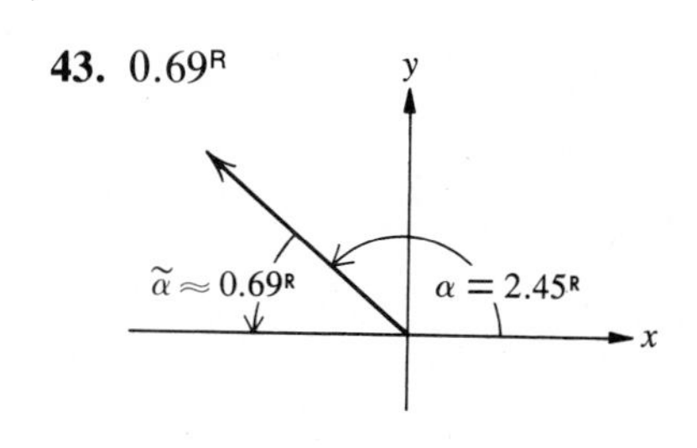

44. 1.52^{R}

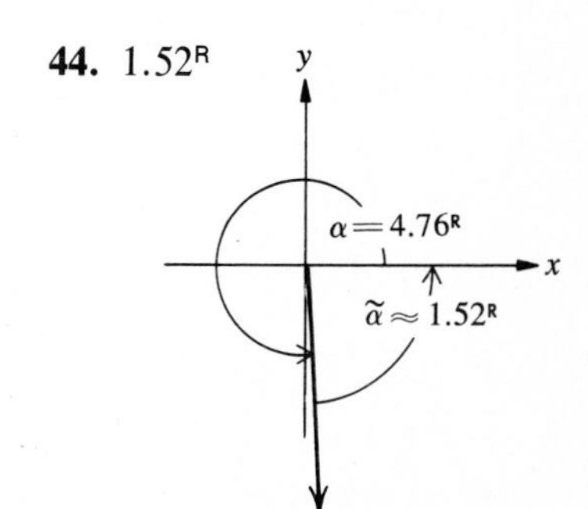

45. 1/2 **46.** -1.8650 **47.** 0.1307

48. -2.395 **49.** {21.0°, 159.0°} **50.** {149.6°, 329.6°}

51. $\{1.14^{R}, 4.28^{R}\}$ **52.** $\{0.99^{R}, 5.29^{R}\}$ **53.** 1

54. 0.9245 **55.** 1 **56.** -19.98

57. $-\sqrt{3}$ **58.** $2/\sqrt{3}$ **59.** $-2/\sqrt{3}$

60. -0.9986 **61.** -0.9259 **62.** 3.916

63. $\{\alpha \mid \alpha \approx 17.7° + k \cdot 360° \text{ or } \alpha \approx 162.3° + k \cdot 360°\}$

64. $\{\alpha \mid \alpha \approx 61.0° + k \cdot 360° \text{ or } \alpha \approx 299.0° + k \cdot 360°\}$

65. $\{\alpha \mid \alpha \approx 27.1° + k \cdot 180°\}$

66. $\{\alpha \mid \alpha \approx 224.6° + k \cdot 360° \text{ or } \alpha \approx 315.4° + k \cdot 360°\}$

Exercise Set 2.1

1. sin 28° **3.** cot 10.7° **5.** sec 1° 48′

7. $\alpha \approx 33°\ 42'$, $\beta \approx 56°\ 18'$, $c \approx 3.61$

9. $\alpha \approx 61°\ 54'$, $\beta \approx 28°\ 6'$, $b \approx 8.00$

11. $\alpha = 39°\ 00'$, $\alpha \approx 14.8$, $b \approx 18.3$

13. $\alpha = 67° 48'$, $b \approx 4.49$, $c \approx 11.9$
15. $\beta \approx 39° 42'$, $a \approx 18.1$, $c \approx 23.5$
17. $\alpha = 48° 24'$, $a \approx 33.8$, $c \approx 45.2$
19. 60.0 **21.** 24.7 **23.** 360 meters
25. $53° 6'$ **27.** 27.1 decimeters **29.** 239 meters
31. $\cos \alpha = \frac{l - h}{l}$, from which $l \cos \alpha = l - h$ and $h = l(1 - \cos \alpha)$.

Exercise Set 2.2

1. $\gamma = 85°$, $b \approx 14.1$, $c \approx 14.6$
3. $\alpha = 42.8°$, $b \approx 52.0$, $c \approx 84.7$
5. $\beta = 52° 54'$, $a \approx 0.307$, $c \approx 0.526$
7. $\gamma = 77° 30'$, $a \approx 7.05$, $b \approx 13.3$
9. One triangle: $\beta \approx 20° 54'$, $\gamma \approx 129° 6'$, $c \approx 10.9$
11. Two triangles: $\gamma \approx 50° 42'$, $\beta \approx 97° 6'$, $b \approx 7.82$ or
$\gamma' \approx 129° 18'$, $\beta' \approx 18° 30'$, $b' \approx 2.50$
13. One right triangle: $\alpha = 60° 00'$, $\gamma = 90° 00'$, $a \approx 27.7$
15. Two triangles: $\alpha \approx 60°24'$, $\gamma \approx 76°54'$, $c \approx 5.60$ or
$\alpha' \approx 119° 36'$, $\gamma' \approx 17° 42'$, $c' \approx 1.75$
17. 1.51 **19.** 3,049 feet
21. 7.90 centimeters; 14.8 centimeters
23. 78.4 meters
25. From the Law of Sines, $\frac{a}{b} = \frac{\sin \alpha}{\sin \beta}$. Adding 1 to both members gives $\frac{a}{b} + 1 = \frac{\sin \alpha}{\sin \beta} + 1$, from which $\frac{a + b}{b} = \frac{\sin \alpha + \sin \beta}{\sin \beta}$.
27. From the results of Exercises 25 and 26 and dividing left members and right members respectively gives

$$\frac{\frac{a - b}{b}}{\frac{a + b}{b}} = \frac{\frac{\sin \alpha - \sin \beta}{\sin \beta}}{\frac{\sin \alpha + \sin \beta}{\sin \beta}},$$

from which

$$\frac{a - b}{a + b} = \frac{\sin \alpha - \sin \beta}{\sin \alpha + \sin \beta}.$$

Exercise Set 2.3

1. $a \approx 4.37$, $\beta \approx 59° 54'$, $\gamma \approx 49° 6'$
3. $c \approx 3.39$, $\alpha \approx 65° 54'$, $\beta \approx 38° 48'$
5. $b \approx 1.42$, $\alpha \approx 17° 54'$, $\gamma \approx 20° 36'$
7. $a \approx 3.83$, $\beta \approx 24° 18'$, $\gamma \approx 55° 18'$

9. $\alpha \approx 95° 42'$, $\beta \approx 50° 42'$, $\gamma \approx 33° 36'$
11. $\alpha \approx 4° 00'$, $\beta \approx 31° 36'$, $\gamma \approx 144° 24'$
13. $\alpha \approx 26° 24'$, $\beta \approx 36° 24'$, $\gamma \approx 117° 12'$
15. $\alpha \approx 41° 24'$, $\beta \approx 55° 48'$, $\gamma \approx 82° 48'$
17. $42° 42'$ **19.** $82° 30'$ **21.** 10.9
23. 5.33 kilometers **25.** 11.7 centimeters **27.** 47.8°

29. Adding 1 to both members of $\cos \alpha = \dfrac{b^2 + c^2 - a^2}{2bc}$ gives

$$1 + \cos \alpha = 1 + \frac{b^2 + c^2 - a^2}{2bc}$$
$$= \frac{2bc + b^2 + c^2 - a^2}{2bc} = \frac{b^2 + 2bc + c^2 - a^2}{2bc}$$
$$= \frac{(b + c)^2 - a^2}{2bc} = \frac{(b + c + a)(b + c - a)}{2bc}.$$

2

31. $a + b + c = 2s$ and $b + c - a = 2s - 2a = 2(s - a)$. Substituting $2s$ for $a + b + c$ and $2(s - a)$ for $b + c - a$ in the results of Exercise 29 gives

$$1 + \cos \alpha = \frac{(2s)(2)(s - a)}{2bc},$$

from which

$$\frac{1 + \cos \alpha}{2} = \frac{s(s - a)}{bc}.$$

33. Multiplying left members and right members respectively of the result of Exercises 31 and 32 gives

$$\left(\frac{1 + \cos \alpha}{2}\right)\left(\frac{1 - \cos \alpha}{2}\right) = \left[\frac{s(s - a)}{bc}\right]\left[\frac{(s - b)(s - c)}{bc}\right],$$
$$\frac{1 - \cos^2 \alpha}{4} = \frac{s(s - a)(s - b)(s - c)}{b^2c^2},$$
$$\frac{\sin^2 \alpha}{4} = \frac{s(s - a)(s - b)(s - c)}{b^2c^2}.$$

Taking the square root of both members gives

$$\tfrac{1}{2} \sin \alpha = \frac{\sqrt{s(s - a)(s - b)(s - c)}}{bc},$$
$$\tfrac{1}{2}bc \sin \alpha = \sqrt{s(s - a)(s - b)(s - c)}.$$

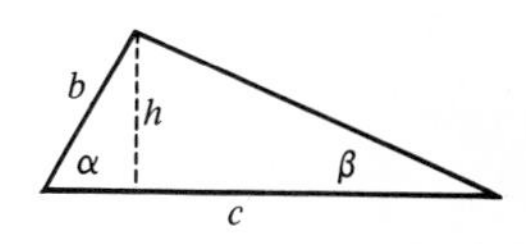

Because $h = b \sin \alpha$ (see the figure), the area $\mathscr{A}$ of a triangle is $\frac{1}{2}$(base $\times$ altitude), or $\frac{1}{2}c(b \sin \alpha)$. Thus,

$$\mathscr{A} = \sqrt{s(s - a)(s - b)(s - c)}.$$

35. 3.14

37. Noting in the figure that the radius r is the altitude of the three inner triangles, we have that the total area is $\frac{1}{2}(ra + rb + rc) = \frac{1}{2}(a + b + c)(r) = sr$. From the results of Exercise 33, we have

$$\mathscr{A} = \sqrt{s(s - a)(s - b)(s - c)} = sr,$$

from which

$$r = \sqrt{\frac{s(s - a)(s - b)(s - c)}{s^2}} = \sqrt{\frac{(s - a)(s - b)(s - c)}{s}}.$$

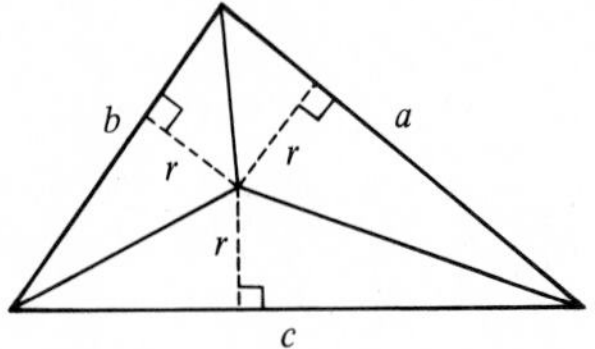

Exercise Set 2.4

1. 2.3; 90°

3. 3.5; −60°

5. 2.4; 90°

7.

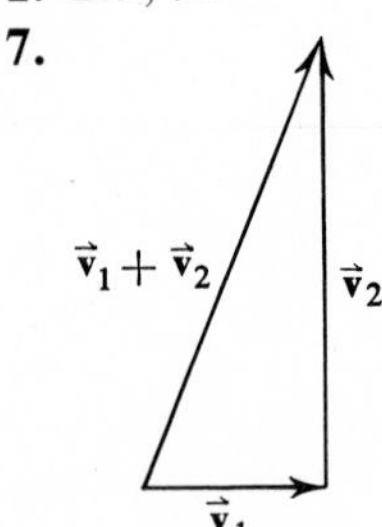

9.

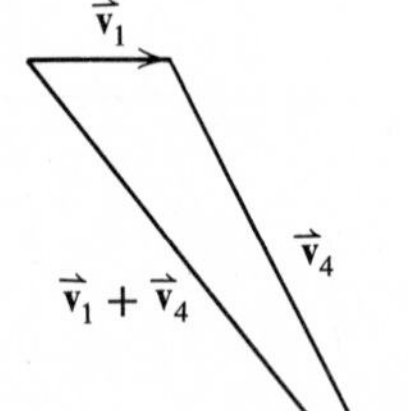

11.

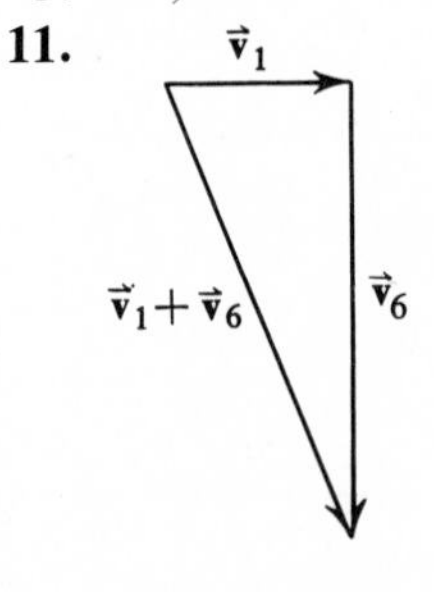

13.

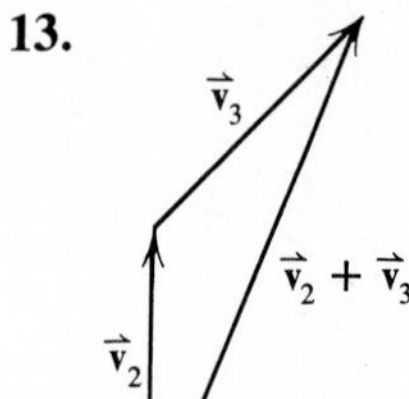

15.

17.

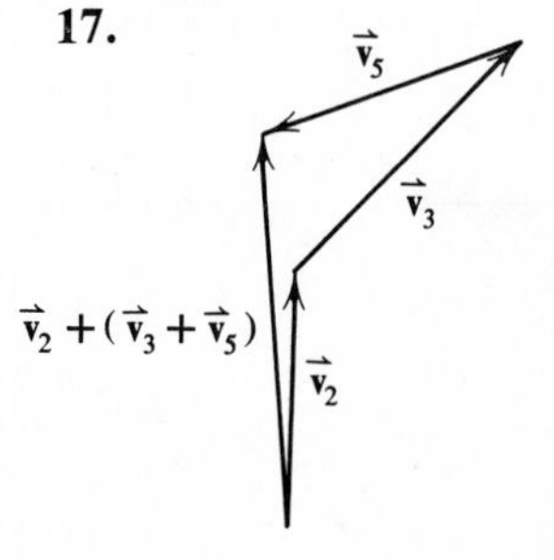

19. $-\vec{\mathbf{v}}_1$

21. **23.**

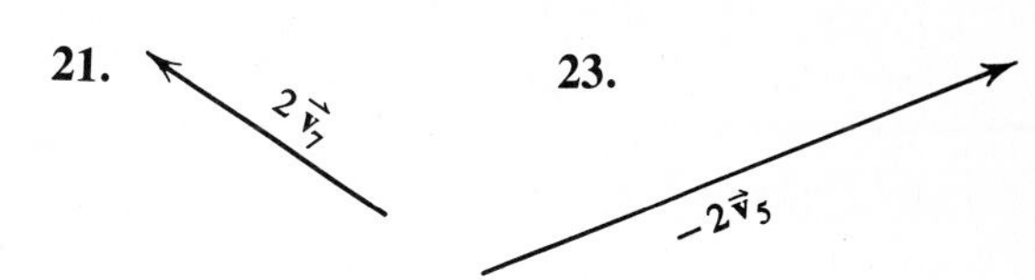

25. $\frac{1}{2}\vec{\mathbf{v}}_3$

27.

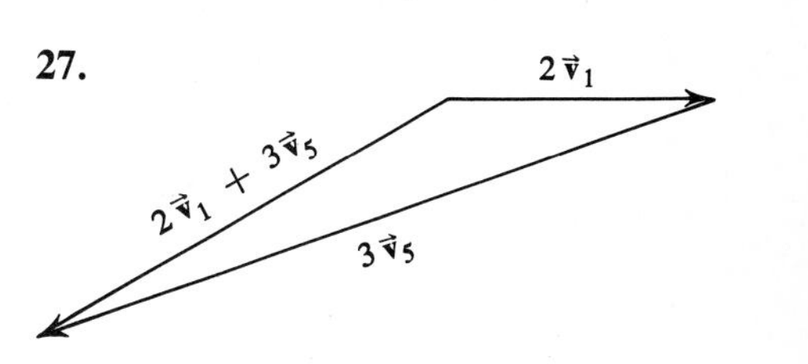

29.

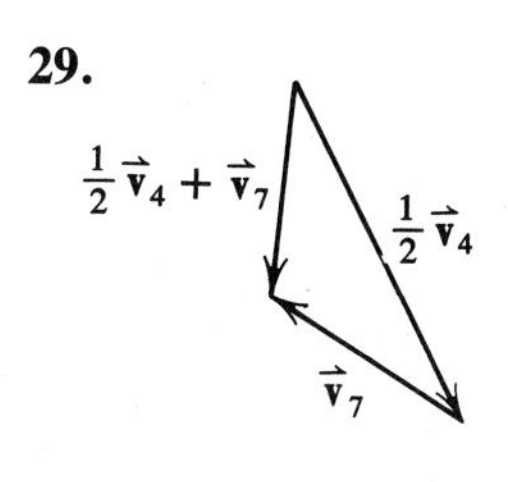

31.

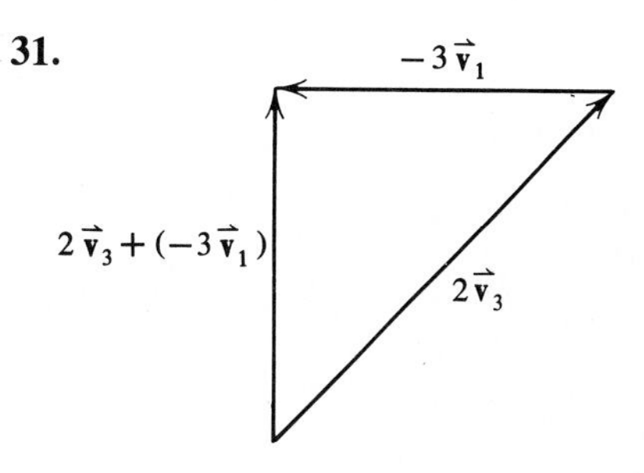

33.

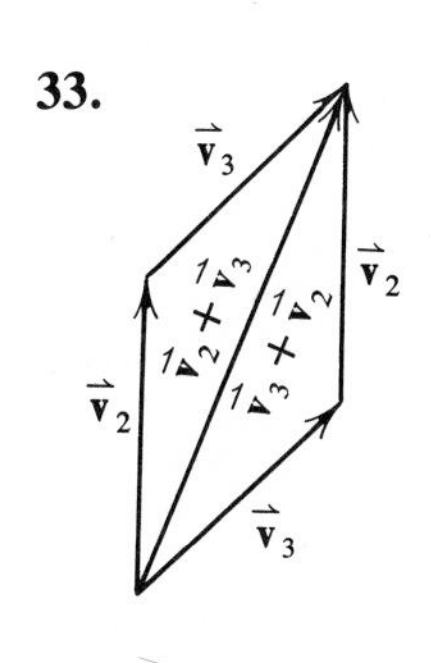

35.

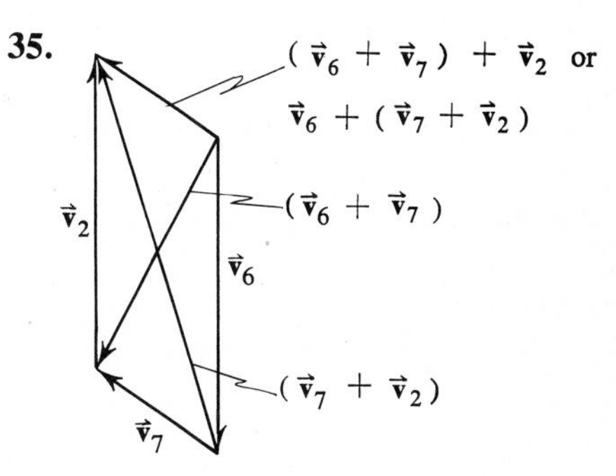

37.

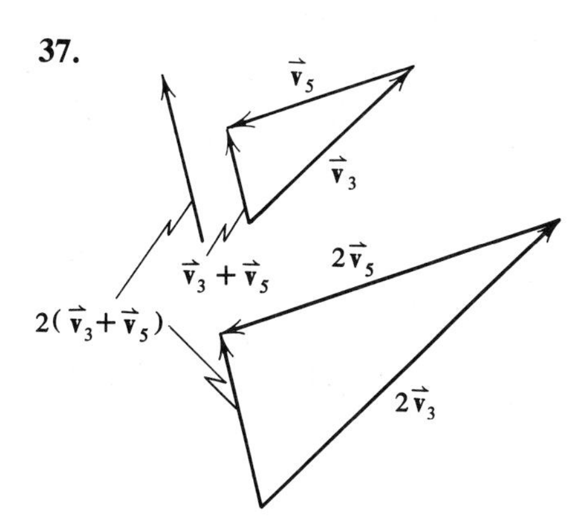

39.

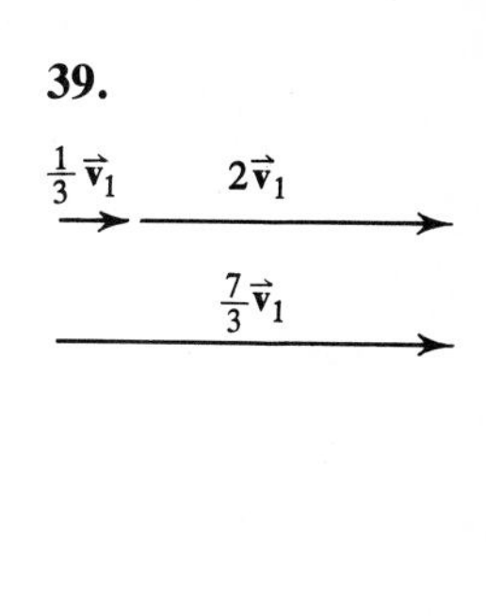

Exercise Set 2.5

1. $\|\vec{v}_x\| = 5, \quad \|\vec{v}_y\| = 5$
3. $\|\vec{v}_x\| = 0, \quad |\vec{v}_y\| = 15$
5. $\|\vec{v}_x\| = 6\sqrt{3}, \quad \|\vec{v}_y\| = 6$
7. $|\vec{v}_x\| = 4, \quad \|\vec{v}_y\| = 4\sqrt{3}$
9. $\|\vec{v}_3\| = 5$ pounds, $\beta \approx 53°\ 6'$, $\alpha \approx 36°\ 54'$
11. $\|\vec{v}_3\| = 27.9$ kilograms, $\beta \approx 43°48'$, $\alpha \approx 31°18'$.
    (*Note:* $\alpha + \beta = 75°\ 6'$ because of rounding off procedures.)
13. 8.72 kilograms, $\alpha \approx 36°\ 36'$, $\beta \approx 83°\ 24'$
15. 40 kilograms, 36° 54′
17. 19.7 pounds, 37.1 pounds
19. 248 kilograms
21. 33° 24′
23. $\|\vec{v}_x\| = 1.64$ meters/sec^2, $\|\vec{v}_y\| = 1.15$ meters/sec^2
25. 0.967 kilometer per hour
27. Drift angle is 11° 18′ to the right; ground speed is 612 miles per hour; course is 146° 18′
29. Drift angle is 12° to the right; wind direction is 318° measured from true north; wind speed is 106.5 miles per hour
31. Heading is 85° 12′; air speed is 602 kilometers per hour
33. 34.6 nautical miles, 265° 24′
35. 70.7 kilograms on each half

Exercise Set 2.6

1. 6.2 meters
3. $5\pi^2/6 \approx 8.22$ decimeters
5. $-9\pi/4 \approx -7.07$ feet
7. 126.9°
9. 280.7°
11. 3.6 meters per second
13. 30 radians per second
15. $-54\pi \approx -169.65$ decimeters per second
17. 104.72 meters
19. 1,745 miles
21. 3.98 centimeters
23. 1600 π meters per minute
25. 9.39 radians per minute
27. 7.17×10^{-4} radians per hour
29. 2.24×10^3 miles per hour
31. Area of the sector is to the area of the circle as the central angle θ is to 2π radians. Thus,

$$\frac{\mathscr{A}(\text{of sector})}{\pi r^2} = \frac{\theta^{\mathrm{R}}}{2\pi},$$

from which

$$\mathscr{A} = \frac{\pi r^2(\theta^{\mathrm{R}})}{2\pi} = \tfrac{1}{2}r^2\theta^{\mathrm{R}}.$$

Exercise Set 2.7

1. 8 centimeters; 8 centimeters; 8 centimeters
3. 4.33 centimeters; -1.12 centimeters
5. $(40/3)\sqrt{3}$ centimeters
7. 20.9 centimeters

9. -106 volts
11. -66.7 volts
13. 85.2 volts
15. -0.0295 ampere
17. 0.002 ampere
19. 0.01 ampere
21. 1.2 volts
23. 1.2 volts
25. 0 volt
27. -5.89 dynes per square centimeter
29. 0 dyne per square centimeter
31. 0.242 foot

Review Exercises, Chapter 2

1. $\alpha = 39°$, $a \approx 8.81$, $b \approx 10.9$
2. $\alpha \approx 53° \, 42'$, $\beta \approx 36° \, 18'$, $b \approx 3.97$
3. $\beta = 57° \, 24'$, $b \approx 20.2$, $c \approx 24.0$
4. $\beta = 18° \, 42'$, $a \approx 28.7$, $c \approx 30.3$
5. $\gamma = 77.5°$, $a \approx 7.03$, $b \approx 13.3$
6. $\beta = 10° \, 00'$, $a \approx 1.73$, $b \approx 0.425$
7. Two triangles: $\alpha \approx 60.4°$, $\gamma \approx 76.9°$, $c \approx 5.60$ or
$\alpha' \approx 119.6°$, $\gamma' \approx 17.7°$, $c' \approx 1.75$
8. One triangle: $\beta \approx 30° \, 36'$, $\gamma \approx 13° \, 54'$, $c \approx 17.5$
9. $\alpha \approx 51° \, 18'$, $\beta \approx 70° \, 30'$, $c \approx 2.07$
10. $\alpha \approx 35° \, 12'$, $\beta \approx 42° \, 12'$, $\gamma \approx 102° \, 36'$

11.

$(\vec{v}_1 + \vec{v}_2) + \vec{v}_3$
$\vec{v}_1$
$\vec{v}_2$
$\vec{v}_3$
$\vec{v}_1 + \vec{v}_2$

12.

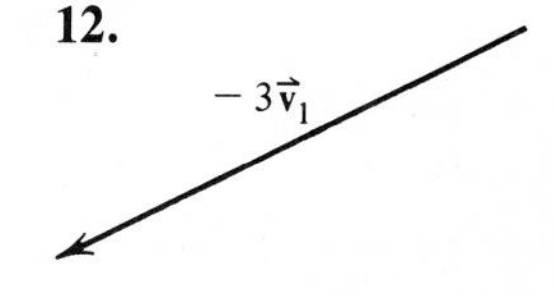

13.

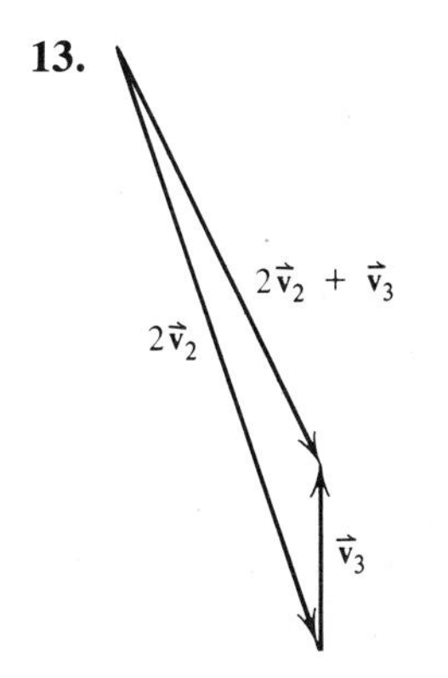

14.

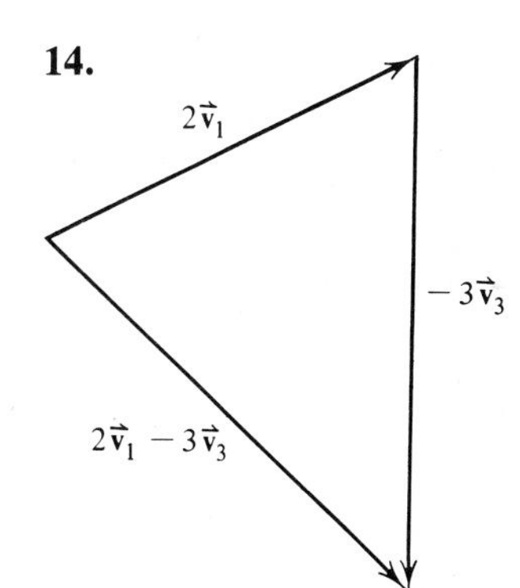

15. $4\sqrt{3}$; 4

16. $\dfrac{6}{\sqrt{2}}, \dfrac{6}{\sqrt{2}}$

17. 120 kilograms; 22° 36′

18. 6.71 miles per hour; drift of 63° 24′

19. Wind bearing is 23° 00′; wind speed is 183 kilometers per hour

20. 36° 54′

21. 6.82 centimeters

22. 13.2°

23. 12 inches per minute

24. 13.1 radians per second

25. 0.0276 ampere

26. −1.5 centimeters

27. 0 volt

28. 9.52 dynes per square centimeter

Exercise Set 3.1

1. $-2\pi \quad -\frac{3\pi}{2} \quad -\pi \quad -\frac{\pi}{2} \quad 0 \quad \frac{\pi}{2} \quad \pi \quad \frac{3\pi}{2} \quad 2\pi$

$-5 \quad 0 \quad 5$

3. $0 \quad \frac{\pi}{4} \quad \frac{\pi}{2} \quad \frac{3\pi}{4} \quad \pi \quad \frac{5\pi}{4} \quad \frac{3\pi}{2} \quad \frac{7\pi}{4}$

$0 \quad \pi \quad 2\pi$

5. $-2\pi \quad -\frac{4\pi}{3} \quad -\frac{2\pi}{3}$

$-\frac{5\pi}{3} \quad -\pi \quad -\frac{\pi}{3} \quad 0 \quad \frac{\pi}{3} \quad \frac{2\pi}{3} \quad \pi \quad \frac{4\pi}{3} \quad \frac{5\pi}{3} \quad 2\pi$

$-2\pi \quad -\pi \quad 0 \quad \pi \quad 2\pi$

7. Zeros: $\{x \mid x = k\pi, k \in J\}$; amplitude: 3

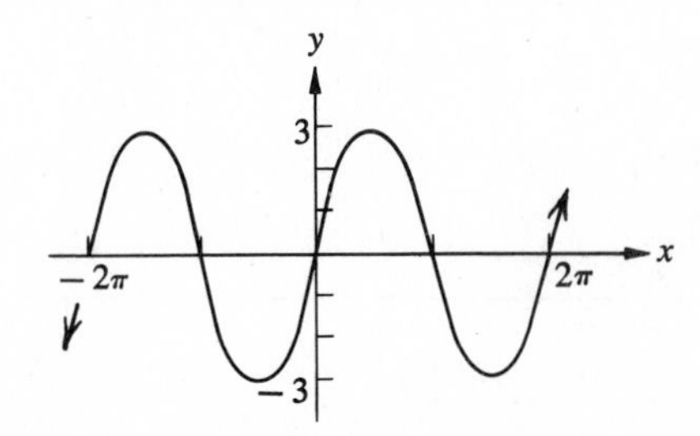

9. Zeros: $\{x \mid x = \pi/2 + k\pi, k \in J\}$; amplitude: 1/2

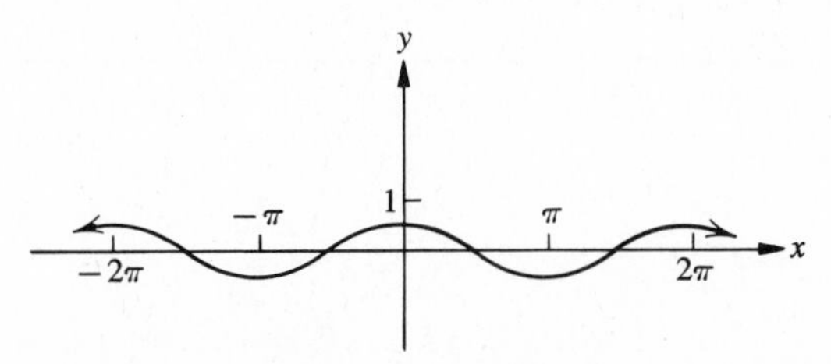

11. Zeros: $\{x \mid x = k\pi,\ k \in J\}$; amplitude: 1

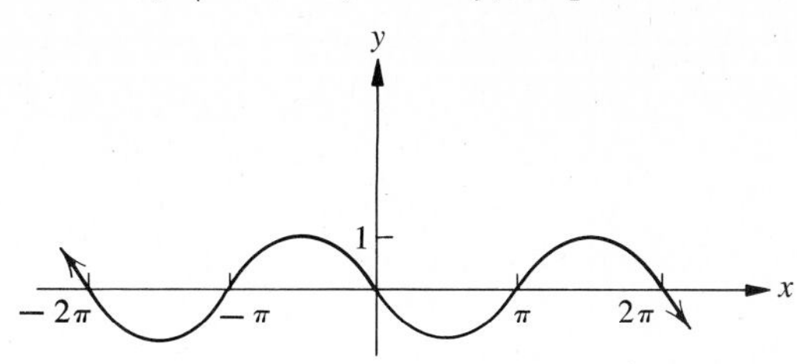

13. Zeros: $\{x \mid x = k\pi/2,\ k \in J\}$; amplitude: 1

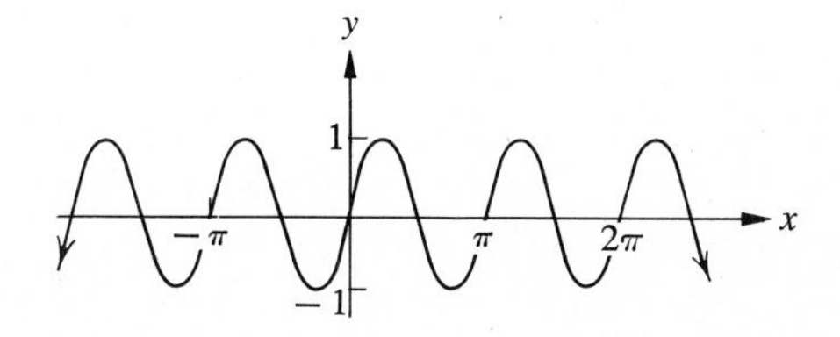

15. Zeros: $\{x \mid x = \pi + k \cdot 2\pi,\ k \in J\}$; amplitude: 1

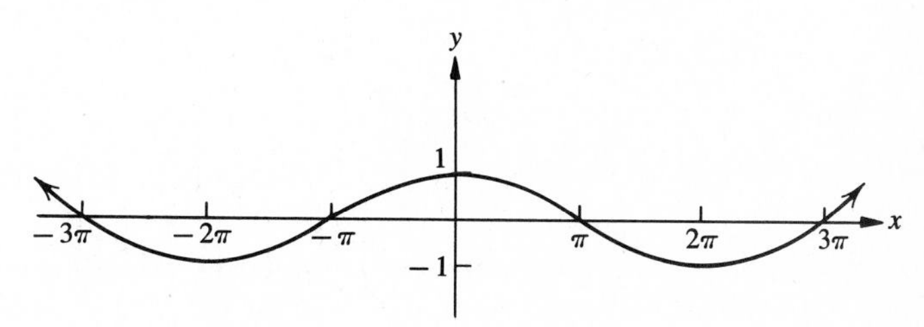

17. Zeros: $\{x \mid x = \pi/2 + k\pi,\ k \in J\}$; amplitude: 1

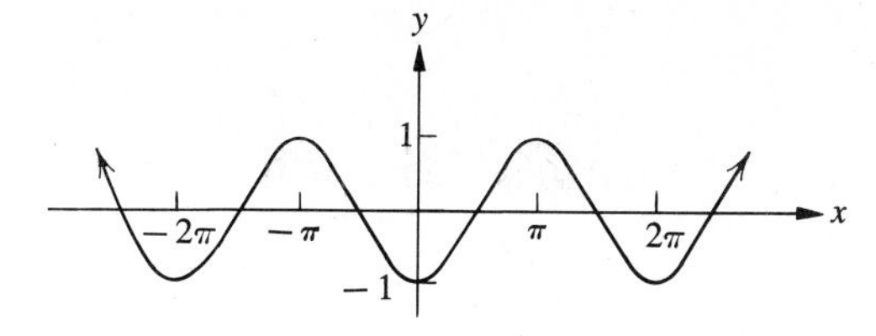

19.

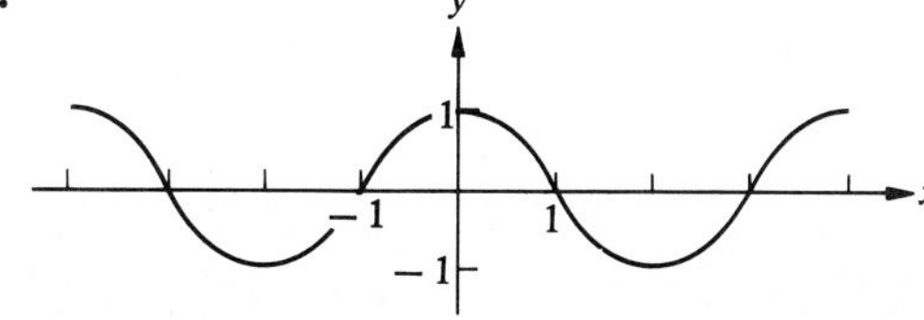

21.

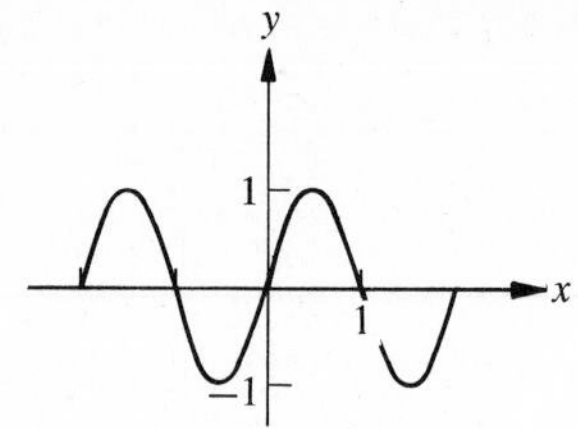

23.

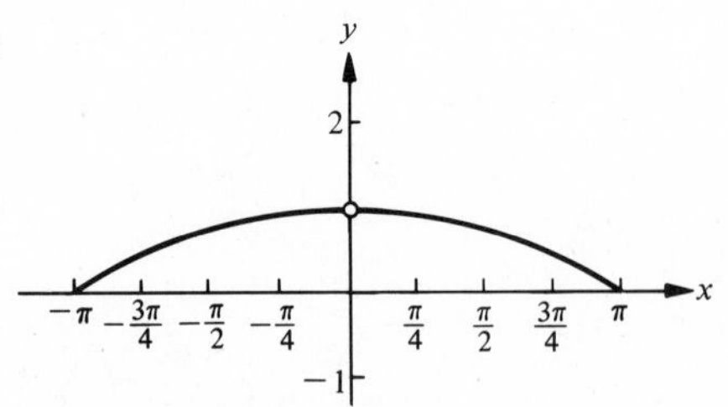

As x approaches zero from the left or right, $\frac{\sin x}{x}$ approaches a value of 1. The graph is not periodic because there is no number a ($a \neq 0$) such that $\frac{\sin(x + a)}{x + a} = \frac{\sin x}{x}$.

25.

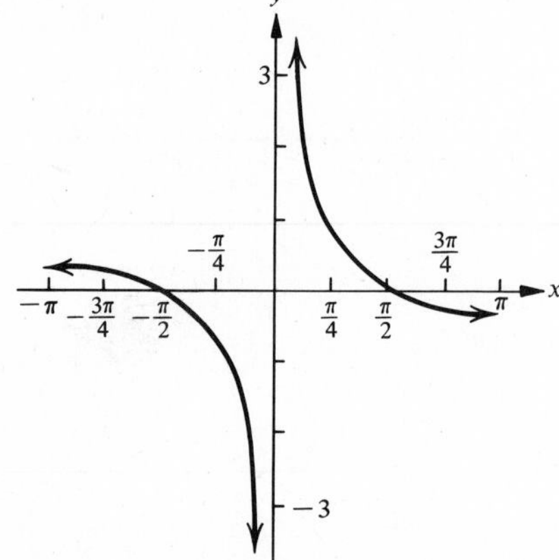

As x approaches zero from the left, $\frac{\cos x}{x}$ is negative and its absolute value becomes greater. As x approaches zero from the right, $\frac{\cos x}{x}$ is positive and its absolute value becomes greater. The graph is not periodic because there is no number a ($a \neq 0$) such that $\frac{\cos(x + a)}{x + a} = \frac{\cos x}{x}$.

Exercise Set 3.2

1.

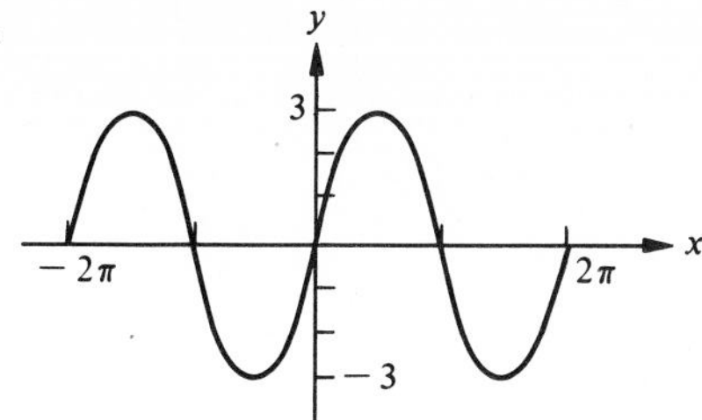

3.

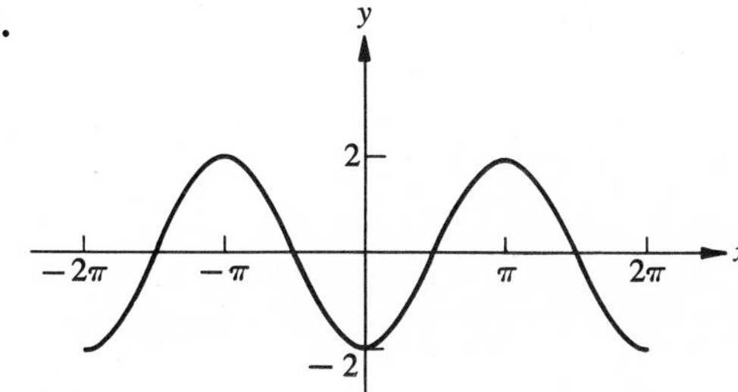

5.

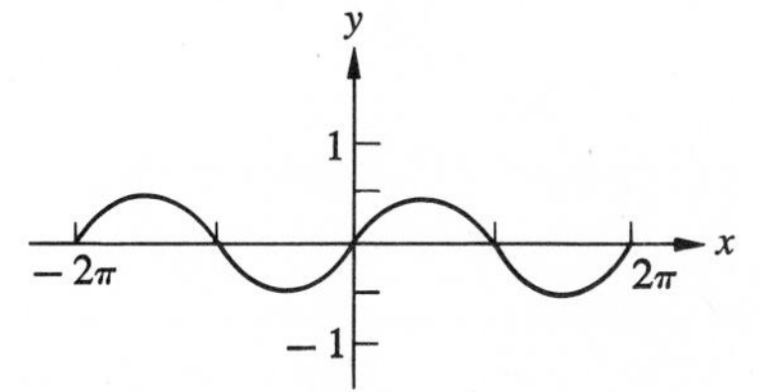

7.

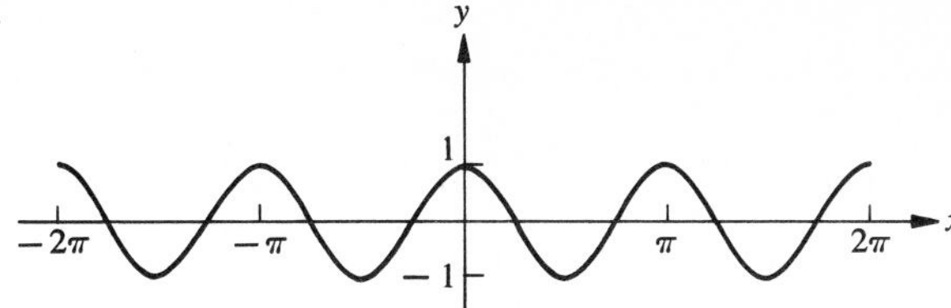

9.

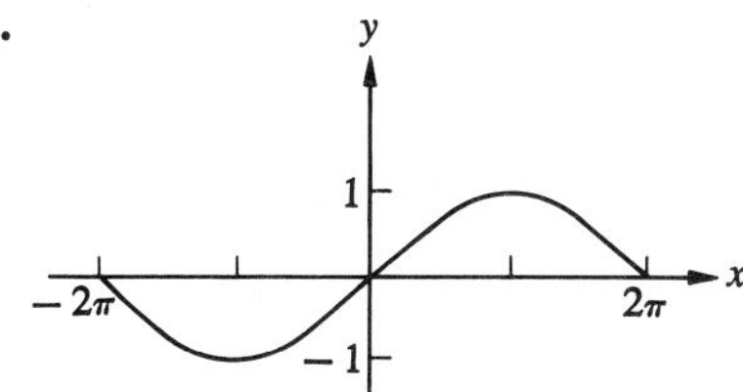

11.

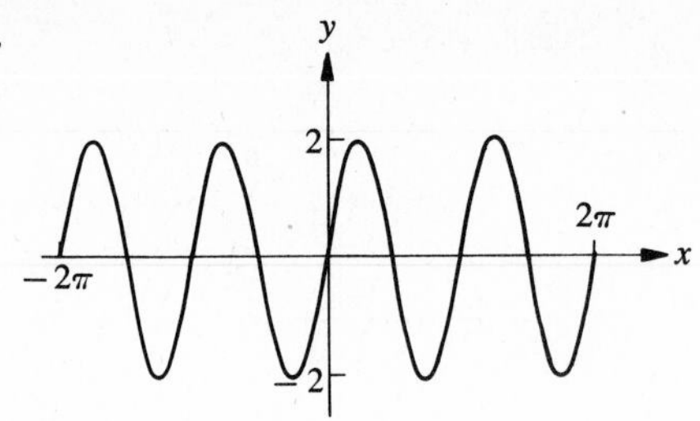

13.

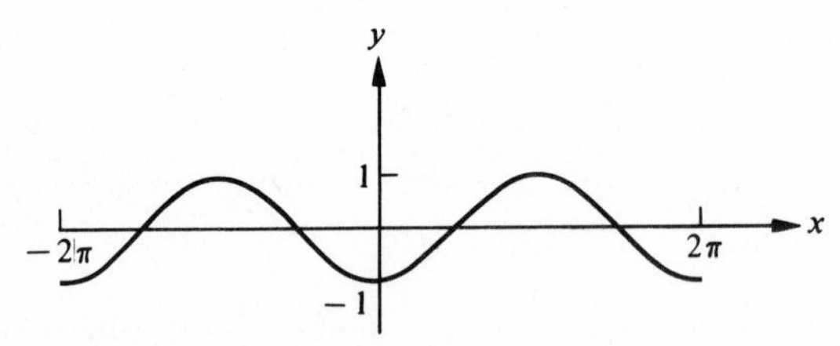

15.

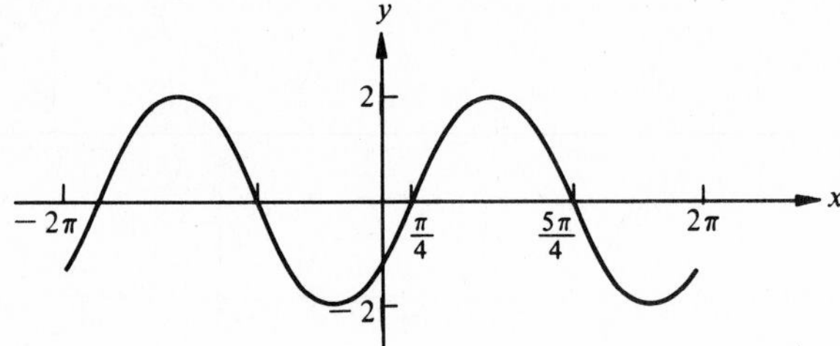

17.

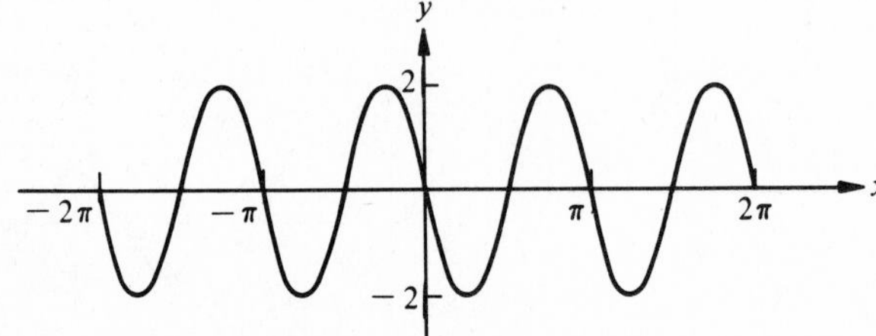

19.

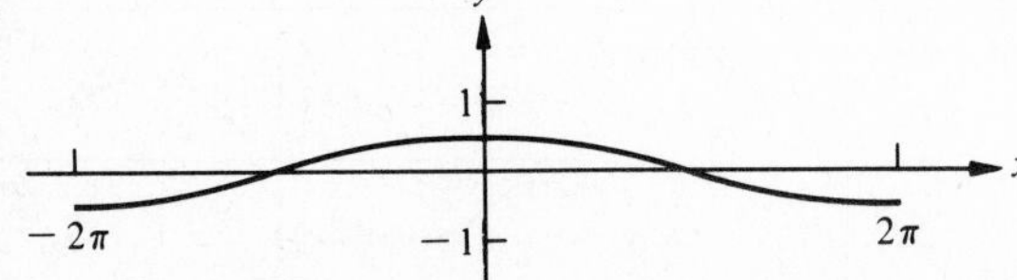

21.

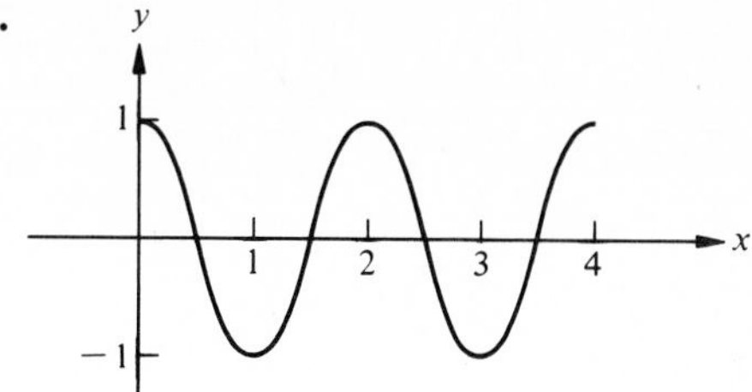

23.

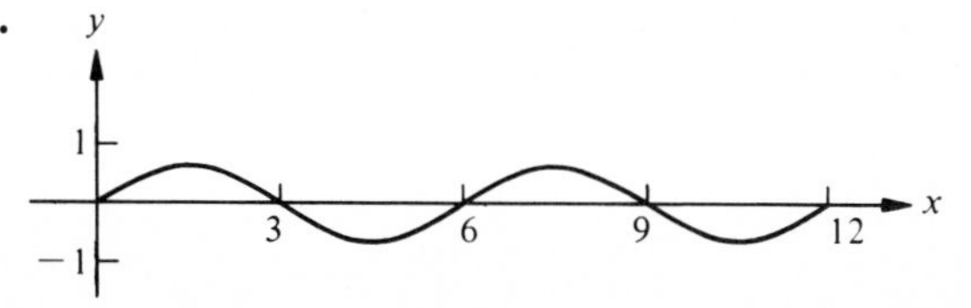

3

Exercise Set 3.3

1.

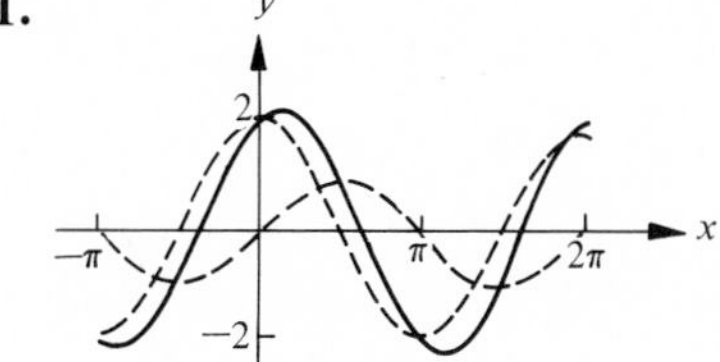

3.

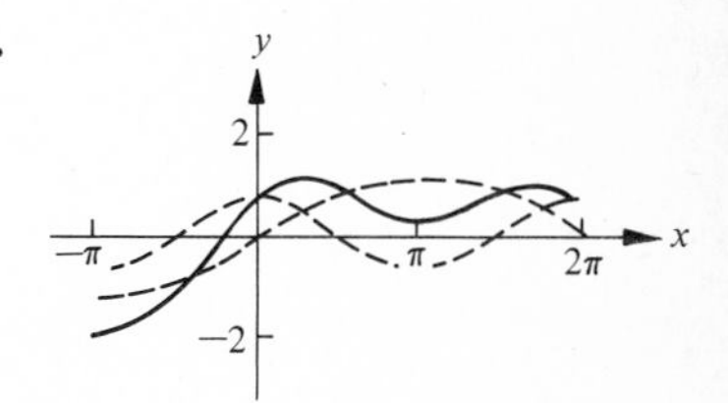

5.

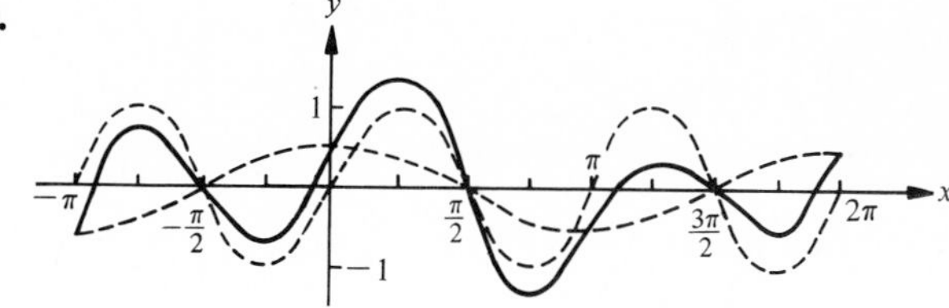

7.

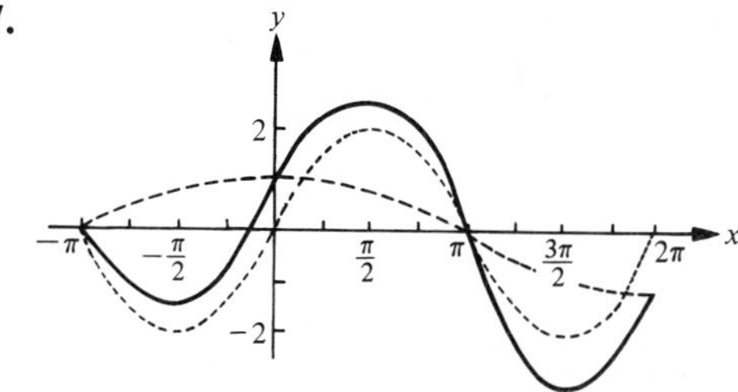

9.

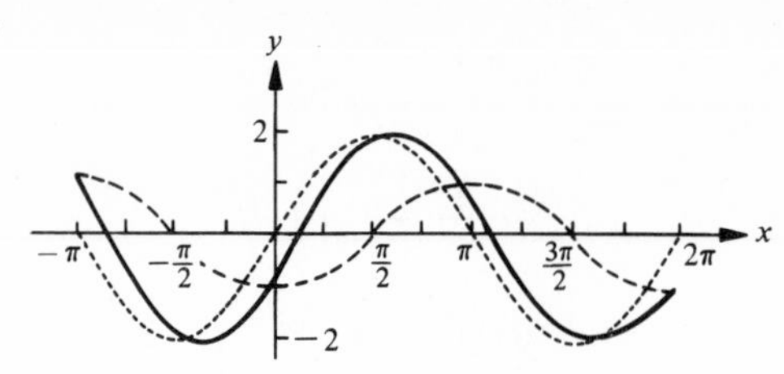

11.

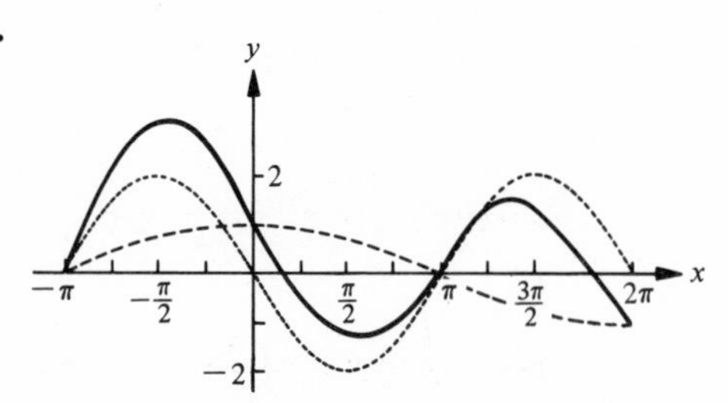

13.

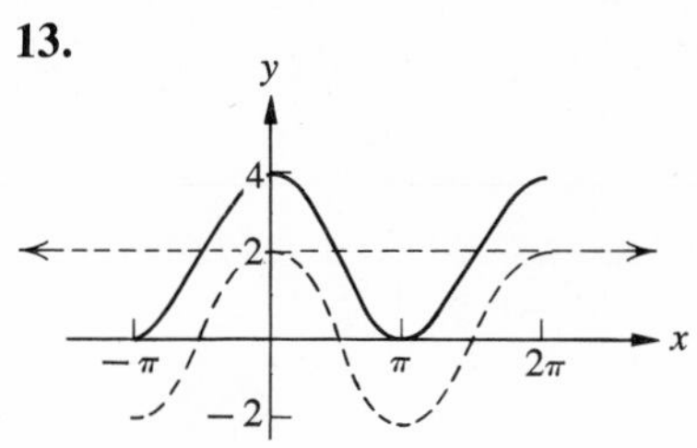

15.

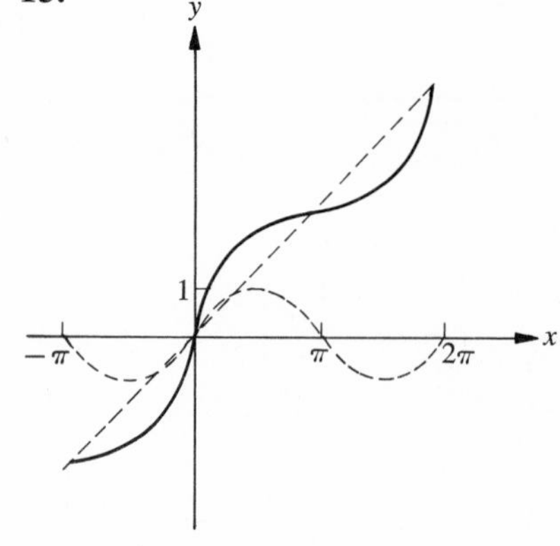

17.

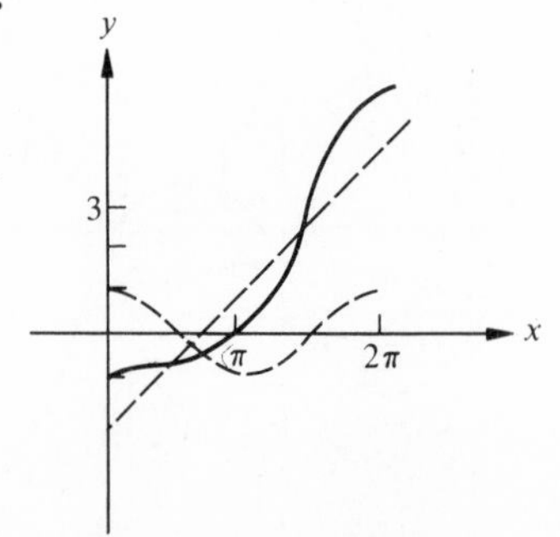

19.

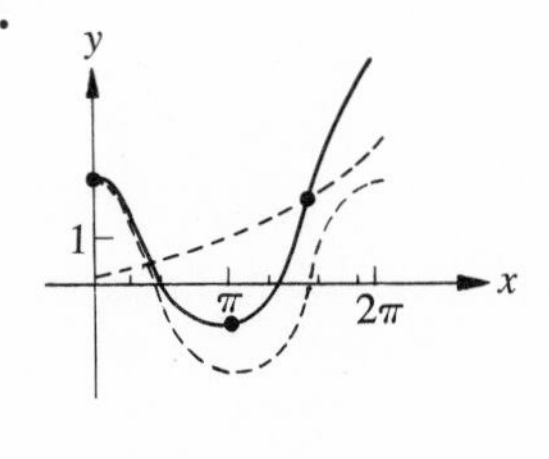

21.

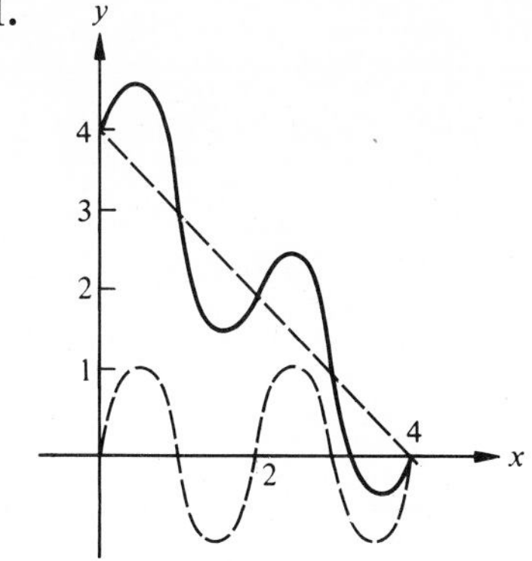

Exercise Set 3.4

1. 

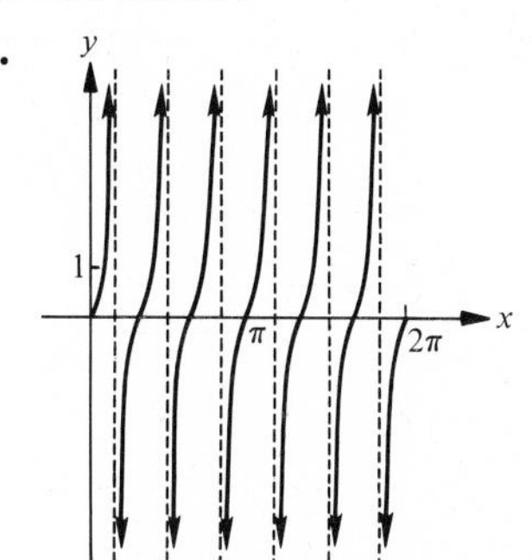

a. $\{y \mid y \in R\}$
b. $x = \dfrac{\pi}{6} + k \cdot \dfrac{\pi}{3},\ k \in J$
c. $\{x \mid x = k \cdot \dfrac{\pi}{3},\ k \in J\}$

3.

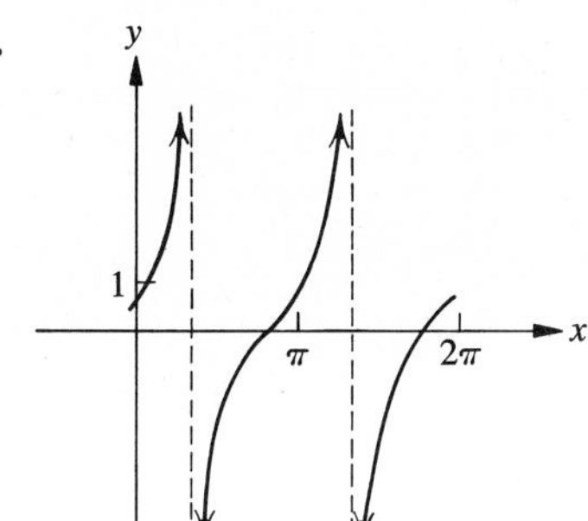

a. $\{y \mid y \in R\}$
b. $x = \pi/3 + k\pi, \quad k \in J$
c. $\{x \mid x = 5\pi/6 + k\pi,\ k \in J\}$

5.

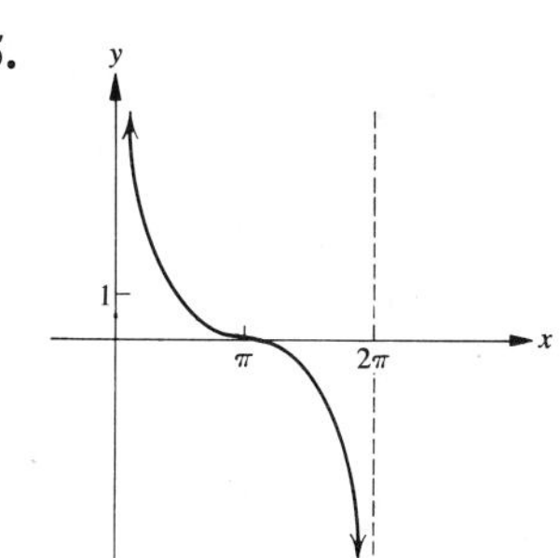

a. $\{y \mid y \in R\}$
b. $x = k \cdot 2\pi, \quad k \in J$
c. $\{x \mid x = \pi + k \cdot 2\pi,\ k \in J\}$

7.

9. $\{x \mid x = k\pi,\ k \in J\}$

11.

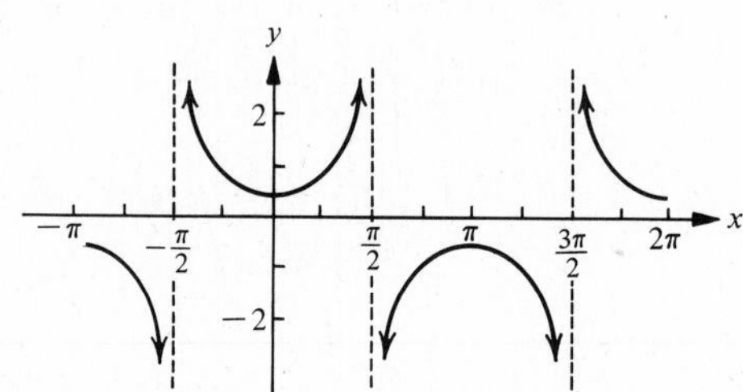

13.

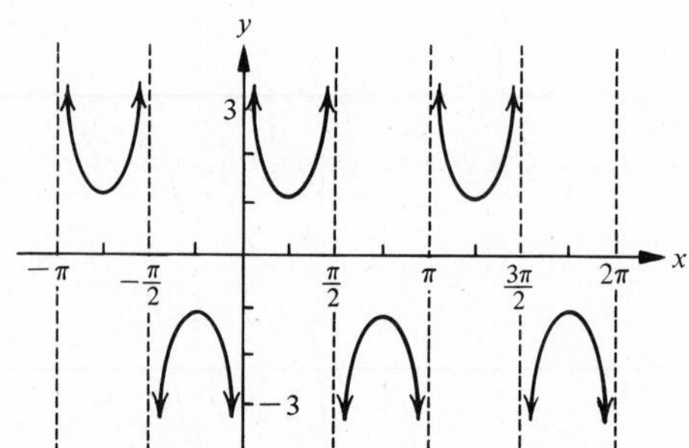

15.

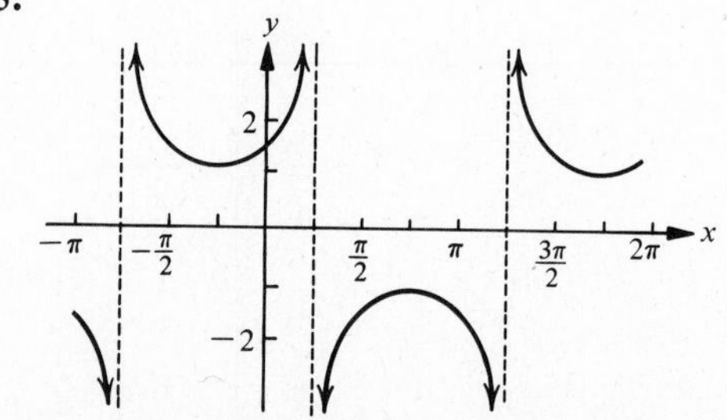

Exercise Set 3.5

1. $\{(0, 1), (6.3, 1)\}$

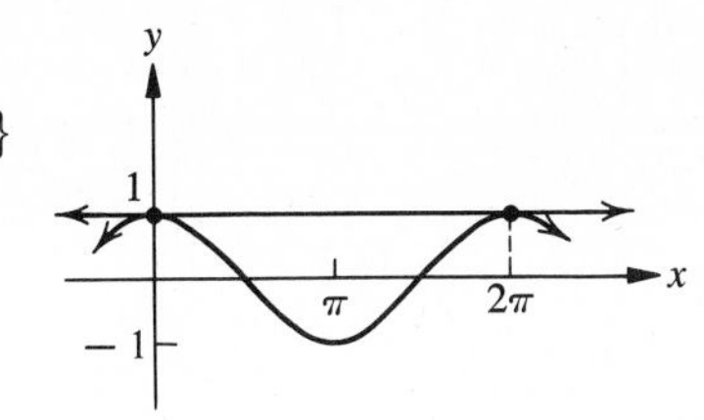

3. $\{(2.1, 0.5)\}$

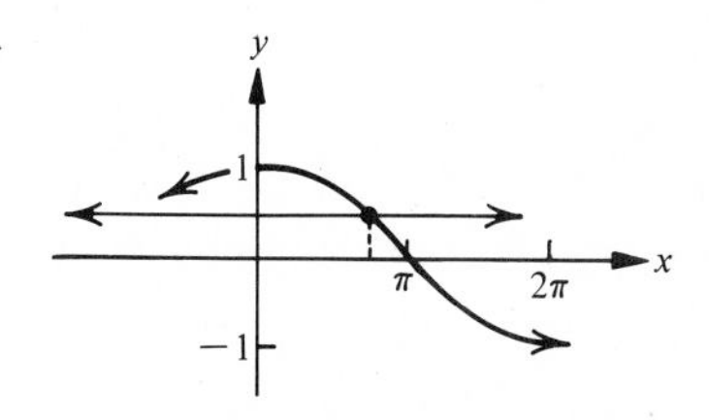

5. $\{(3\pi/4, -1), (7\pi/4, -1)\}$

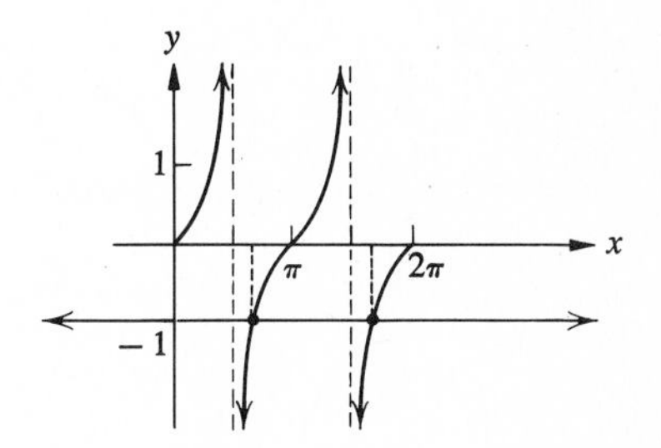

7. $\{0, 1.3, \pi, 5.0, 2\pi\}$

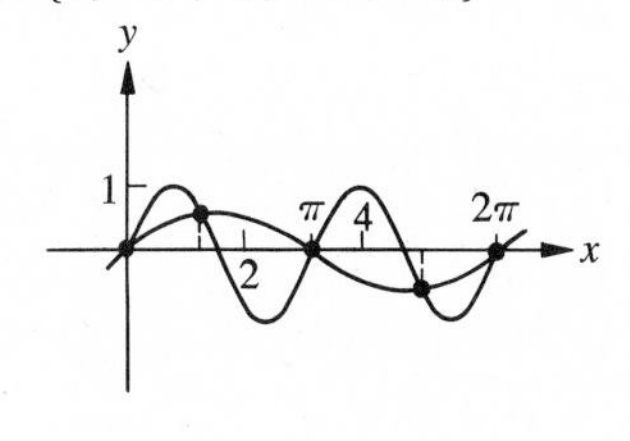

9. $\{0.7, 2.6, 3.7, 5.6\}$

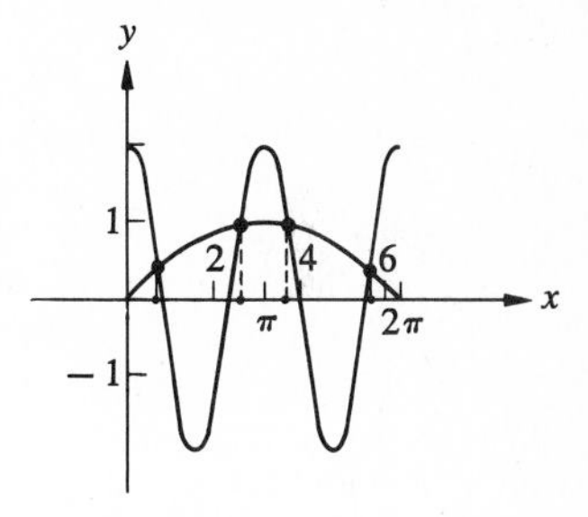

11. a. Some ordered pairs are $(1, 0)$, $(\sqrt{3}/2, 1/2)$, $(1/\sqrt{2}, 1/\sqrt{2})$, $(0, 1)$, $(-\sqrt{3}/2, 1/2)$, $(-1/\sqrt{2}, 1/\sqrt{2})$, $(-1, 0)$, $(-\sqrt{3}/2, -1/2)$, $(-1/\sqrt{2}, -1/\sqrt{2})$, $(0, -1)$, $(\sqrt{3}/2, -1/2)$, $(1/\sqrt{2}, -1/\sqrt{2})$

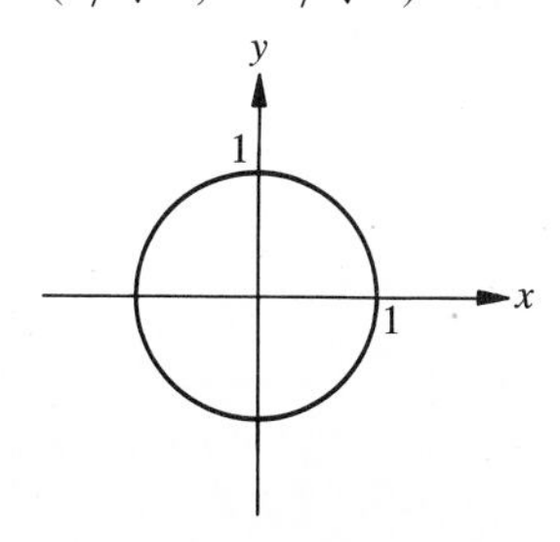

b. $x^2 = \cos^2 \alpha$, $y^2 = \sin^2 \alpha$,
$x^2 + y^2 = \sin^2 \alpha + \cos^2 \alpha$,
$x^2 + y^2 = 1$

13. a. Some ordered pairs are $(4, 0)$, $(2\sqrt{2}, 3/\sqrt{2})$, $(0, 3)$, $(-2\sqrt{2}, 3/\sqrt{2})$, $(-4, 0)$, $(-2\sqrt{2}, -3/\sqrt{2})$, $(0, -3)$, $(2\sqrt{2}, -3/\sqrt{2})$

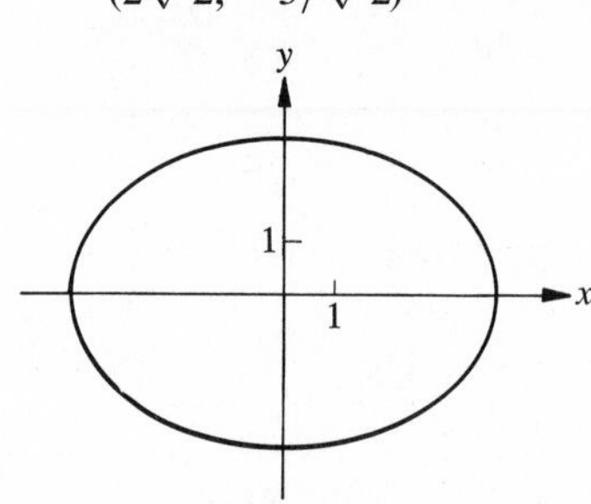

b. $(x/4)^2 = \cos^2 \alpha$, $(y/3)^2 = \sin^2 \alpha$,
$x^2/16 + y^2/9 = \cos^2 \alpha + \sin^2 \alpha$,
$x^2/16 + y^2/9 = 1$

15. a. Some ordered pairs are $(5, 0)$, $(5\sqrt{2}/2, \sqrt{2})$, $(0, 2)$, $(-5\sqrt{2}/2, \sqrt{2})$, $(-5, 0)$, $(-5\sqrt{2}/2, -\sqrt{2})$, $(0, -2)$, $(5\sqrt{2}/2, -\sqrt{2})$

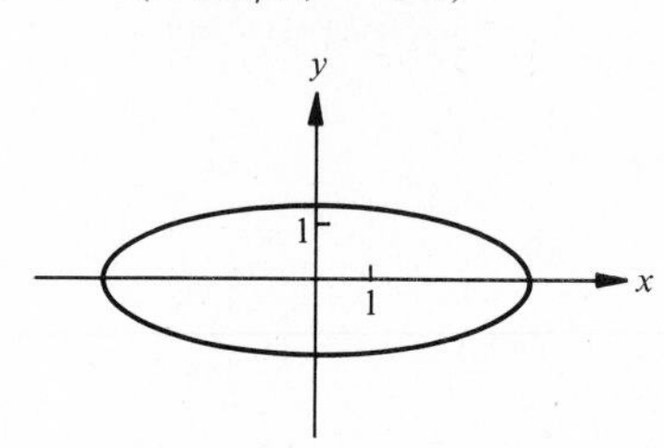

b. $(x/5)^2 = \cos^2 \alpha$, $(y/2)^2 = \sin^2 \alpha$,
$x^2/25 + y^2/4 = \cos^2 a + \sin^2 \alpha$,
$x^2/25 + y^2/4 = 1$

17. a. Some ordered pairs are $(-1, 0)$, $(-\sqrt{2}/2, 3\sqrt{2}/2)$, $(0, 3)$, $(\sqrt{2}/2, 3\sqrt{2}/2)$, $(1, 0)$, $(\sqrt{2}/2, -3\sqrt{2}/2)$, $(0, -3)$, $(-\sqrt{2}/2, -3\sqrt{2}/2)$

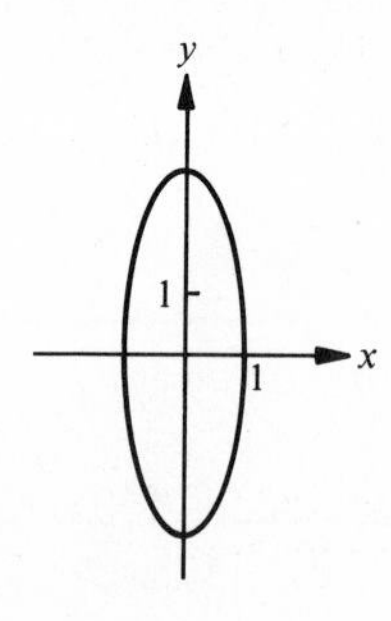

b. $(x/-1)^2 = \cos^2 \alpha$, $(y/3)^2 = \sin^2 \alpha$,
$x^2/1 + y^2/9 = \cos^2 \alpha + \sin^2 \alpha$,
$x^2/1 + y^2/9 = 1$

19. $y = \dfrac{8}{x^2 + 4}$

21. y 2 5 x

Exercise Set 3.6

1. $\cos y = \sqrt{3}/2;\ \pi/6$
3. $\sin y = 1;\ \pi/2$
5. $\cot y = 1/\sqrt{3};\ \pi/3$
7. $\pi/4$
9. $\pi/4$
11. $\pi/3$
13. 0
15. $\pi/6$
17. $\pi/4$
19. $-\pi/6$
21. 0.52
23. 0.46
25. 0.36
27. 1.24
29. 1.45
31. -0.50
33. 2.92
35. $x = \text{Cos}^{-1}\frac{3}{4},\ 0 \le x \le \pi$
37. $x = \frac{1}{3}\text{Sin}^{-1}\frac{1}{4},\ -\frac{\pi}{6} \le x \le \frac{\pi}{6}$
39. $x = \frac{1}{2}\text{Tan}^{-1}\frac{5}{2},\ -\frac{\pi}{4} < x < \frac{\pi}{4}$
41. $x = \frac{1}{3}\text{Tan}^{-1}\frac{y}{2},\ -\frac{\pi}{6} < x < \frac{\pi}{6}$
43. $\sqrt{3}/2$
45. $-1/\sqrt{2}$
47. 0.7151
49. 1.073
51. $\pi/4$
53. 0
55. π
57. 4/5
59. 2/5
61. 3/4
63. $\frac{\sqrt{9 - x^2}}{3}$
65. $\sqrt{1 - x^2}$

Review Exercises, Chapter 3

1.

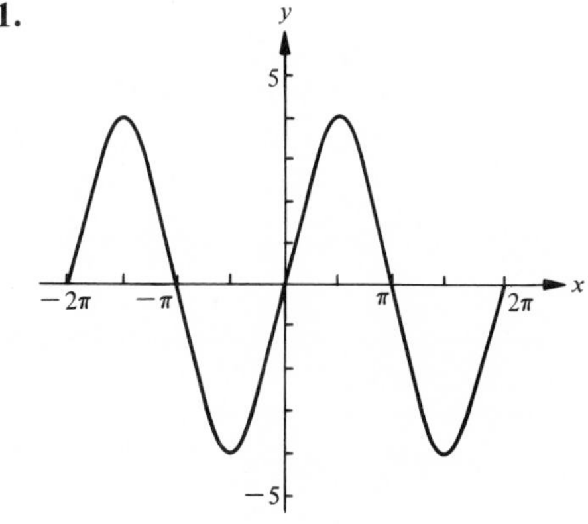

a. amplitude: 4
b. period: 2π
c. no phase shift

2.

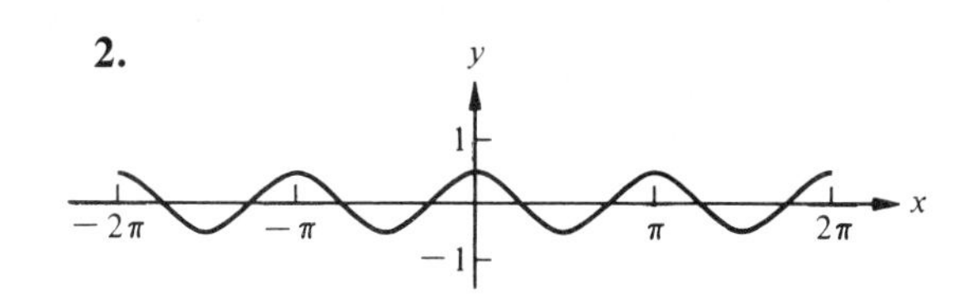

a. amplitude: 1/2
b. period: π
c. no phase shift

3.

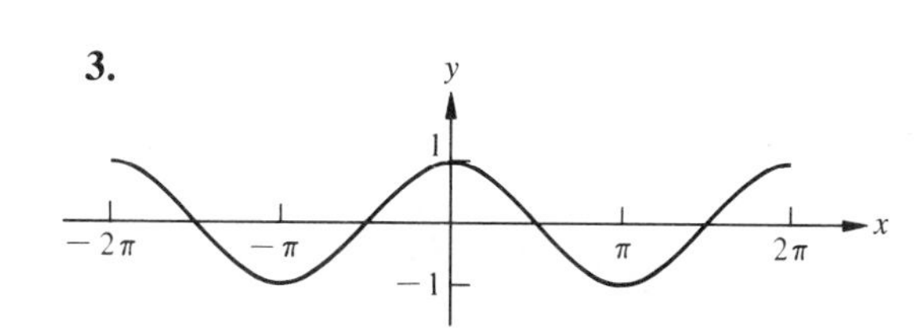

a. amplitude: 1
b. period: 2π
c. phase shift: $\pi/2$ to the left

4.

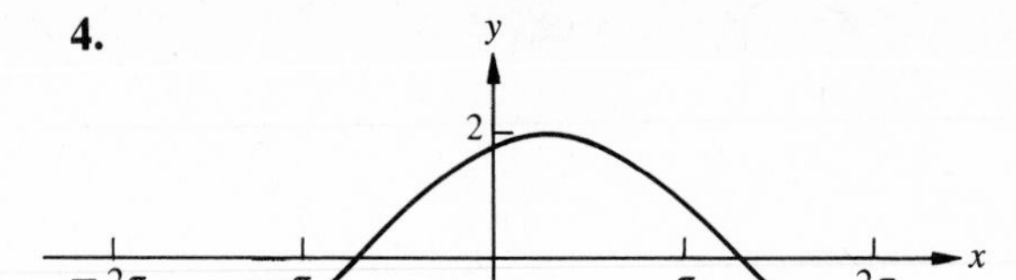

a. amplitude: 2
b. period: 4π
c. phase shift: $\pi/3$ to the right

5.

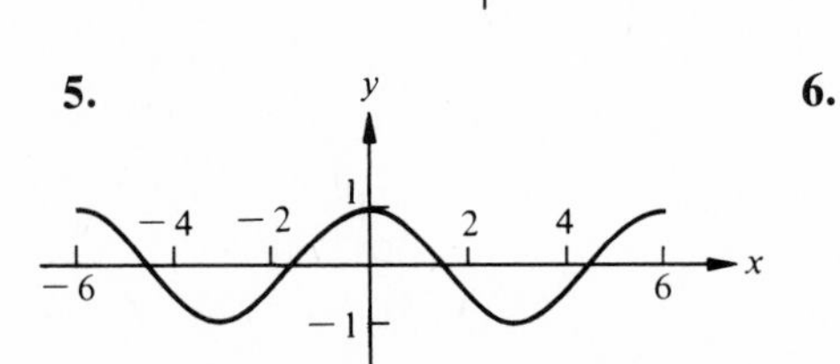

6.

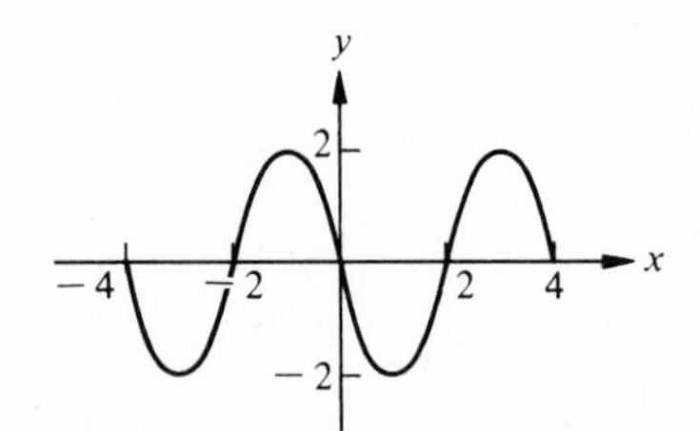

7.

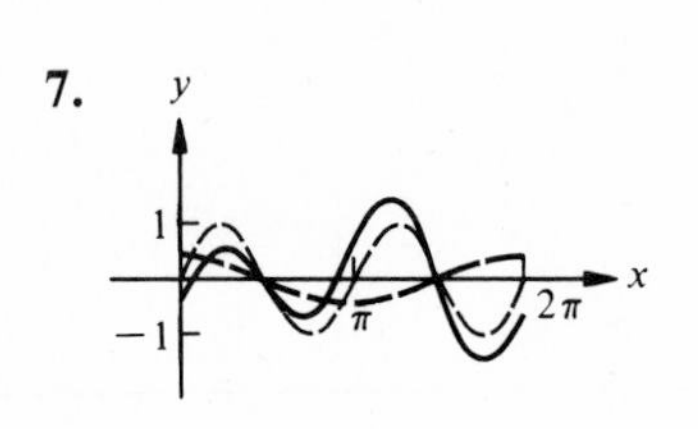

8.

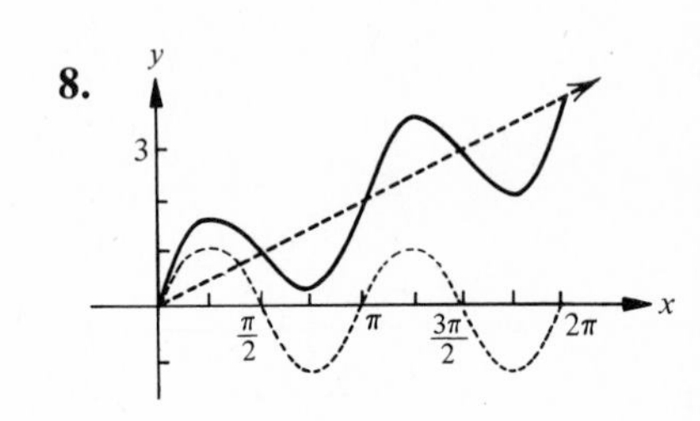

9.

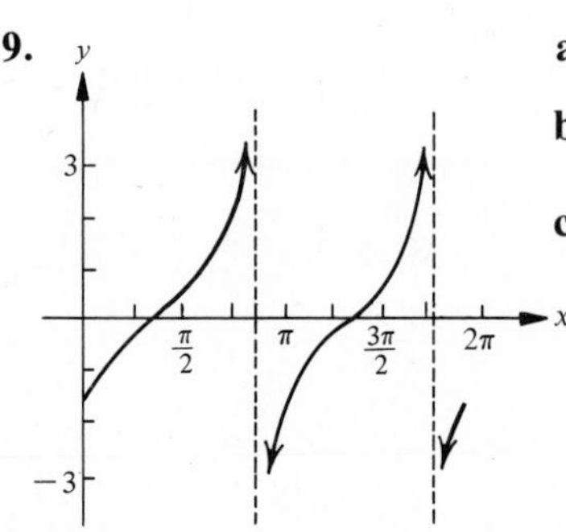

a. range: $\{y \mid y \in R\}$
b. asymptotes: $x = \dfrac{5\pi}{6} + k \cdot \pi,\ k \in J$
c. zeros: $\{x \mid x = \dfrac{\pi}{3} + k \cdot \pi,\ k \in J\}$

10.

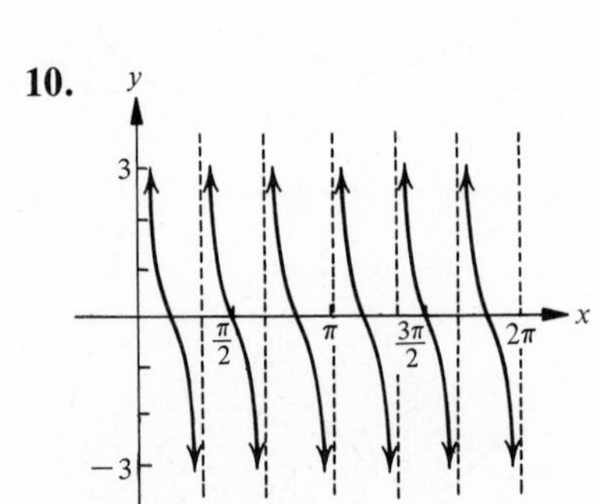

a. range: $\{y \mid y \in R\}$
b. asymptotes: $x = k \cdot \dfrac{\pi}{3},\ k \in J$
c. zeros: $\left\{x \mid x = \dfrac{\pi}{6} + k \cdot \dfrac{\pi}{3},\ k \in J\right\}$

11.

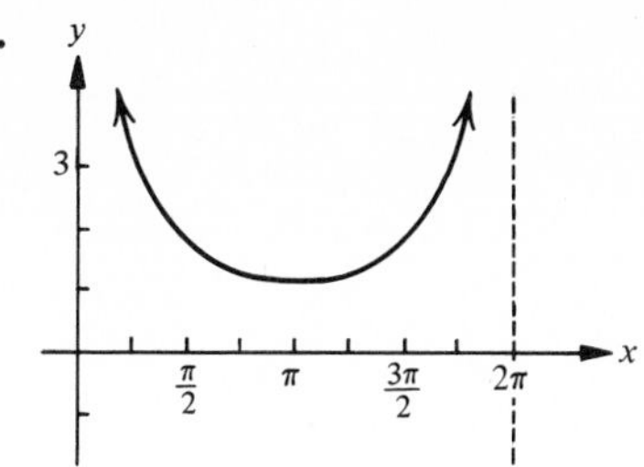

a. range: $\{y \mid y \leq -1 \text{ or } y \geq 1\}$
b. asymptotes: $x = k \cdot 2\pi,\ k \in J$
c. no zeros

12.

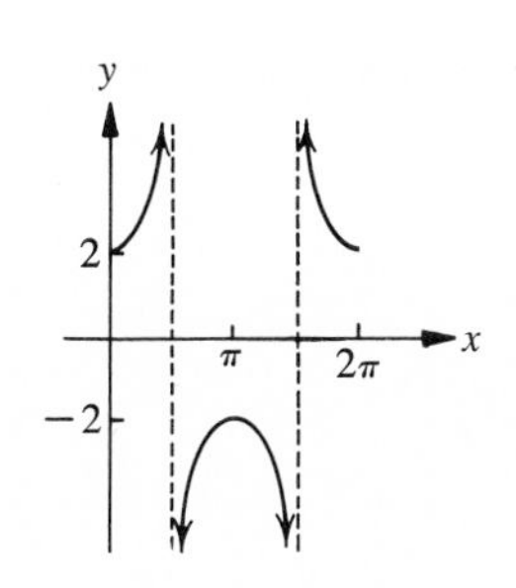

a. range: $\{y \mid y \leq -2 \text{ or } y \geq 2\}$
b. asymptotes: $x = \pi/2 + k\pi,\ k \in J$
c. no zeros

3

13.

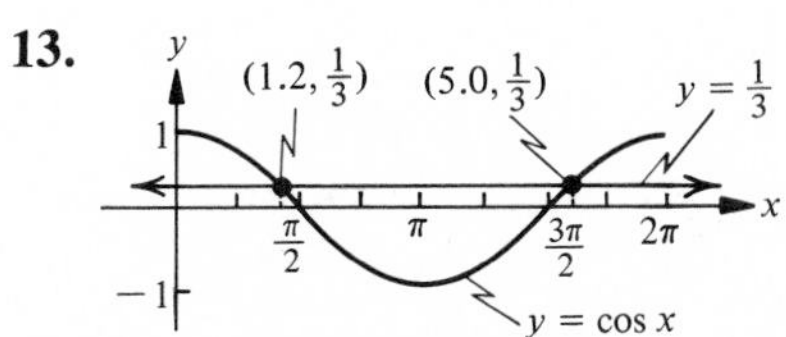

Solution set is
$\{(1.2, \frac{1}{3}), (5.0, \frac{1}{3})\}$

14.

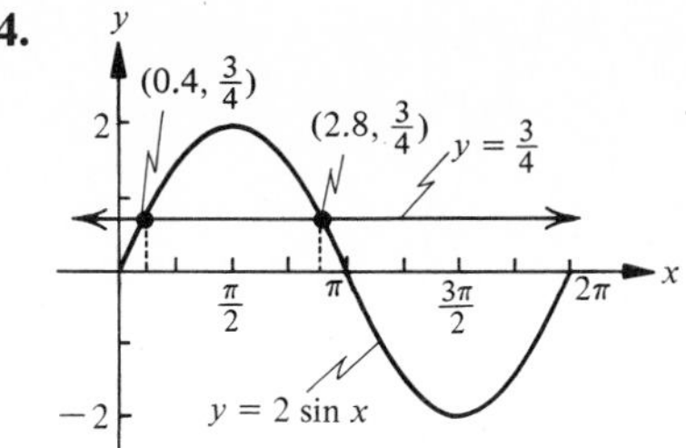

Solution set is
$\{(0.4, \frac{3}{4}), (2.8, \frac{3}{4})\}$

15.

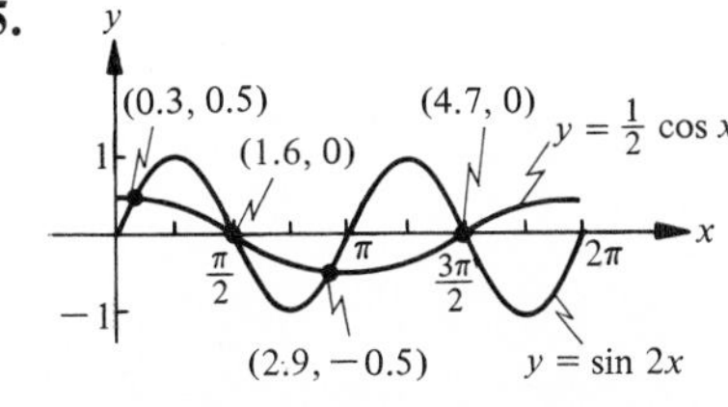

Solution set is
$\{(0.3, 0.5), (1.6, 0),$
$(2.9, -0.5), (4.7, 0)\}$

16. 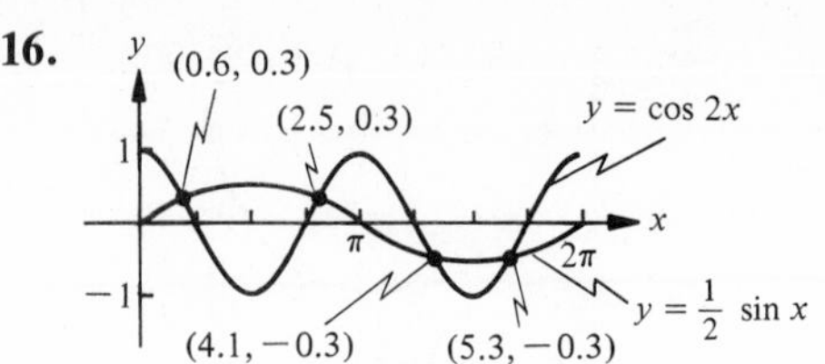

Solution set is $\{(0.6, 0.3), (2.5, 0.3), (4.1, -0.3), (5.3, -0.3)\}$

17.

18. $\dfrac{x^2}{1} + \dfrac{y^2}{16} = 1$ or $16x^2 + y^2 = 16$

19. $\cos y = 1/\sqrt{2}$, $\pi/4$
20. $\cot y = -1$; $-\pi/4$
21. $\csc y = -2$; $-\pi/6$
22. $\sec y = 2/\sqrt{3}$; $\pi/6$
23. $\pi/2$ **24.** $-\pi/4$ **25.** $3\pi/4$ **26.** $\pi/6$
27. 1.45 **28.** 0.95 **29.** 0.47 **30.** -1.02

31. $x = \operatorname{Cos}^{-1}\dfrac{5}{3}$, $0 \le x \le \pi$

32. $x = \dfrac{1}{2}\operatorname{Sin}^{-1}\dfrac{1}{5}$, $-\dfrac{\pi}{4} \le x \le \dfrac{\pi}{4}$

33. $x = \dfrac{1}{2}\operatorname{Tan}^{-1}\dfrac{y}{3}$, $-\dfrac{\pi}{4} < x < \dfrac{\pi}{4}$

34. $x = 2\operatorname{Cos}^{-1}\dfrac{y}{2}$, $0 \le x \le 2\pi$

35. $\sqrt{3}$ **36.** 0.8870 **37.** $\dfrac{\pi}{2}$ **38.** $\dfrac{\pi}{4}$

Exercise Set 4.1

1. $\dfrac{1}{\sin\alpha}$ **3.** $\dfrac{1}{\cos^2\alpha}$ **5.** $\sin^2\alpha$

7. $\dfrac{1}{\sin\alpha}$ **9.** $\dfrac{1}{\cos\alpha}$ **11.** $\dfrac{1}{\sin\alpha\cos\alpha}$

59. If $x = \pi/4$, then $\sin\dfrac{\pi}{4} + \left(\cos\dfrac{\pi}{4}\right)\left(\tan\dfrac{\pi}{4}\right) = 2; \dfrac{1}{\sqrt{2}} + \left(\dfrac{1}{\sqrt{2}}\right)\cdot 1 \ne 2$

61. If $y = \pi/2$, then $2\left(\sin \frac{\pi}{2}\right)\left(\cos \frac{\pi}{2}\right) + \sin \frac{\pi}{2} = 0$; $2(1)(0) + 1 \neq 0$

63. $\csc(-\alpha) = \dfrac{1}{\sin(-\alpha)} = \dfrac{1}{-\sin \alpha} = -\csc \alpha$

65. $\cot(-\alpha) = \dfrac{1}{\tan(-\alpha)} = \dfrac{1}{-\tan \alpha} = -\cot \alpha$

Exercise Set 4.2

1. $\dfrac{\sqrt{6} - \sqrt{2}}{4}$ **3.** $-\dfrac{\sqrt{2} + \sqrt{6}}{4}$ **5.** $\dfrac{\sqrt{2} + \sqrt{6}}{4}$ **13.** $\cos 9x$

15. $\cos 2x$ **17.** $\cos x$ **19.** $\cos 2x$ **21.** $16/65$

23. $13/\sqrt{170}$ **25.** $-\cos 60°$ **27.** $-\cos 30°$ **29.** $\cos 45°$

37. If $\alpha = 0°$ and $\beta = 60°$, then

$$\cos(\alpha + \beta) = \cos(0° + 60°) = \cos 60° = \tfrac{1}{2};$$

$$\cos \alpha + \cos \beta = \cos 0° + \cos 60° = 1 + \tfrac{1}{2} = \tfrac{3}{2} \neq \tfrac{1}{2}.$$

39. Using the difference formula and substituting $\pi/2$ for α and α for β, we obtain

$$\cos\left(\frac{\pi}{2} = \alpha\right) = \cos \frac{\pi}{2} \cos \alpha + \sin \frac{\pi}{2} \sin \alpha$$

$$= 0(\cos \alpha) + 1(\sin \alpha) = \sin \alpha.$$

41. 0 **43.** $\dfrac{\sqrt{3} + 1}{2\sqrt{2}}$ **45.** $\dfrac{\sqrt{3} - 1}{2\sqrt{2}}$

Exercise Set 4.3

1. $\dfrac{\sqrt{6} + \sqrt{2}}{4}$ **3.** $\dfrac{\sqrt{6} - \sqrt{2}}{4}$ **5.** $\dfrac{\sqrt{2} + \sqrt{6}}{4}$ **13.** $\sin 3x$

15. $\sin 4x$ **17.** $\sin 2x$ **19.** $\sin 2x$ **21.** $-44/125$

23. $5/\sqrt{754}$ **25.** $-\sin 60°$ **27.** $\sin 30°$ **29.** $-\sin 30°$

37. If $\alpha = 90°$ and $\beta = 60°$, then

$$\sin(\alpha - \beta) = \sin(90° - 60°) = \sin 30° = \frac{1}{2};$$

$$\sin \alpha - \sin \beta = \sin 90° - \sin 60° = 1 - \frac{\sqrt{3}}{2} \neq \frac{1}{2}.$$

39. 0 **41.** $-\dfrac{1}{2}$ **43.** $\dfrac{\sqrt{3} - 1}{2\sqrt{2}}$

Exercise Set 4.4

1. $\dfrac{1+\sqrt{3}}{1-\sqrt{3}}$ **3.** $\dfrac{\sqrt{3}-1}{1+\sqrt{3}}$ **5.** $\dfrac{\sqrt{3}-1}{1+\sqrt{3}}$ **7.** $\cot 54°$

9. $\tan 24°$ **11.** $\cot 77° 30'$ **13.** $\cot \dfrac{\pi}{6}$ **15.** $\tan \dfrac{\pi}{4}$

17. $\tan \dfrac{5\pi}{12}$ **25.** $\tan 9x$ **27.** $\tan 2x$ **29.** $\tan x$

31. $-\tan 60°$ **33.** $\tan 30°$ **35.** $-\tan 45°$

Exercise Set 4.5

1. $\sin 2x$ **3.** $\cos 2x$ **5.** $\tan 2x$

7. $\frac{1}{2} \sin x$ **9.** $\cos 10\alpha$ **11.** $2 \cos 12\alpha$

13. $\sin \alpha$ **15.** $\cos 2x$ **33.** $24/25$

35. $5/\sqrt{34}$ **37.** $-24/7$

Exercise Set 4.6

1. $\sqrt{\dfrac{1-\sqrt{3}/2}{2}}$ **3.** $\dfrac{1-\sqrt{3}/2}{1/2}$ **5.** $\sqrt{\dfrac{1-\sqrt{3}/2}{2}}$

7. $\sqrt{\dfrac{1+1/\sqrt{2}}{2}}$ **9.** $\dfrac{1+\sqrt{3}/2}{1/2}$ **11.** $-\sqrt{\dfrac{1-\sqrt{3}/2}{2}}$

19. $\dfrac{1}{\sqrt{3}}$ **21.** $\sqrt{\dfrac{2}{3}}$ **23.** $\sqrt{\dfrac{1+1/\sqrt{2}}{2}}$

25. $\frac{1}{2}(\sin 17° - \sin 7°)$ **27.** $\frac{1}{2}(\sin 3x - \sin x)$ **29.** $\frac{1}{2}(\cos 7x + \cos 3x)$

31. $2 \cos 12° \cos 8°$ **33.** $2 \sin 4\alpha \cos \alpha$ **35.** $2 \cos 3\alpha \sin 2\alpha$

37.
$$\begin{aligned}\cos\alpha\cos\beta &= \tfrac{1}{2}[\cos(\alpha+\beta)+\cos(\alpha-\beta)]\\ &= \tfrac{1}{2}[\cos\alpha\cos\beta - \sin\alpha\sin\beta + \cos\alpha\cos\beta\\ &\qquad + \sin\alpha\sin\beta]\\ &= \tfrac{1}{2}[2\cos\alpha\cos\beta] = \cos\alpha\cos\beta\end{aligned}$$

39.
$$\begin{aligned}\sin\alpha\cos\beta &= \tfrac{1}{2}[\sin(\alpha+\beta)+\sin(\alpha-\beta)]\\ &= \tfrac{1}{2}(\sin\alpha\cos\beta + \cos\alpha\sin\beta + \sin\alpha\cos\beta\\ &\qquad - \cos\alpha\sin\beta)\\ &= \tfrac{1}{2}(2\sin\alpha\cos\beta) = \sin\alpha\cos\beta\end{aligned}$$

41. Let $\alpha = x + y$ and $\beta = x - y$. Then,

$$\begin{aligned}\cos\alpha + \cos\beta &= \cos(x+y) + \cos(x-y)\\ &= \cos x\cos y - \sin x\sin y + \cos x\cos y\\ &\qquad + \sin x\sin y\\ &= 2\cos x\cos y.\end{aligned}$$

Solving $\alpha = x + y$ and $\beta = x - y$ for α and β, respectively, gives $x = \dfrac{\alpha + \beta}{2}$ and $y = \dfrac{\alpha - \beta}{2}$. Thus,

$$2 \cos x \cos y = 2 \cos \frac{\alpha + \beta}{2} \cos \frac{\alpha - \beta}{2}.$$

43. Let $\alpha = x + y$ and $\beta = x - y$. Then,

$$\begin{aligned}\sin \alpha + \sin \beta &= \sin(x + y) + \sin(x - y) \\ &= \sin x \cos y + \cos x \sin y + \sin x \cos y \\ &\qquad - \cos x \sin y \\ &= 2 \sin x \cos y.\end{aligned}$$

As in Exercise 41, $x = \dfrac{\alpha + \beta}{2}$ and $y = \dfrac{\alpha - \beta}{2}$. Thus,

$$2 \sin x \cos y = 2 \sin \frac{\alpha + \beta}{2} \cos \frac{\alpha - \beta}{2}.$$

4

Exercise Set 4.7

1. $\{\pi/6\}$ **3.** $\{\pi/4\}$ **5.** $\{\pi/2\}$
7. $\{\pi/4, \pi/2\}$ **9.** $\{0, \pi/2\}$ **11.** $\{\pi/4, \pi/2\}$
13. $\{0\}$ **15.** $\{0, 2\pi/3, 4\pi/3\}$ **17.** $\{1.91, 4.37\}$
19. a. $\{270°\}$ **b.** $\{x \mid x = 270° + k \cdot 360°\}, k \in J$
21. a. $\{60°, 180°, 300°\}$
b. $\{x \mid x = 60° + k \cdot 360°\} \cup \{x \mid x = 180° + k \cdot 360°\} \cup \{x \mid x = 300° + k \cdot 360°\}, k \in J$
23. a. $\{30°, 90°, 150°\}$
b. $\{x \mid x = 30° + k \cdot 360°\} \cup \{x \mid x = 90° + k \cdot 360°\} \cup \{x \mid x = 150° + k \cdot 360°\}, k \in J$

Exercise Set 4.8

1. $\{0, \pi/3\}$ **3.** $\{\pi/6, 5\pi/12\}$ **5.** $\{\pi/15, \pi/3\}$ **7.** $\{\pi/6\}$
9. $\{0\}$ **11.** $\{\pi/8\}$ **13.** $\{0\}$
15. $\{10°, 50°, 130°, 170°, 250°, 290°\}$
17. $\{15°, 75°, 135°, 195°, 255°, 315°\}$
19. $\{45°, 135°, 225°, 315°\}$
21. $\{15°, 45°, 75°, 105°, 135°, 165°, 195°, 225°, 255°, 285°, 315°, 345°\}$
23. $\{180°, 360°\}$ **25.** $\{60°, 180°, 300°\}$
27. $\{x \mid x = \pi/2 + k \cdot 2\pi\} \cup \{x \mid x = \pi/6 + k \cdot 2\pi\} \cup \{x \mid x = 5\pi/6 + k \cdot 2\pi\}, k \in J$

29. $\{x \mid x = k\pi\} \cup \{x \mid x = \pi/6 + k\pi/3\}, k \in J$
31. $\{x \mid x \approx 1.28 + k\pi/2\} \cup \{x \mid x = \pi/8 + k\pi/2\}, k \in J$
33. $\{x \mid x = \pi/12 + k(2\pi/3)\} \cup \{x \mid x = \pi/4 + k(2\pi/3)\}, k \in J$
35. $\{0.5\}$

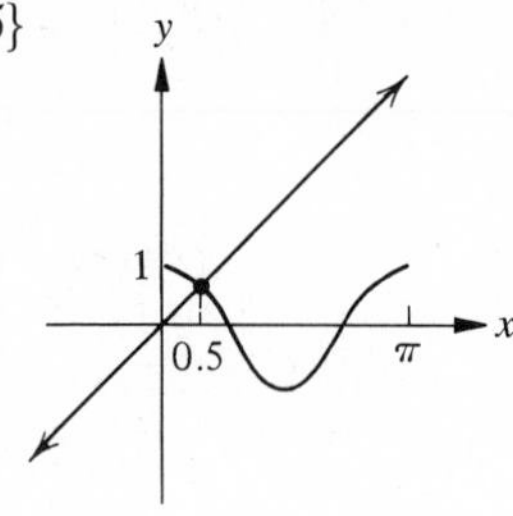

Review Exercises, Chapter 4

1. $1/\cos\alpha$ **2.** $1/\sin^2\alpha$ **3.** $\cos^2\alpha$ **4.** $1/\sin\alpha$

23. If $\alpha = \frac{\pi}{4}$, then $\sin\frac{\pi}{4}\sec\frac{\pi}{4} = 2\tan\frac{\pi}{4}$; $\frac{1}{\sqrt{2}}\cdot\sqrt{2} \neq 2\cdot 1$

24. If $\alpha = \frac{\pi}{4}$, then $\frac{1-\cos\pi/4}{\sin\pi/4} = \frac{1+\sin\pi/4}{\cos\pi/4}$;
$\frac{1-1\sqrt{2}}{1/\sqrt{2}} \neq \frac{1+1/\sqrt{2}}{1/\sqrt{2}}$; $\sqrt{2}-1 \neq \sqrt{2}+1$

25. $-\sqrt{2-\sqrt{3}}/2$
29. $\frac{1}{2}(\cos 6x + \cos 4x)$ **30.** $2\sin 5x\cos x$
31. $\{\pi/3\}$ **32.** $\{\pi/6\}$ **33.** $\{\pi/4\}$ **34.** $\{\pi/2\}$
35. $\{30°, 150°, 210°, 330°\}$ **36.** $\{45°, 135°, 225°, 315°\}$
37. $\{15°, 105°, 135°, 225°, 255°, 345°\}$
38. $\{15°, 30°, 105°, 120°, 195°, 210°, 285°, 300°\}$
39. $\{0°, 45°, 135°, 180°, 225°, 315°\}$ **40.** $\{0°, 180°\}$

Exercise Set 5.1

1. $x = 1, y = -3$ **3.** $x = -\frac{1}{2}, y = 8$
5. $1 + 10i$ **7.** $-10 - 11i$ **9.** $-1 + 6i$ **11.** $0 + 0i$
13. $12 - 12i$ **15.** $-36 - 12i$ **17.** $2 + 3i$ **19.** $-19 + 2i$
21. $62 - 39i$ **23.** $26 - 19i$ **25.** $a^2 + b^2$ **27.** i
29. -1 **31.** -1 **33.** $-i$ **35.** $5i$
37. $2\sqrt{3}\,i$ **39.** $1 + 3i$ **41.** $3 - 2\sqrt{5}\,i$ **43.** $8 + 2i$
45. $-8 + 8i$ **47.** $-10\sqrt{5} + 2\sqrt{10}\,i$
49. $23 + 2i$ **51.** $9 + \sqrt{3}\,i$
53. If $x = 2i$, then $(2i)^2 + 4 = 4i^2 + 4 = -4 + 4 = 0$;
if $x = -2i$, then $(-2i)^2 + 4 = 4i^2 + 4 = -4 + 4 = 0$

55. **a.** $\left(-\frac{1}{2}+i\frac{\sqrt{3}}{2}\right)^2 = \frac{1}{4} - i\frac{\sqrt{3}}{2} + i^2\left(\frac{3}{4}\right) = -\frac{1}{2} - i\frac{\sqrt{3}}{2};$

$$\left(-\frac{1}{2}+i\frac{\sqrt{3}}{2}\right)^3 = \left(-\frac{1}{2}+i\frac{\sqrt{3}}{2}\right)^2\left(-\frac{1}{2}+i\frac{\sqrt{3}}{2}\right)$$
$$= \left(-\frac{1}{2}-i\frac{\sqrt{3}}{2}\right)\left(-\frac{1}{2}+i\frac{\sqrt{3}}{2}\right) = \frac{1}{4} - i^2\left(\frac{3}{4}\right) = 1$$

b. $\left(-\frac{1}{2}-i\frac{\sqrt{3}}{2}\right)^2 = \frac{1}{4} + i\frac{\sqrt{3}}{2} + i^2\left(\frac{3}{4}\right) = -\frac{1}{2} + i\frac{\sqrt{3}}{2};$

$$\left(-\frac{1}{2}-i\frac{\sqrt{3}}{2}\right)^3 = \left(-\frac{1}{2}-i\frac{\sqrt{3}}{2}\right)^2\left(-\frac{1}{2}-i\frac{\sqrt{3}}{2}\right)$$
$$= \left(-\frac{1}{2}+i\frac{\sqrt{3}}{2}\right)\left(-\frac{1}{2}-i\frac{\sqrt{3}}{2}\right) = \frac{1}{4} - i^2\left(\frac{3}{4}\right) = 1$$

57. $a = 0, b = 1$ or $a = 0, b = -1$

59. $e^{\pi i} = \cos \pi + i \sin \pi = -1 + i(0) = -1$

61.
$$\frac{e^{ix} - e^{-ix}}{2i} = \frac{\cos x + i \sin x - [\cos(-x) + i \sin(-x)]}{2i}$$
$$= \frac{\cos x + i \sin x - \cos x + i \sin x}{2i}$$
$$= \frac{2i \sin x}{2i} = \sin x$$

63. Using the results of Exercises 61 and 62,

$$\sin^2 x + \cos^2 x = \left(\frac{e^{ix} - e^{-ix}}{2i}\right)^2 + \left(\frac{e^{ix} + e^{-ix}}{2}\right)^2$$
$$= \frac{e^{2ix} - 2e^0 + e^{-2ix}}{4i^2} + \frac{e^{2ix} + 2e^0 + e^{-2ix}}{4}$$
$$= \frac{-e^{2ix} + 2 - e^{-2ix}}{4} + \frac{e^{2ix} + 2 + e^{-2ix}}{4}$$
$$= \frac{4}{4} = 1.$$

5

Exercise Set 5.2

1. $13 - 4i$
3. $2 + 2i$
5. $5 - i$
7. $6 - 3i$
9. $-3 + 3i$
11. $-3i$
13. $\frac{3}{2} - \frac{9}{4}i$
15. $\frac{8}{85} + \frac{36}{85}i$
17. $\frac{9}{17} + \frac{2}{17}i$
19. $-\frac{6}{5} + \frac{11}{10}i$
21. $\frac{-i}{\sqrt{5}}$
23. $-\frac{2i}{\sqrt{2}}$
25. $\frac{4}{7} - \frac{\sqrt{5}}{7}i$
27. $\frac{18}{11} - \frac{6\sqrt{2}}{11}i$
29. $\frac{6\sqrt{2}}{7} - \frac{6\sqrt{5}}{7}i$
31. $\frac{2\sqrt{5}}{13} - \frac{4\sqrt{2}}{13}i$

Exercise Set 5.3

1. $(-1, 6)$ **3.** $(8, 1)$ **5.** $(0, 6)$ **7.** $-8 + 7i$
9. $-4 + 0i$ **11.** $7 - \sqrt{2}i$

13.

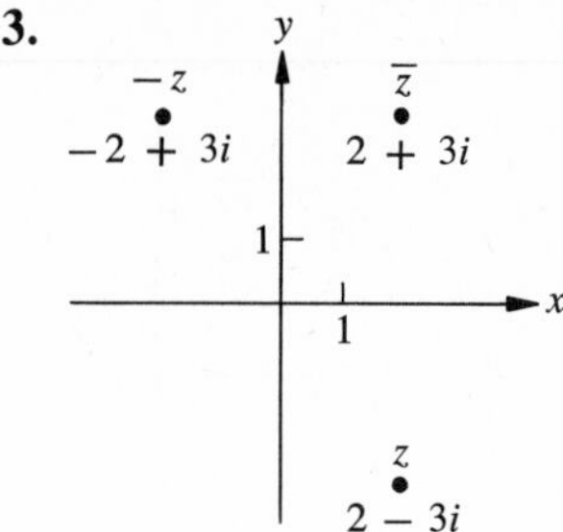

15.

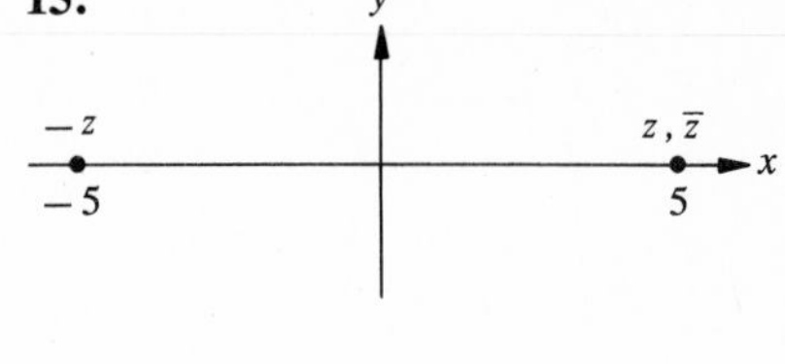

17.

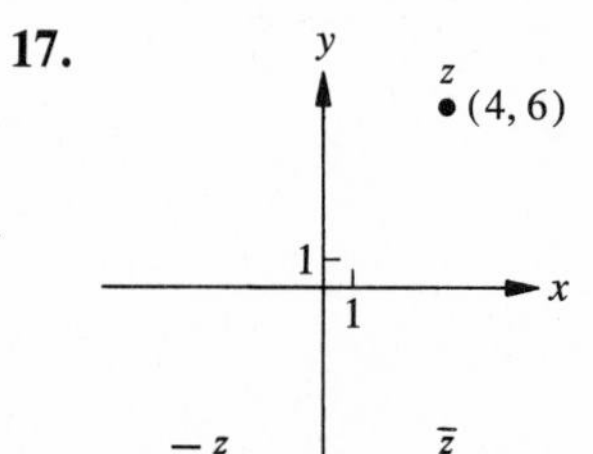

19.

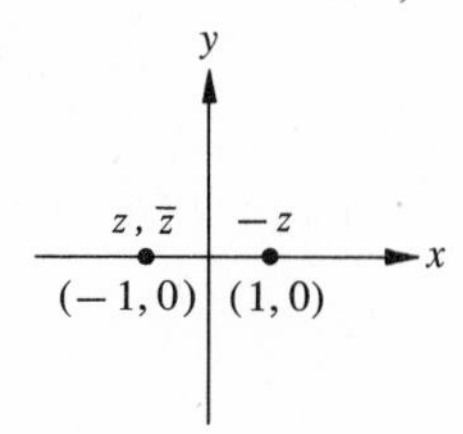

21. $-1 + 6i$

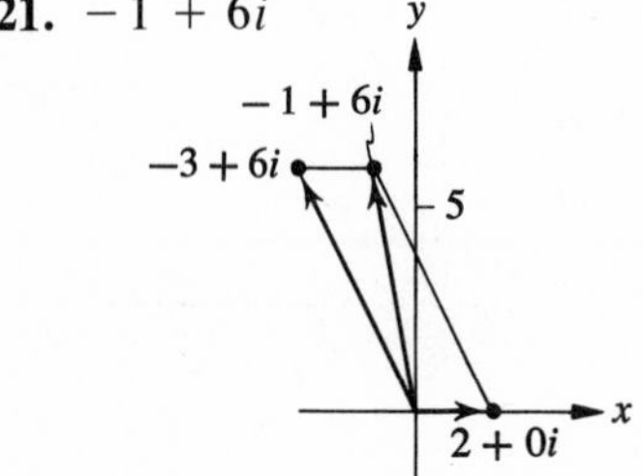

23. $-4 - 5i$

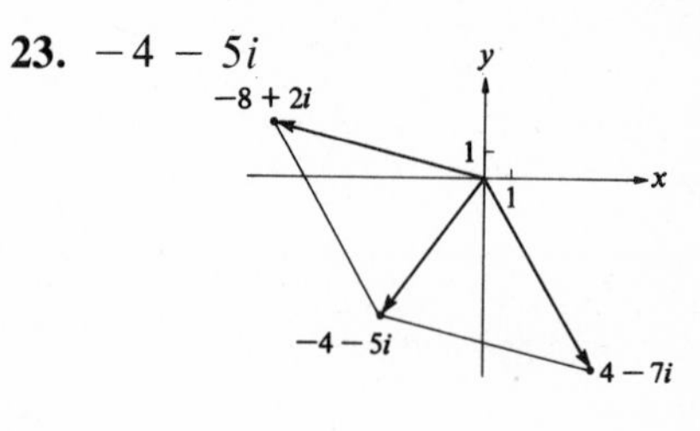

25. $-1 + 4i$

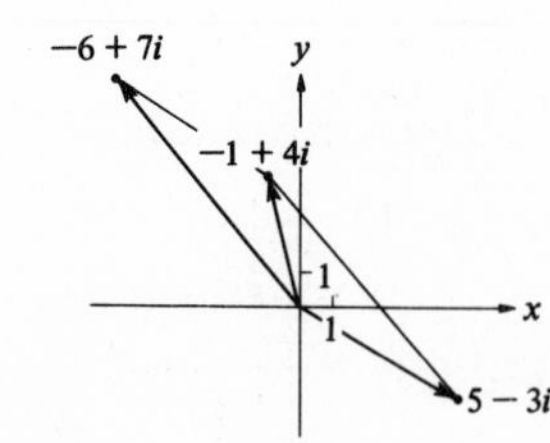

27. $\sqrt{29}$; 21° 48′ **29.** $2\sqrt{2}$; 225°
31. 7; 0° **33.** 5; 270°
35. $3\sqrt{10}$; 288° 24′ **37.** 4; 0°
39. a. $a \neq 0, b = 0$ **b.** $a = 0, b \neq 0$ **c.** $b > 0$ **d.** $b < 0$

41.

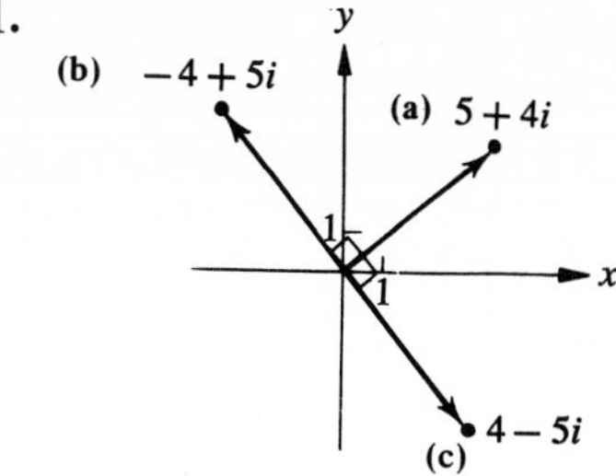

43.

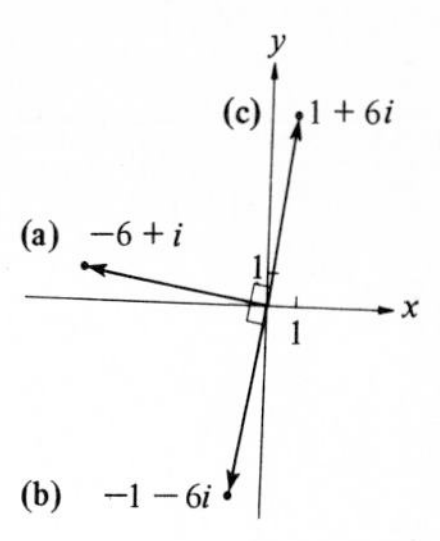

45.

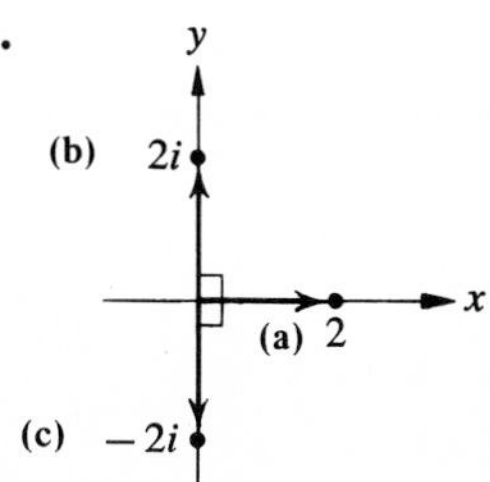

Exercise Set 5.4

1. $4(\cos 120° + i \sin 120°)$ **3.** $10(\cos 210° + i \sin 210°)$

5. $2\sqrt{2}(\cos 315° + i \sin 315°)$ **7.** $8(\cos 180° + i \sin 180°)$

9. $-1 - \sqrt{3}i$ **11.** $-\sqrt{2} + \sqrt{2}i$

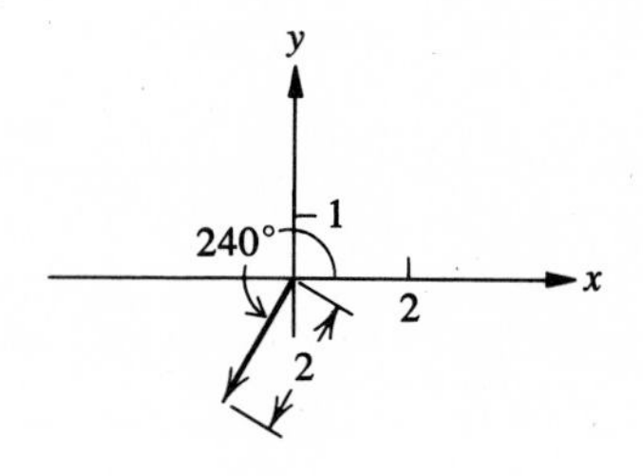

13. $\dfrac{1}{\sqrt{2}} - \dfrac{1}{\sqrt{2}}i$ **15.** $-\dfrac{\sqrt{3}}{4} + \dfrac{1}{4}i$

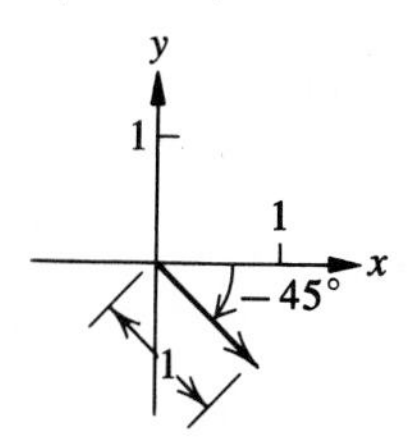

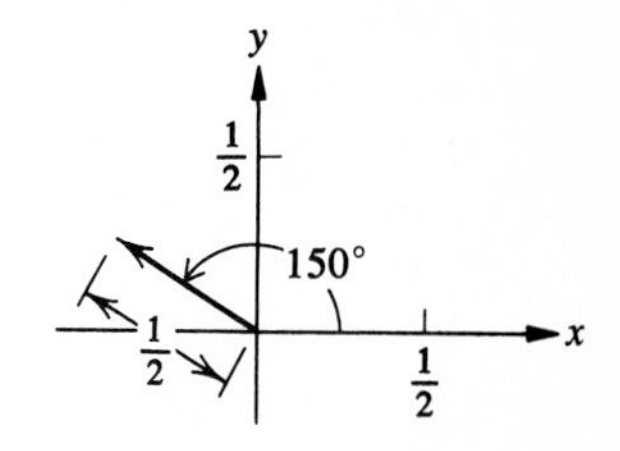

17. $-5\sqrt{2} + 5\sqrt{2}i$ **19.** $\frac{3}{2}\sqrt{3} - \frac{3}{2}i$ **21.** $-\sqrt{2} + \sqrt{2}i$

23. $-4 + 0i$ **25.** $-\dfrac{1}{5} - \dfrac{\sqrt{3}}{5}i$ **27.** $\dfrac{3}{2} - \dfrac{3\sqrt{3}}{2}i$

29. $4 + 4\sqrt{3}i$

31. $(a + bi)^2 = \rho(\cos\theta + i\sin\theta) \cdot \rho(\cos\theta + i\sin\theta)$; using Equation (1) in this section gives

$$\begin{aligned}(a + bi)^2 &= \rho \cdot \rho[\cos(\theta + \theta) + i\sin(\theta + \theta)]\\ &= \rho^2[\cos 2\theta + i\sin 2\theta].\end{aligned}$$

33. $(a + bi)^4 = (a + bi)^3(a + bi)$; using the results of Exercise 32 and Equation (1) in this section, gives

$$\begin{aligned}&\rho^3(\cos 3\theta + i\sin 3\theta) \cdot \rho(\cos\theta + i\sin\theta)\\ &\qquad = \rho^3 \cdot \rho[\cos(3\theta + \theta) + i\sin(3\theta + \theta)]\\ &\qquad = \rho^4(\cos 4\theta + i\sin 4\theta).\end{aligned}$$

35. From Section 4.1, $\sin(-\alpha) = -\sin\alpha$ and $\cos(-\alpha) = \cos\alpha$. Thus, $\cos(-\theta) + i\sin(-\theta) = \cos\theta - i\sin\theta$, the conjugate of $\cos\theta + i\sin\theta$.

Exercise Set 5.5

1. $\frac{27}{2} + \frac{27\sqrt{3}}{2}i$ **3.** $0 + 8i$ **5.** $-\frac{1}{8} + 0i$

7. $0 + \frac{1}{81}i$ **9.** $-\frac{1}{32} + \frac{1}{32}i$ **11.** $-\frac{1}{64} - \frac{\sqrt{3}}{64}i$

13. **a.** $\cos 40° + i\sin 40°$, $\cos 220° + i\sin 220°$
b. $\cos 30° + i\sin 30°$, $\cos 210° + i\sin 210°$

15. **a.** $2(\cos 40° + i\sin 40°)$, $2(\cos 160° + i\sin 160°)$, $2(\cos 280° + i\sin 280°)$
b. $\cos 20° + i\sin 20°$, $\cos 140° + i\sin 140°$, $\cos 260° + i\sin 260°$

17. **a.** $\cos 60° + i\sin 60°$, $\cos 150° + i\sin 150°$, $\cos 240° + i\sin 240°$, $\cos 330° + i\sin 330°$
b. $\cos 75° + i\sin 75°$, $\cos 165° + i\sin 165°$, $\cos 255° + i\sin 255°$, $\cos 345° + i\sin 345°$

19. $\left\{\frac{1}{2} + \frac{\sqrt{3}}{2}i, -1, \frac{1}{2} - \frac{\sqrt{3}}{2}i\right\}$ **21.** $\left\{1, -\frac{1}{2} + \frac{\sqrt{3}}{2}i, -\frac{1}{2} - \frac{\sqrt{3}}{2}i\right\}$

23. $\left\{\frac{1}{2^{1/3}}(\cos 110° + i\sin 110°), \frac{1}{2^{1/3}}(\cos 230° + i\sin 230°), \frac{1}{2^{1/3}}(\cos 350° + i\sin 350°)\right\}.$

25. In all cases, the graphs are spaced equally on the unit circle.

a.

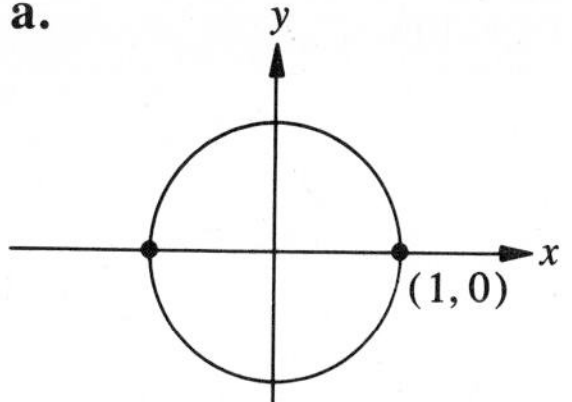

b.

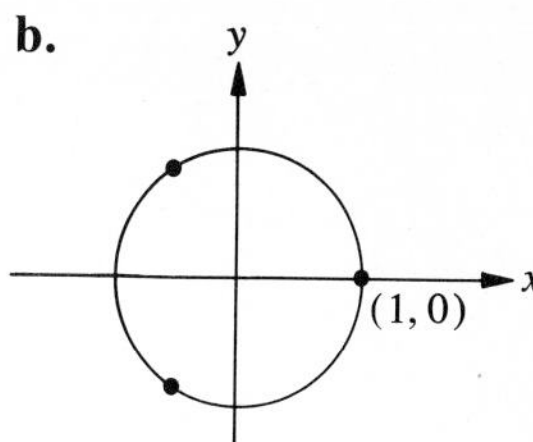

c.

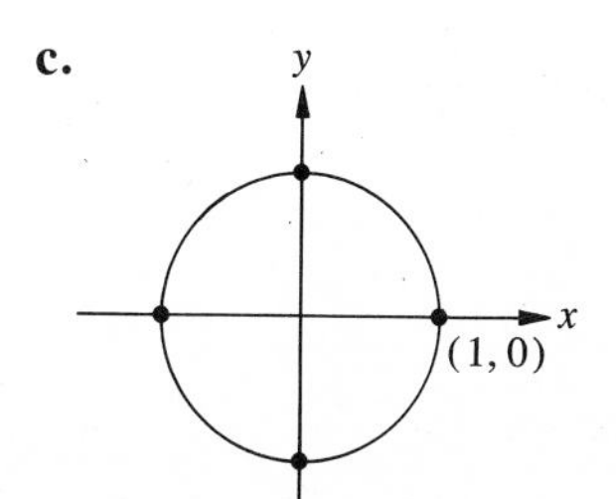

d.

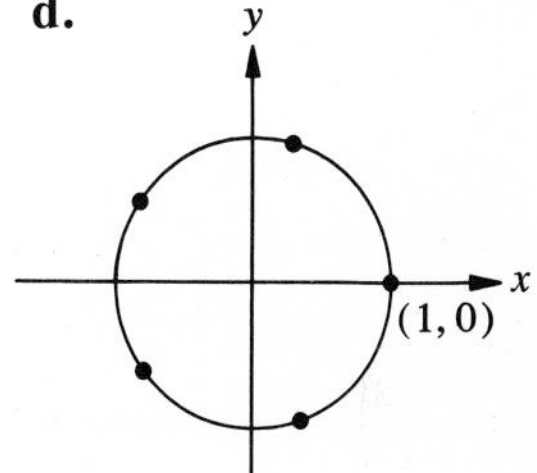

e.

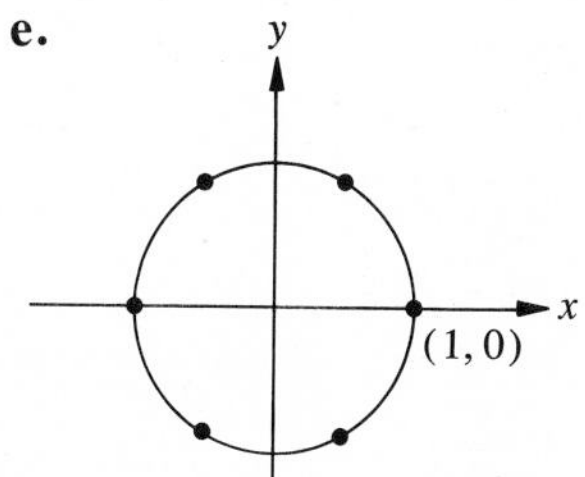

f.

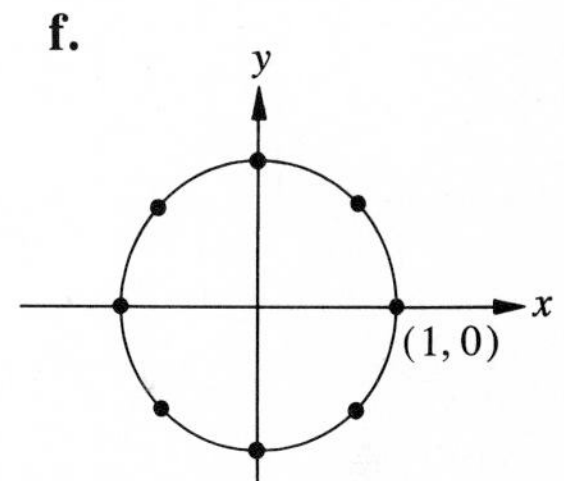

Exercise Set 5.6

1. $(3, 20°)$, $(-3, 200°)$, $(-3, -160°)$, $(3, -340°)$
3. $(5, -90°)$, $(-5, -270°)$, $(-5, 90°)$, $(5, 270°)$
5. $(-4, 60°)$, $(4, 240°)$, $(4, -120°)$, $(-4, -300°)$
7. $(2\sqrt{3}, 2)$ **9.** $(-3/\sqrt{2}, -3/\sqrt{2})$ **11.** $(3\sqrt{3}/2, -3/2)$
13. $(4, 0^R)$ **15.** $\left(2\sqrt{2}, \frac{\pi^R}{4}\right)$ **17.** $\left(1, \frac{5\pi^R}{6}\right)$
19. $\rho \cos \theta = 4$ **21.** $\rho = 4$ **23.** $\rho = 3 \sin \theta$
25. $y = -1$ **27.** $x^2 + y^2 = 4y$ **29.** $x^2 - 8y - 16 = 0$

31.

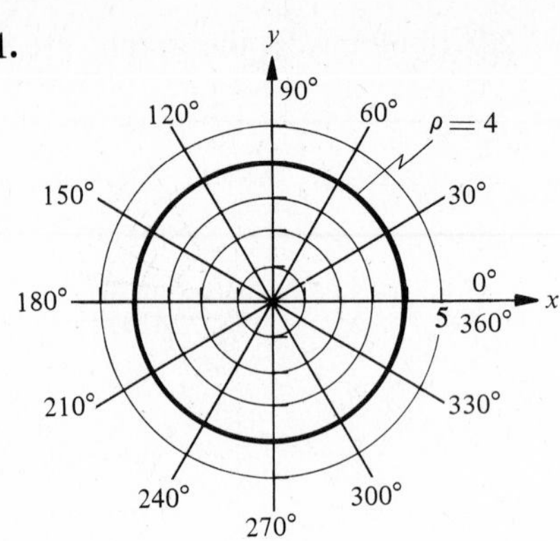
y
90°
120°
60°
$\rho = 4$
150°
30°
180°
0°
5
360°
x
210°
330°
240°
300°
270°

33.

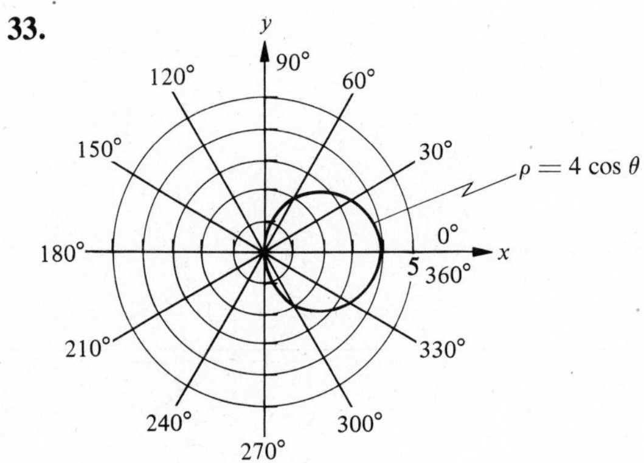
y
90°
120°
60°
150°
30°
$\rho = 4\cos\theta$
180°
0°
5
360°
x
210°
330°
240°
300°
270°

35.

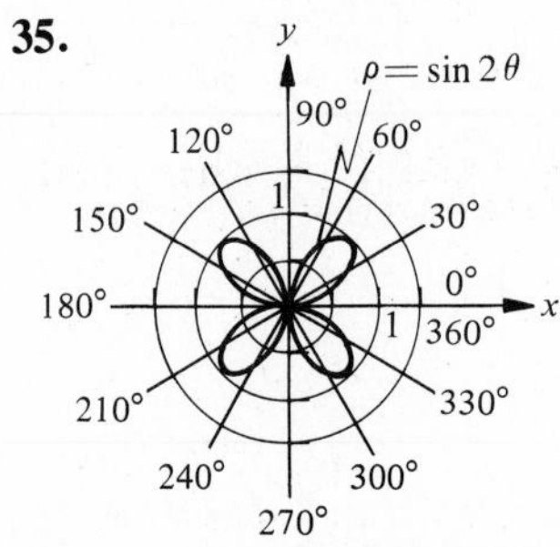
y
$\rho = \sin 2\theta$
90°
120°
60°
1
150°
30°
180°
0°
1
360°
x
210°
330°
240°
300°
270°

37.

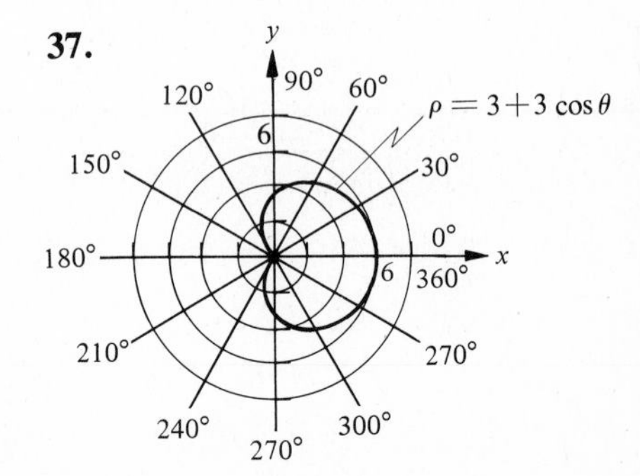
y
90°
120°
60°
$\rho = 3 + 3\cos\theta$
6
150°
30°
180°
0°
6
360°
x
210°
270°
240°
300°
270°

39.

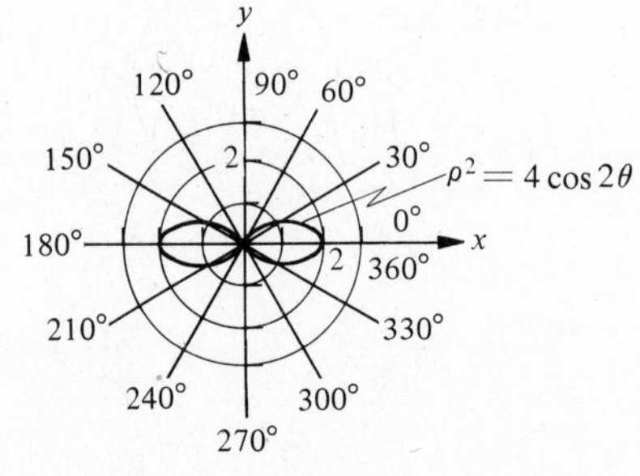
y
120°
90°
60°
150°
2
30°
$\rho^2 = 4\cos 2\theta$
0°
180°
2
360°
x
210°
330°
240°
300°
270°

41.

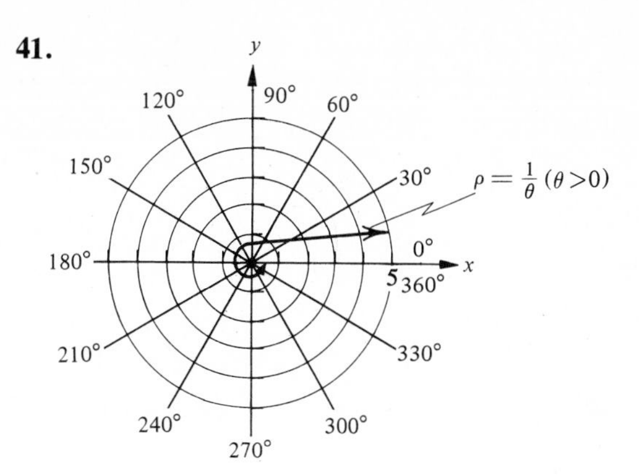

43.

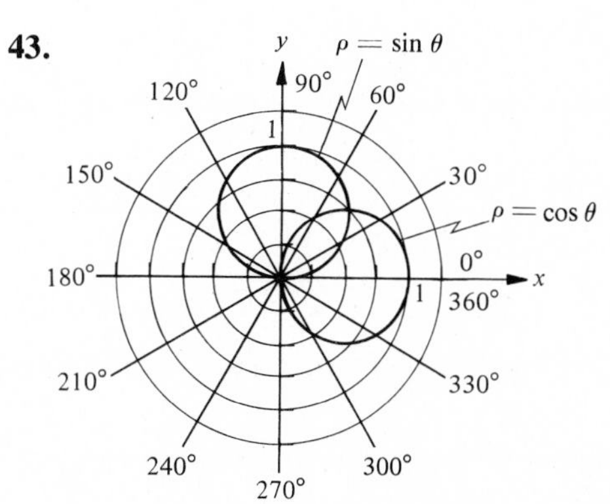

(0.7, 45°) and (0, θ) where θ has any measure

45.

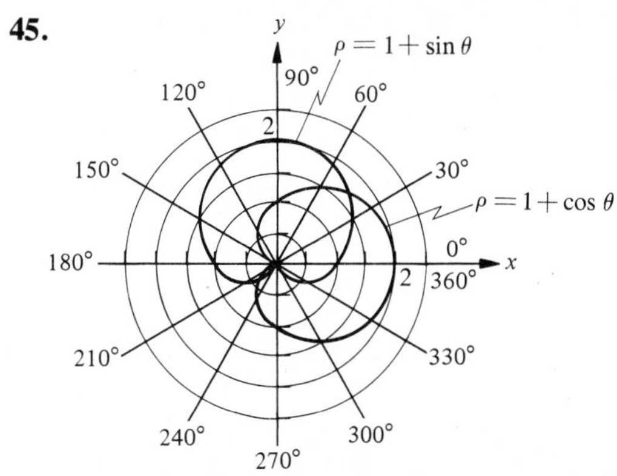

(1.7, 45°), (0.3, 225°) and (0, θ) where θ has any measure

47. $\{(1/\sqrt{2}, 45°)\}$; since ρ is a distance measured from the origin, it may be zero for more than one value of θ. In this case, $\rho = 0$ when $\theta = 0°$ in $\rho = \sin\theta$, and $\rho = 0$ when $\theta = 90°$ in $\rho = \cos\theta$.

49. $\left\{\left(\frac{\sqrt{2}+1}{\sqrt{2}}, 45°\right), \left(\frac{\sqrt{2}-1}{\sqrt{2}}, 225°\right)\right\}$

51. ρ_1 and ρ_2 are the lengths of sides with included angle $\theta_1 - \theta_2$. Using the Law of Cosines, $d^2 = \rho_1^2 + \rho_2^2 - 2\rho_1\rho_2\cos(\theta_1 - \theta_2)$,

$$d = \sqrt{\rho_1^2 + \rho_2^2 - 2\rho_1\rho_2\cos(\theta_1 - \theta_2)}.$$

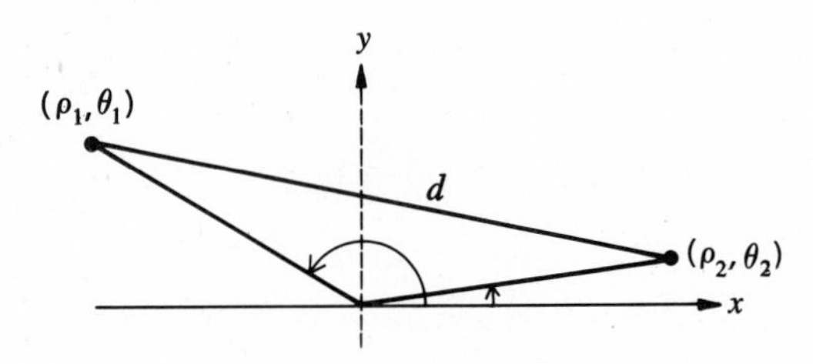

Review Exercises, Chapter 5

1. $x = -\frac{1}{3}, \quad y = -\frac{2}{3}$
2. $x = 1, \quad y = 4; \quad x = -1, \quad y = -4$
3. $-8 + 8i$ **4.** $0 - i$ **5.** $-46 + 28i$ **6.** $22 - 14i$
7. i **8.** $-i$ **9.** $\sqrt{10} + 3\sqrt{5}i$ **10.** 14
11. If $x = 3i$, then $(3i)^2 + 9 = 9i^2 + 9 = -9 + 9 = 0$;
if $x = -3i$, then $(-3i)^2 + 9 = 9i^2 + 9 = -9 + 9 = 0$.
12. If $x = 1 + 3i$, then

$$(1 + 3i)^2 - 2(1 + 3i) + 10 = 1 + 6i + 9i^2 - 2 - 6i + 10 = 0;$$

if $x = 1 - 3i$, then

$$(1 - 3i)^2 - 2(1 - 3i) + 10 = 1 - 6i + 9i^2 - 2 + 6i + 10 = 0.$$

13. $-15 - i$ **14.** $3 - 5i$ **15.** $\frac{7}{5} - \frac{13}{10}i$
16. $-\frac{13}{34} - \frac{8}{17}i$ **17.** $\frac{2i}{\sqrt{6}}$ **18.** $\frac{8}{9} - \frac{4\sqrt{5}}{9}i$
19. $(-4, 1)$ **20.** $(2\sqrt{3}, -2)$ **21.** $9 - 6i$
22. $3 + 4i$

23.

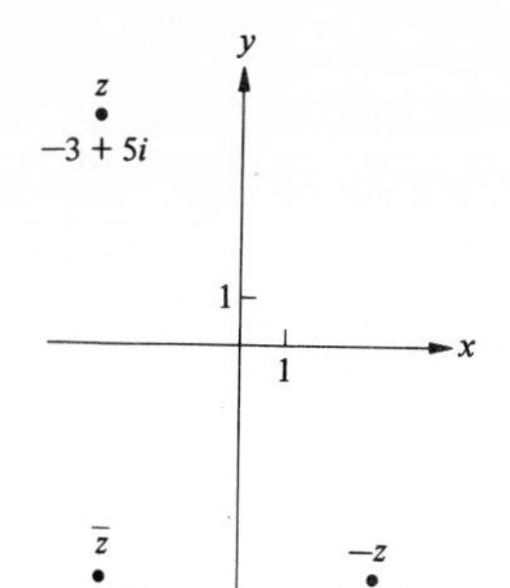

24.

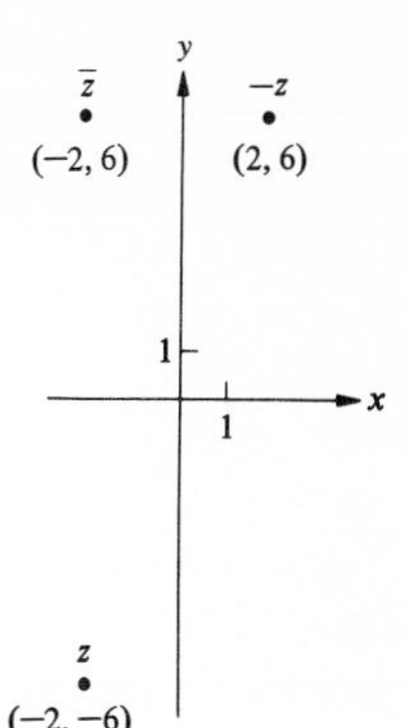

25. $4 + i$

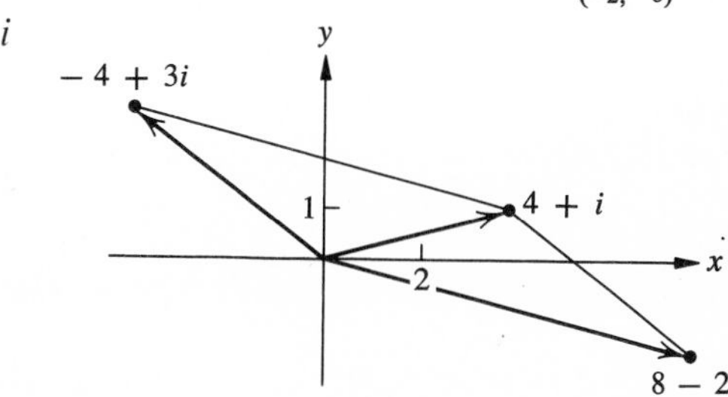

26. $(5, -1)$

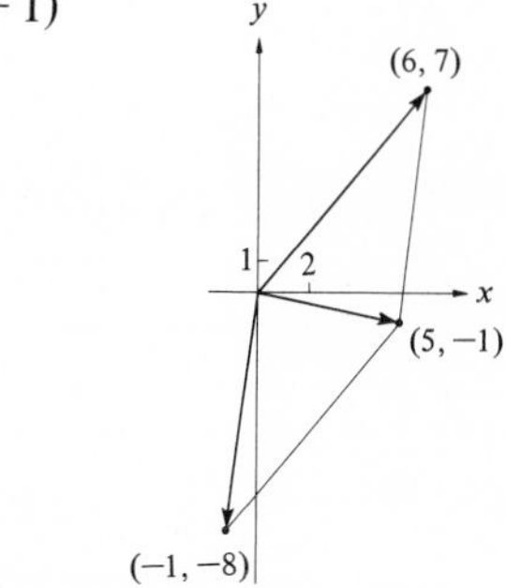

27. $2\sqrt{5}$, $26°\ 30'$

28. $\sqrt{29}$, $248°\ 10'$

29. $\sqrt{2}(\cos 45° + i \sin 45°)$

30. $6(\cos 120° + i \sin 120°)$

31. $-\dfrac{7}{2} + \dfrac{7\sqrt{3}}{2}i$

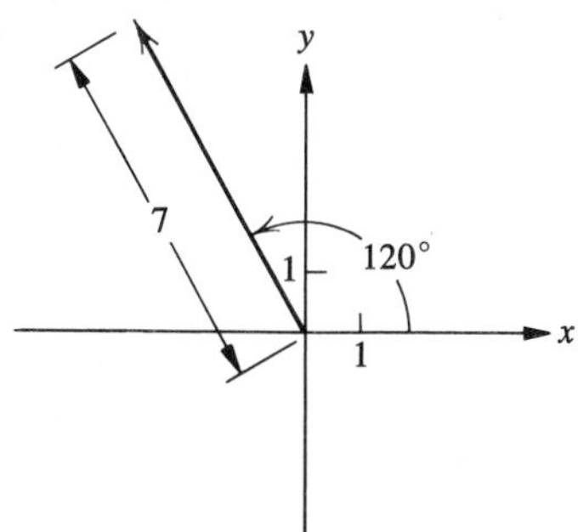

5

32. $\sqrt{2} + \sqrt{2}i$

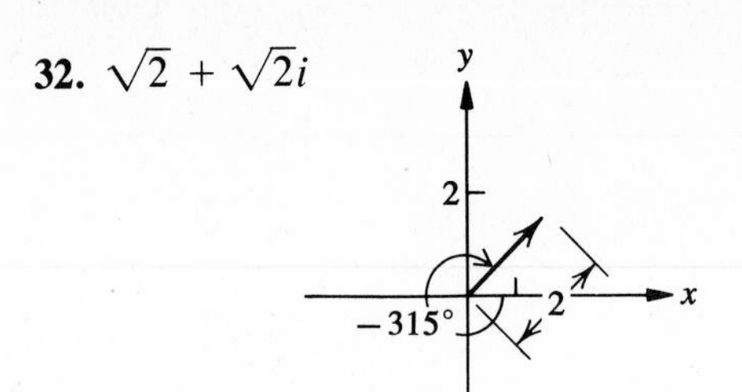

33. $-3\sqrt{2} - 3\sqrt{2}i$

34. $3\sqrt{3} + 3i$

35. $-\frac{8}{9} + 0i$

36. $-2\sqrt{2} + 2\sqrt{2}i$

37. $\frac{81}{2} + \frac{81\sqrt{3}}{2}i$

38. $0 - 8i$

39. $-\frac{1}{32} - \frac{\sqrt{3}}{32}i$

40. $-\frac{1}{256} + \frac{1}{256}i$

41. $\cos 15° + i \sin 15°, \cos 195° + i \sin 195°$

42. $5(\cos 110° + i \sin 110°), 5(\cos 230° + i \sin 230°), 5(\cos 350° + i \sin 350°)$

43. $\{\cos 45° + i \sin 45°, \cos 135° + i \sin 135°, \cos 225° + i \sin 225°, \cos 315° + i \sin 315°\}$

44. $\{10^{1/5}(\cos 48° + i \sin 48°), 10^{1/5}(\cos 120° + i \sin 120°), 10^{1/5}(\cos 192° + i \sin 192°), 10^{1/5}(\cos 264° + i \sin 264°), 10^{1/5}(\cos 336° + i \sin 336°)\}$

45. $(2, 250°), (-2, 70°), (2, -110°), (-2, -290°)$

46. $(-3, -330°), (-3, 30°), (3, 210°), (3, -150°)$

47. $\left(\frac{3}{2}, \frac{3\sqrt{3}}{2}\right)$ **48.** $(\sqrt{2}, -\sqrt{2})$ **49.** $\left(2\sqrt{2}, \frac{7\pi^R}{4}\right)$ **50.** $\left(2, \frac{7\pi^R}{6}\right)$

51. $\rho \cos \theta = 3$

52. $\rho^2(\cos^2 \theta + 3 \sin^2 \theta) = 5$

53. $x^2 + y^2 = 5y$

54. $y^2 - 4x - 4 = 0$

55.

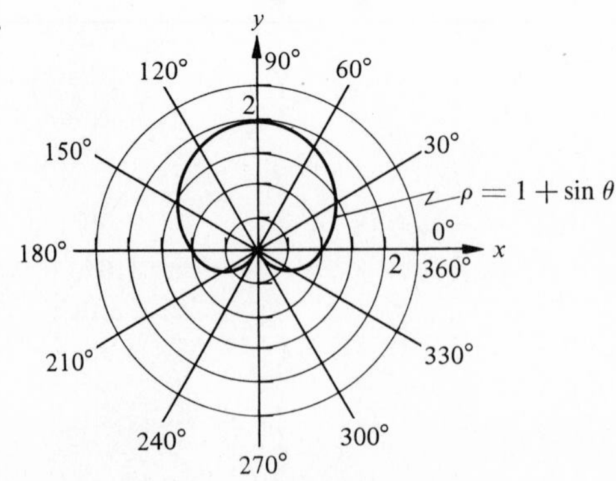

56.

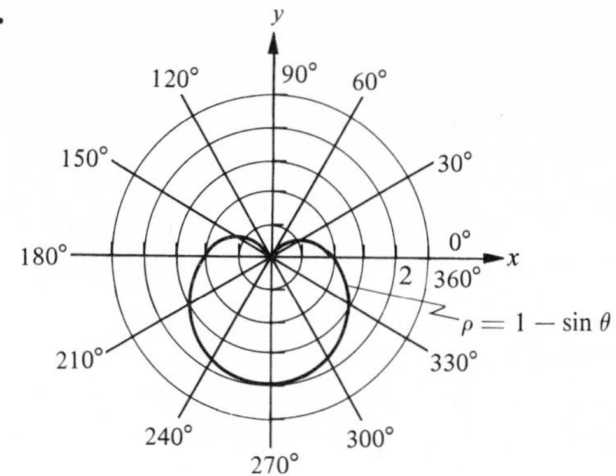

57.

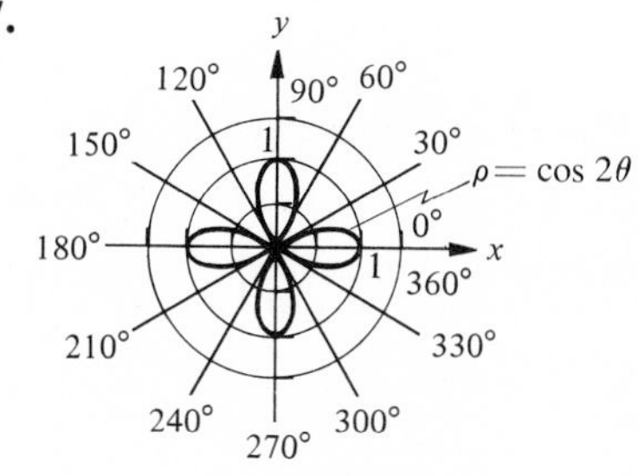

58.

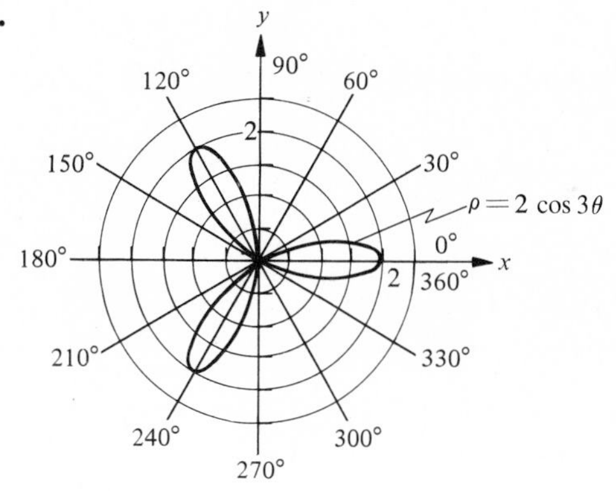

Exercise Set B.1

1. 0.0634 **3.** 0.5716 **5.** 0.9105 **7.** 0.9631
9. 0.2523 **11.** 1.617 **13.** 15.93° **15.** 25.73°
17. 37.85° **19.** 52.06° **21.** 0.315^{R} **23.** 1.006^{R}
25. 0.803^{R} **27.** 0.969^{R}

Index

Summary

1.1 Conversion Formulas

$$\frac{\alpha^\circ}{360} = \frac{\alpha^R}{2\pi} \qquad \alpha^\circ = \frac{180}{\pi}\alpha^R \qquad \alpha^R = \frac{\pi}{180}\alpha^\circ$$

1.2 Trigonometric Ratios ($r = \sqrt{x^2 + y^2}$, $r \neq 0$)

$\sin \alpha = \dfrac{y}{r}$ $\qquad$ $\csc \alpha = \dfrac{r}{y}$ $(y \neq 0)$

$\cos \alpha = \dfrac{x}{r}$ $\qquad$ $\sec \alpha = \dfrac{r}{x}$ $(x \neq 0)$

$\tan \alpha = \dfrac{y}{x}$ $(x \neq 0)$ $\qquad$ $\cot \alpha = \dfrac{x}{y}$ $(y \neq 0)$

Signs of Trigonometric Ratios

Quadrant:	**I**	**II**	**III**	**IV**
	$x > 0$, $y > 0$	$x < 0$, $y > 0$	$x < 0$, $y < 0$	$x > 0$, $y < 0$
$\sin \alpha$ or $\csc \alpha$	+	+	−	−
$\cos \alpha$ or $\sec \alpha$	+	−	−	+
$\tan \alpha$ or $\cot \alpha$	+	−	+	−

1.3 Functions for Special Angles

α°	α^R	$\sin \alpha$	$\cos \alpha$	$\tan \alpha$	$\csc \alpha$	$\sec \alpha$	$\cot \alpha$
0°	0^R	0	1	0	undef.	1	undef.
30°	$\frac{\pi^R}{6}$	$\frac{1}{2}$	$\frac{\sqrt{3}}{2}$	$\frac{1}{\sqrt{3}}$	2	$\frac{2}{\sqrt{3}}$	$\sqrt{3}$
45°	$\frac{\pi^R}{4}$	$\frac{1}{\sqrt{2}}$	$\frac{1}{\sqrt{2}}$	1	$\sqrt{2}$	$\sqrt{2}$	1
60°	$\frac{\pi^R}{3}$	$\frac{\sqrt{3}}{2}$	$\frac{1}{2}$	$\sqrt{3}$	$\frac{2}{\sqrt{3}}$	2	$\frac{1}{\sqrt{3}}$
90°	$\frac{\pi^R}{2}$	1	0	undef.	1	undef.	0
180°	π^R	0	−1	0	undef.	−1	undef.
270°	$\frac{3\pi^R}{2}$	−1	0	undef.	−1	undef.	0

$\frac{1}{2} = 0.500$

$\sqrt{2} \approx 1.414$

$\frac{1}{\sqrt{2}} \approx 0.707$

$\sqrt{3} \approx 1.732$

$\frac{\sqrt{3}}{2} \approx 0.866$

$\frac{1}{\sqrt{3}} \approx 0.577$

$\frac{2}{\sqrt{3}} \approx 1.155$

1.4 For $y > 0$ and $0^\circ \le \alpha \le 90^\circ$,

$\text{Trig}^{-1}\, y = \alpha \;\leftrightarrow\; \text{trig}\, \alpha = y$

2.1 Trigonometric Ratios in a Right Triangle

$$\sin \alpha = \frac{\text{length of side opposite } \alpha}{\text{length of hypotenuse}}$$

$$\cos \alpha = \frac{\text{length of side adjacent to } \alpha}{\text{length of hypotenuse}}$$

$$\tan \alpha = \frac{\text{length of side opposite } \alpha}{\text{length of side adjacent to } \alpha}$$

$$\csc \alpha = \frac{\text{length of hypotenuse}}{\text{length of side opposite } \alpha}$$

$$\sec \alpha = \frac{\text{length of hypotenuse}}{\text{length of side adjacent to } \alpha}$$

$$\cot \alpha = \frac{\text{length of side adjacent to } \alpha}{\text{length of side opposite } \alpha}$$

2.2 Solution of Oblique Triangles

2∠'s and 1 side; 2 sides and an ∠ opposite one of the given sides — Law of Sines:

$$\frac{\sin \alpha}{a} = \frac{\sin \beta}{b} = \frac{\sin \gamma}{c}$$

2.3 Solution of Oblique Triangles

2 sides and the included ∠; 3 sides — Law of cosines:

$$c^2 = a^2 + b^2 - 2ab \cos \gamma$$
$$b^2 = a^2 + c^2 - 2ac \cos \beta$$
$$a^2 = b^2 + c^2 - 2bc \cos \alpha$$

2.4 Geometric Vector

Norm (or magnitude):

$\|\vec{\mathbf{v}}\|$, the length of the line segment $\vec{\mathbf{v}}$.

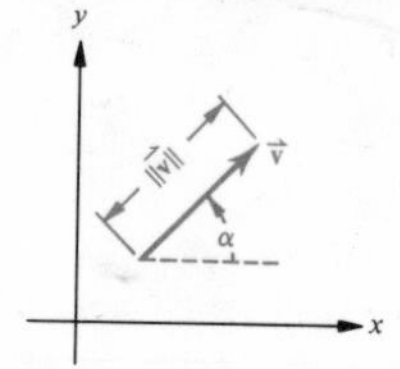

Direction angle:

α, where $-180^\circ < \alpha \le 180^\circ$

Equivalent vectors:

$\vec{\mathbf{v}}_1 = \vec{\mathbf{v}}_2$ if $\|\vec{\mathbf{v}}_1\| = \|\vec{\mathbf{v}}_2\|$ and $\alpha_1 = \alpha_2$

Sum (or resultant):

$\mathbf{v}_1 + \mathbf{v}_2$

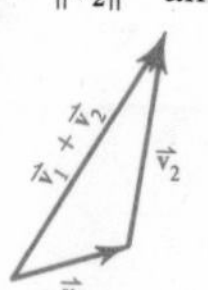

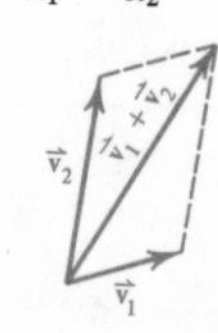

Product of a scalar and a vector:

$c\vec{\mathbf{v}}$, where $\|c\vec{\mathbf{v}}\| = |c| \cdot \|\vec{\mathbf{v}}\|$:
same direction as $\vec{\mathbf{v}}$ if $c > 0$ and
same direction as $-\vec{\mathbf{v}}$ if $c < 0$.

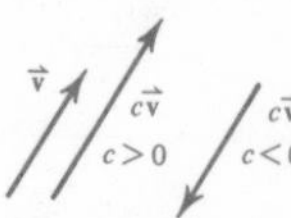

Rectangular components of $\mathbf{v}$:

$\|\vec{\mathbf{v}}_y\| = \|\vec{\mathbf{v}}\| \cdot |\sin \alpha|$

$\|\vec{\mathbf{v}}_x\| = \|\vec{\mathbf{v}}\| \cdot |\cos \alpha|$

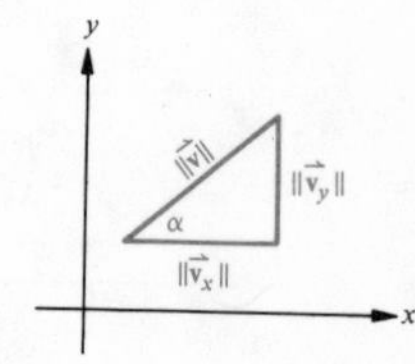

2.6 Circular Motion

Length of arc: $\quad s = r \cdot \alpha^R$

Linear velocity: $\quad v = \dfrac{s}{t} = \dfrac{r \cdot \alpha^R}{t} = r\omega$

Angular velocity: $\quad \omega = \dfrac{\alpha^R}{t}$